Unbottled

Unbottled

THE FIGHT AGAINST PLASTIC WATER
AND FOR WATER JUSTICE

Daniel Jaffee

UNIVERSITY OF CALIFORNIA PRESS

University of California Press
Oakland, California

Cataloging-in-Publication Data is on file at the Library of Congress.

ISBN 978-0-520-30661-5 (cloth : alk. paper)
ISBN 978-0-520-30662-2 (pbk. : alk. paper)
ISBN 978-0-520-97371-8 (ebook)

Manufactured in the United States of America

32 31 30 29 28 27 26 25 24 23
10 9 8 7 6 5 4 3 2 1

In Memory of Iris Jaffee

Contents

List of Figures and Tables

FIGURES

TABLES

Preface

FLINT, MICHIGAN

The lines began to form every day before dawn—long lines of cars, backed up for over a mile, their engines running in the winter cold for hours. Waiting for bottled water. For several years, this was the primary source of drinking water for most of the ninety-five thousand residents of Flint, who were exposed to toxic lead- and bacteria-laden tap water in an egregious case of environmental injustice. Today, thousands of people in Flint still rely on bottled water. "Why make Flint residents wait four hours in line for two cases of water?" Gina Luster, organizer and cofounder of the group Flint Rising, asks me rhetorically. "I have video and pictures where literally, if the water giveaway starts at 10, there's people already in line at 5 a.m. for two cases of water. All they want to do is rinse their fruits and vegetables off, cook, brush their teeth and bathe. Oh yeah, people are still bathing in bottled water."[1] Luster began experiencing severe health problems shortly after Flint's state-appointed emergency manager approved a switch to the polluted Flint River as the city's water source in April 2014, but it was nearly a year before she discovered the cause was her own tap water.

The most salient public image of the Flint water crisis has been plastic water bottles: hundreds of millions of them, donated by celebrities, bottling firms, and individuals—and from 2016 to 2018, distributed free to residents by the state of Michigan. "All of the water being provided by the state was Nestlé water," says Luster. Nestlé also faced heavy criticism for pumping hundreds of millions of gallons of Michigan groundwater per year for its Ice Mountain spring water brand, for most of which it paid only an annual permit fee of $200.[2] "So we found out that this company was really banking on the Flint water crisis, and making a profit off of it. . . . Mind you, this water was coming from [springs] one hour away from us. . . . So that's why I would say, we were buying our own water back!" As of late 2022, Flint residents continued to line up for bottled water, much of it donated by Nestlé's successor firm, BlueTriton.[3]

"If you go to the emergency room and you're sick, you're in triage," Luster continues. "That's [how] I look at this bottled water. . . . It's a triage kind of thing. It's not going to heal us. It's not going to make us necessarily better. It's just something to hold us captive, basically, until we figure this out. But there was nothing to figure out—fix the damn pipes and the infrastructure, and we won't need this bottled water."

BRASILIA, BRAZIL

A rock band has been playing for at least half an hour in the enormous pavilion located on the edge of Brasilia's main city park. Thousands of empty white plastic chairs sit in long orderly rows, facing a massive stage lined with amplifiers and two huge video screens. Overhead hang bright cloth banners painted with messages including the forum's slogan, "Agua É Direito, Não Mercadoría" (Water Is a Right, Not a Commodity). A large crowd makes its way toward the seats, some playing batucada drums and many waving big flags identifying the mass social movements they represent—the most visible the Landless Workers' Movement (MST), the Movement of Dam-Affected People (MAB), and the Brazilian Federation of Agricultural Workers' Unions (CONTAG). I follow them to find a colorful ocean of people dancing energetically in the midday heat. This is the opening ceremony of the Forum Alternativo Mundial da Agua (FAMA)—the Alternative World Water Forum, which has gathered over seven thousand attendees, including grassroots peasant, urban, labor, environmental, and water justice activists from Brazil, Latin America, and the rest of the world.

Suddenly the music and dancing stop. All eyes turn to a theatrical event down in front of the stage, magnified on the monitors overhead. Two four-meter-high puppets with huge painted papier-maché heads and cloth bodies stand in the wings: one with a twisted face and pursed lips, representing Brazilian president Michel Temer, and the other with the face and hat of Uncle Sam, a blue tie, and a white suit emblazoned with the logos of Pepsi, Coca-Cola, Nestlé, and Danone—the four biggest global bottled water and beverage corporations. Three groups of people in costumes stroll into view, representing indigenous people, *quilombo* dwellers of African descent, and peasant farmers. An upbeat samba tune plays as they walk and dance around a raised, winding river of blue fabric signifying water. Suddenly the music turns dark and threatening. The monster puppets move inward, swaying and circling around each other, as the actors huddle on the ground in fear, trying to protect their water. Ten minutes later, the performance ends with the people rising up in victory over the forces that attempted to take their water. The music returns to a major key, and the huge puppets disappear.

This is a *mística*—a ritualized theater performance typically used to open meetings of the MST and other Brazilian activist groups, drawing on the movement's roots in radical liberation theology. The effect is powerful, and it leaves the crowd buzzing with energy.

At the same time, only a few kilometers north in the heart of this planned city but encircled by a tight security cordon with heavy police and military presence, the Eighth World Water Forum is meeting in a large convention center and a former Olympic stadium. This "official forum" is attended by about ten thousand representatives of private water corporations, bottled water and beverage firms, the World Bank and other international financial bodies, UN agencies, national and local government officials, academics, and representatives of nongovernmental organizations (NGOs) from around the world. Military helicopters hover overhead. Limousine caravans with police escorts, sirens blaring, announce the arrival and departure of several heads of state, grinding city traffic to a halt. After passing through police lines and two security checks, I enter the convention center, turn down a long crowded hallway, and encounter a large sign displaying the logos of the Forum's top financial sponsors, which include Coca-Cola, Nestlé, and AmBev, the Brazil-based Latin American bottler of Pepsi. The World Water Forum—which the activists at FAMA refer to as the "corporate forum"—has been held every three years around the world since its first public meeting in The Hague in 2000. At nearly every World Water Forum since then, groups representing the global water justice movement, who oppose water privatization, have staged a simultaneous alternative gathering in the same city. The juxtaposition between these two opposing events in Brasilia could not be more stark.

The next morning, I learn that overnight, six hundred women from the Landless Workers' Movement, their faces covered by bandanas, and clad in the MST's trademark red baseball caps and shirts, have seized a controversial Nestlé water bottling plant one thousand kilometers away in São Lourenço, in protest of the World Water Forum and in support of FAMA. Their press statement accuses President Temer of allowing Nestlé to exploit Brazil's groundwater, pointing to alleged meetings between Temer and company officials to discuss exploration of the enormous Guaraní Aquifer.[4] The occupiers are detained by military police but released a few hours later.

CAPE TOWN, SOUTH AFRICA

The municipal authorities in Cape Town made a stunning announcement: by April 12, 2018, the city would reach "Day Zero"—the moment its public water supply would effectively run dry and water service would be cut off to over one million homes, making it the world's first major city to run out of drinking water.[5] After three years of extreme drought, the city's reservoir had been virtually depleted. Officials implemented severe water rationing—with residents waiting hours to fill containers at two hundred collective taps with a maximum of twenty-five liters (6.6 gallons) per person per day—but acknowledged they were preparing for "anarchy" if the taps ran dry. In this quasi-apocalyptic context, the wealthy dug personal wells and built water storage tanks, while many residents attempted to stock up on bottled water to fill the void.[6] Members of the Cape Town Water Crisis Coalition protested outside Coca-Cola Peninsula Beverages, the region's leading bottler of water and soft drinks, which draws 530 million liters annually from the municipal water supply. Accusing the company of exploiting the crisis for profit, the activists demanded that Coke cut its water extraction by half immediately and provide free water to Capetonians. "The water they have access to should be made available to the communities where water has been limited unfairly," argued the Coalition's Shaheed Mohamed.[7] While poor households faced substantial debt from soaring water rates and heavy fines for excessive water use, Coca-Cola was allowed to continue pumping unrestricted.[8] Although heavy rains postponed Cape Town's reckoning at the last minute, in 2022 the Eastern Cape city of Gqeberha, population one million, was on the verge of its own Day Zero, with those able to afford it buying bottled water and a member of the local Water Crisis Committee calling the situation a case of "water apartheid." Other large cities across the global South, including Monterrey, Mexico, and Delhi and Chennai, India, are now also facing similar fates.[9]

. . .

In the space of only four decades, bottled water has transformed from a luxury niche item into a ubiquitous global consumer good. This relatively new commodity sits at the convergence of at least three major struggles: a mounting social crisis of affordable access to safe drinking water; a severe ecological crisis of plastic waste, climate change, and increasing fresh water scarcity; and a battle over the future of our public water systems. The snapshots above embody the tensions that arise between the goal of ensuring public access to drinking water—a substance essential for life—and the private provision of water for profit, in the form of bottled water. These tensions map onto a long-standing clash between two opposing visions of water: on the one hand, as an economic good, a commodity that should be provided by the market in order to ensure its efficient use; and on the other, as a public good, a public trust, and a human right. Bottled water represents a newer, underemphasized front in this ongoing conflict.

Consumption of bottled water has grown with startling speed in the rich nations of the global North and is now expanding even faster in the global South, where access to clean drinking water is not as widespread. In the United States, bottled water surpassed soft drinks in 2016 to become the nation's most consumed beverage. Nearly nine in ten consumers now buy some bottled water, four in ten drink mostly or entirely bottled water, and one-sixth consume it exclusively, shunning the tap for drinking altogether.[10] Yet the United States is only fourth in per capita consumption of bottled water, behind Mexico, Thailand, and Italy. Worldwide bottled water consumption surpassed 120 billion gallons in 2021, two-thirds of it in single-serving plastic containers.[11] The worldwide bottled water market—with revenues of $300 billion in 2021 and projected to reach $509 billion by 2030—has long been led by four global food and beverage corporations: Nestlé, Coca-Cola, PepsiCo, and Danone Group.[12]

Several forces have combined to drive this dramatic growth in bottled water consumption. In the global North, an increased focus on health, fitness, diet, and obesity, the advent of lightweight plastic bottles, and a desire to consume "on the go" have hastened a shift away from soft drink consumption, but also away from tap water. The result has been the normalization of bottled water across society in a remarkably short period of time. The bottled water industry argues that it is in competition not with tap water but rather with soft drinks and other packaged beverages. Critics respond that the industry's implicit or explicit disparagement of the quality of public tap water has stoked distrust and played a major role in altering public perceptions and behavior.[13] Ironically, in the United States almost two-thirds of bottled water is now actually filtered tap water, drawn from already treated municipal water supplies, rather than springs or groundwater.[14]

But it is bottled water's environmental footprint that has generated the greatest negative publicity, particularly the global plastic waste crisis created by the disposal of over six hundred billion single-use plastic beverage bottles every year—a problem described by *The Guardian* as being "as dangerous as climate change."[15] Fewer than 27 percent of the single-use plastic bottles consumed in the United States are recycled, and the global recycling rate is a mere 7 percent.[16] Since China halted the importation of most recyclables in 2018, wealthy nations have been swamped by their own accumulating plastic waste, forcing local governments to grapple with the by-products of overconsumption in new ways.

These impacts of bottled water have not gone uncontested. Since the turn of the century, they have spawned oppositional social movements around the world that are challenging both the need for this commodity and the local effects of the water extraction it entails. At the consumer end, a growing number of campaigns by NGOs, city governments, and others are reasserting the value and purity of tap water and the public infrastructure that delivers it and are questioning the necessity of bottled water in places where access to safe, virtually free water is almost universal. Vehement struggles are also being waged at the source, by residents angry at the bottled water industry's efforts to extract water from local springs and groundwater, which have often divided neighbors and public officials alike.

While a good deal of media coverage has examined the rapid growth of bottled water as a consumer good, the cultural changes it has produced, and the major plastic waste problems it has created, there is far more to the story. The intense and persistent conflicts over bottled water suggest that this commodity is linked to much broader struggles over human rights and social justice, the question of who owns nature, and the future of the public sphere.

Unbottled examines both the causes and the environmental, social, cultural, and political consequences of the proliferation of bottled water, with an emphasis on the vibrant opposition movements it has generated at both ends of the commodity chain: where the water is extracted, and where it is consumed. The latter part of the book centers on two in-depth case studies of conflict over groundwater extraction by global beverage firms in the United States and Canada. These contentious struggles involve sustained opposition by coalitions of community residents and activists, supported by national and international advocacy groups. The book draws on an extensive set of interviews with local residents, public officials, activists, bottling firm staff, water experts, and a wide range of other participants on all sides of the controversies around the role and impacts of bottled water in and on society.

In analyzing these issues and conflicts, I attempt to answer a series of thorny questions. Is the continued expansion of bottled water compatible with the human right to water, a right codified by the United Nations in 2010? Is bottled water an important part of the solution to the global crisis of water access, as the industry and even some international institutions claim, or is it a dire threat to water justice, as numerous critics and opponents insist? What does the growth of this commodity—in places both with and without widespread access to safe tap water—mean for ecological sustainability and social inequality? What are, and what will be, the effects of bottled water's massive growth on our vital but increasingly underfunded public water systems?

The book also poses a broader question: Should access to sufficient safe, affordable, and reliable drinking water for all people—a key element of what some term *water justice*[17]—be an inviolable part of the social contract? If so, what does the continued expansion of bottled water as a market commodity mean for the possibility of fulfilling that contract?

Introduction

It's a question of whether we should privatize the normal water supply for the population. . . . The NGOs . . . bang on about declaring water a public right. That means that as a human being you should have a right to water. That's an extreme solution. And the other view says that water is a foodstuff like any other, and like any other foodstuff it should have a market value. Personally I believe it's better to give a foodstuff a value so that we're all aware that it has its price.

—Nestlé CEO Peter Brabeck, 2005

Forty years ago, when I was in grade school, the prospect of a large segment of the population shunning tap water, or families spending hundreds or even thousands of dollars per year on heavy multipacks of plastic water bottles for drinking, which they would lug from the store to the car to the kitchen, would have struck virtually everyone as a ludicrous vision or perhaps a dystopian fantasy.

Yet here we are. In 1980, U.S. consumption of bottled water barely reached two gallons per person per year, mainly imported Perrier in heavy glass bottles. In 2016, bottled water surpassed soft drinks to become the most consumed beverage in the United States, and by the end of 2021 Americans were swilling 47 gallons per year of it on average, for a total of 15.7 billion gallons, 70 percent of that in single-use plastic bottles (see figure 1).[1] One study found that among U.S. adults, bottled water accounted for a stunning 44 percent of *total* drinking water intake.[2] The bottled water industry's annual sales in 2021 reached $40 billion in the United States and $300 billion worldwide.[3] As figure 2 shows, it is far and away the world's most consumed packaged drink. China is now the biggest consumer of bottled water, guzzling one-fourth of the global total of 120 billion gallons in 2021. That total has increased

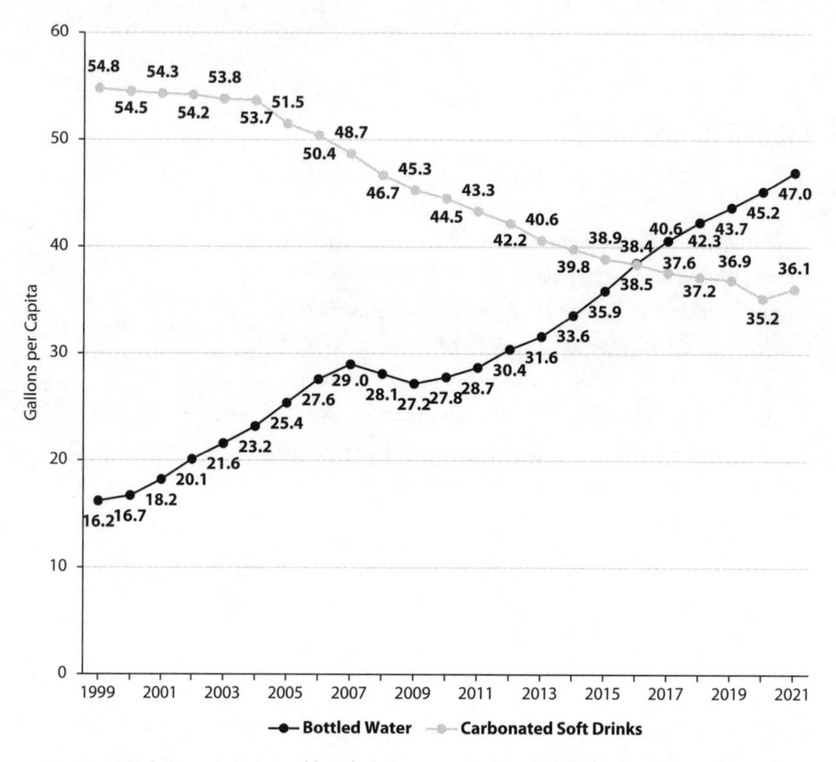

FIGURE 1. U.S. consumption of bottled water and carbonated soft drinks, gallons per capita, 1999–2021. Sources: Rodwan 2019; Beverage Marketing Corporation 2013, 2017, 2021; IBWA 2022; Statista 2022b.

by an average of 6 to 7 percent per year, with the fastest growth in East and Southeast Asia.[4] Clearly this is no minor phenomenon.

All of this water has to come from somewhere. Just under half of bottled water worldwide is extracted from groundwater, via natural springs, boreholes, or wells.[5] This requires gaining access to those sources, which in many cases are already in use by local communities and farmers, and certainly by natural ecosystems.[6] Much of the remainder—including nearly two-thirds of the bottled water sold in the United States—is instead drawn from public tap water supplies, a process that is ironically far less visible to the public.

Who is selling us this water? The biggest players in the bottled water industry are four huge multinational corporations: two of the largest food giants, Nestlé and Danone Group, and the two top beverage behemoths, Coca-Cola and PepsiCo. These companies went on a worldwide buying spree after the turn of the century, snapping up regional and

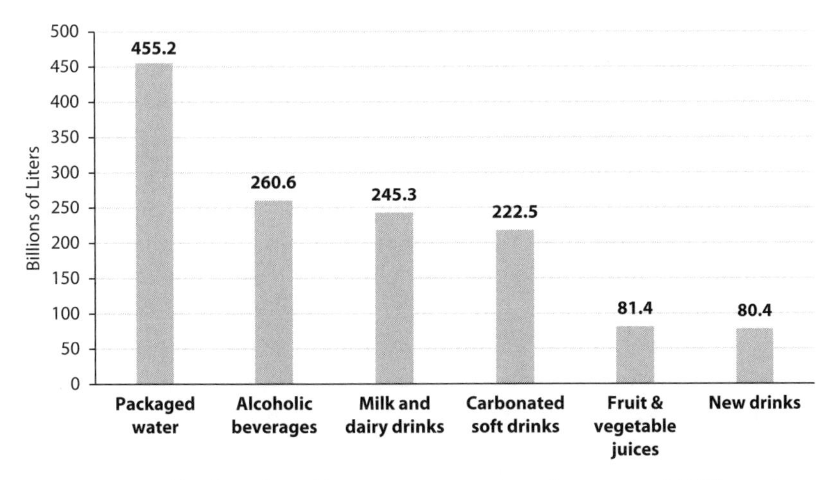

FIGURE 2. Worldwide packaged beverage consumption volume, 2021 (in billions of liters). Source: Adapted from Statista 2022a.

national bottlers along with their water sources and bottling plants. Most of these firms (which also use enormous quantities of water in their food products) work to influence global water policy as well, either through involvement in the World Water Forum and its sponsor the World Water Council or via the 2030 Water Resources Group, an industry-dominated body created by the World Bank to advise the United Nations, whose governing board includes the chairman of Nestlé and the CEO of Coca-Cola.[7] At the same time, many smaller water bottlers continue to survive and thrive, especially across the global South.

• • •

The vignettes in the preface, and Peter Brabeck's quote in the epigraph above, are scenes from a particular kind of war over water. This is a conflict in which bottled water and beverage firms are squaring off not only against many residents of communities whose water they are extracting or want to bottle, but also, it would seem, against public tap water itself—or at least our reliance on the tap as a trustworthy source of drinking water. This raises challenging issues for public water utilities, whose central purpose is to provide the very same substance—drinking water—which they do for an infinitesimally small fraction of the cost and environmental impact of bottled water. The bottled water industry, however, insists that its product is in competition not with tap water but rather with soft drinks, beer, and other beverages.[8]

But this battle over bottled water is only one facet of a much larger global conflict that has been raging in its current form since at least the 1980s: the struggle over whether water should be primarily a market commodity or a public good. These linked conflicts revolve around a simple truth: in a highly unequal world, when access to safe drinking water is premised on the ability to pay, some people will inevitably go without. This obvious yet fundamental fact lies at the root of the deep objections by critics to both the privatization of tap water systems and the commodity of bottled water. It also explains why the tensions over bottled water—a substance dramatically more expensive per unit volume than tap water—are closely related to, and often as intense as, the major battles over water utility privatization that have erupted around the world in recent decades.

Unbottled examines the social movements that are increasingly contesting the commodity of bottled water and the social, cultural and environmental consequences of its growth, both in North America and worldwide. It explores the implications of a profound and ongoing shift, in which the world's people are getting an increasing share of the water they drink from private corporations in plastic containers, rather than from a household or shared faucet served by a public water utility. It addresses the fraught question of where our next drink of water should come from—a tap or a bottle—and what the answer means for human rights, the natural environment, and the future of public water systems.

The dynamics behind bottled water's meteoric growth differ by world region. In significant parts of the global South (or Majority World),[9] because of colonial legacies, debt, austerity, and other factors, many governments have been unable to extend tap water systems fast enough to keep pace with rapid urbanization, even where the political will exists to do so. In this context, corporations, consumers, and governments are increasingly turning to packaged water—in single-serving bottles, multigallon jugs, plastic sachets, cartons, and other forms—as a solution to the actual or perceived scarcity of safe drinking water.[10] In this book I use the term *packaged water* to refer to this wider range of forms of commodified water, of which the various types of bottled water are the biggest subset. In most large cities in the South, there is a two-tier packaged water market: the transnational firms and their subsidiaries target middle- and upper-income consumers with higher-priced branded water, while local vendors and refillers supply poor and working-class residents with lower-cost water of often uncertain origin and quality. Yet for the poorest residents, even the cheapest options can be prohibitively expensive, highlighting the concerning implications of this commodity for the human right to water.[11]

In the rich countries of the global North (or Minority World), where access to clean tap water in the home is nearly universal, the reasons for bottled water's rise are different. Bottled water firms promote their product by appealing to consumer concerns with social status, purity, fitness, and health. Their advertising campaigns have sometimes also disparaged tap water, both capitalizing on and contributing to public fears about water quality.[12] News coverage of disasters of unsafe tap water, such as in Flint, Michigan, or Walkerton, Ontario,[13] further increases demand for bottled water. However, bottled water on average is no safer than tap water, is less strictly regulated, contains much higher levels of microplastics,[14] and at least in the United States largely consists of refiltered municipal water—including Coke's Dasani, Pepsi's Aquafina, and Nestlé Pure Life brands.[15]

These dynamics have provoked resistance in a wide range of forms and places. The oppositional movements fall into two broad categories: those contesting bottled water consumption and those resisting the bottling industry's water extraction. On the consumption end, campaigns to "reclaim the tap" have succeeded in pushing hundreds of city governments, schools and universities, and other institutions to promote the high quality of local tap water, reinvest in public water infrastructure including drinking fountains, and ban the purchase and sale of bottled water. One major impetus for these campaigns is bottled water's major negative environmental effects, which include an energy footprint up to two thousand times higher than tap water, major greenhouse gas emissions, substantial water waste in manufacturing, and the immense worldwide plastic pollution problems generated by the disposal of over half a trillion plastic beverage bottles annually.[16] At the extraction end, proposals to site or expand high-volume pumping and bottling facilities have generated fervent opposition, with local residents mobilizing around concerns including depletion or pollution of local groundwater, harm to fisheries, increased truck traffic, minimal water fees paid by bottlers, and negligible economic returns to communities. Drought and climate change–related water scarcity tend to supercharge these conflicts. Bottled water extraction is also a target of activism across the global South, with conflicts in Mexico, Brazil, Pakistan, Indonesia, India, and other nations.[17] Many of these struggles have received support from a cluster of national and international advocacy groups, some of which also facilitate local and global campaigns against tap water privatization.

Academic attention to the riddles posed by the rapid growth of bottled water has been surprisingly sparse. A substantial body of work examines the privatization of municipal tap water, particularly the efforts by international financial institutions and the global water services industry to open public water utilities to private management and ownership, and the vibrant movements that have arisen in opposition, from Bolivia to Indonesia to Indianapolis to Italy. However, this work has largely neglected the other major avenue of drinking water commodification: the growth of bottled and packaged water and its transformation into a global industry. Much of the published research that does address bottled water places it as a minor addendum to discussions of tap water privatization, and vanishingly little centers primarily on the social movements that are contesting this commodity. Yet the bottled water industry is dominated by a different group of multinational companies, and it has generated distinct opposition movements. This industry's continued rapid expansion, along with the explosion of public concern around the linked crises of climate change, drought, groundwater depletion, plastic waste, and decaying public infrastructure, makes a current examination of this commodity and its countermovements especially necessary.

In the following chapters I explore the causes, as well as the social, environmental, and cultural consequences, of bottled water's dramatic growth. I examine how the soaring fortunes of bottled and packaged water are connected to the growing global crisis of fresh water access, and I assess the implications of this commodity for realizing the human right to water. I also chart a range of social movements around packaged water, considering the parallels and divergences in their tactics and strategy, and asking how they are situated in relation to the broader international water justice movement, which fights water utility privatization and defends public and community water systems. Finally, I analyze the repercussions of bottled water's expansion for the future provision of safe public tap water—which many view as a quintessential public good[18]—and for social justice and sustainability more broadly.

Chapter 1 examines struggles over public versus private provision of water, focusing on the past four decades, during which a wave of privatization of public water utilities has taken place around the world, pushed by international financial institutions and private water firms and often abetted by governments. It briefly describes the opposition movements that have arisen in response to this privatizing trend and evaluates their outcomes. This chapter also explores several conceptual lenses for understanding privatization, commodification, and capital accumulation. It employs those ideas to develop a set of arguments about how bottled water is distinct—both in the way that it commodifies water and in the kinds of challenges it poses to the public provision of drinking water. Chapter 2 turns to the global bottled water industry, examining the reasons for its rapid growth and consolidation, and documenting how the industry has promoted its product by contrasting it with tap water. It asks why people in the United States and other wealthy nations have increasingly come to fear their tap water, explores how justified those concerns are, and traces efforts by the industry to cast doubt on the quality of public tap water. It scrutinizes the environmental, economic, and social effects of this commodity, including how the shift away from the tap and toward bottled water has exacerbated existing social inequalities and more recently has contributed to a growing backlash against single-use plastics. It also addresses the spread of bottled and packaged water in the global South, examining the role it plays in settings where tap water does not represent a safe or reliable drinking water source, and explores tensions over the role of packaged water in meeting international goals for improving clean water access.

But what are the actual practices of the movements taking on packaged water—what do they look like on the ground? Beginning with the toxic water disaster in Flint, Michigan, chapter 3 investigates the relationship between threats to tap water safety, environmental injustice, neoliberal austerity, and disinvestment in public infrastructure, and how the bottled water industry has benefited from these trends. It examines how the Flint crisis has spawned a highly diverse coalition that connects urban tap water crises with the bottled water industry's groundwater extraction in rural communities. It also addresses the implications of packaged water's growth for the future of municipal tap water, asking how we can restore trust in our public water infrastructure. Chapter 4 traces the history of bottled water movements in North America, focusing on a handful of key organizations and their shared roots in earlier and broader activism. It then examines the organized pushback against packaged water from the consumption side: a constellation of campaigns by city governments, public and private institutions, university students, community organizations, consumer and environmental NGOs, and others to increase tap water consumption and access, problematize the commodity of bottled water, and often ban its sale.

The following two chapters focus in depth on two major regional conflicts over bottled water extraction in North America. Chapter 5 travels to Cascade Locks, Oregon, the site of a decade-long struggle over Nestlé Waters' proposal to pump and bottle state-owned spring water in the scenic Columbia River Gorge, which culminated in a precedent-setting vote on a ballot measure to ban water bottling. Chapter 6 moves to Canada, where an alliance of water advocacy groups and Indigenous activists in southwestern Ontario is engaged in the nation's most sustained and visible conflict over bottled water. They have squared off against Nestlé Waters (and now its successor, BlueTriton) over its ongoing water extraction in a wholly groundwater-dependent region and its efforts to expand to new sites, substantially reshaping provincial water policy in the process.

In chapter 7, I step back to take stock of this range of oppositional efforts. This chapter assesses the parallels and divergences among the movements, communities, and organizations covered in the book, identifying the lessons they offer regarding water commodification and asking to what extent they represent an effective force for *de*commodification. It considers the implications of their varying degrees of success for the prospect of ensuring the human right to water and for the future of public water and public goods more broadly. Finally, the Conclusion considers where all of this leaves us and offers a series of concrete recommendations for curtailing the negative impacts of this commodity's global spread, regulating the industry's practices, strengthening and expanding access to public drinking water, and more.

While the book delves into a wide range of intersecting issues that lend themselves to multiple interpretations, I make five main arguments in these pages. First, the unique characteristics of bottled water differentiate it from tap water—a substance that has posed major obstacles to profitable privatization—and render it a more ideal commodity for capital accumulation. Bottled water's plastic packaging and far greater mobility allow it to bypass costly and elaborate tap water networks, disconnecting it from any shared public endeavor and hastening the commodification of water on a global scale. Second, these traits make the growth of bottled and packaged water a threat to the future provision of high-quality drinking water by public water systems—a threat potentially even more serious than that posed by tap water privatization. As this commodity increasingly displaces water consumed from public (and community-managed) sources, it is helping to erode the century-old project of universal public provision of safe drinking water that has brought incalculable health benefits to many parts of the world. Third, bottled and packaged water both illuminate and exacerbate the racial, class, and geographic divides between the water "haves" and "have nots." Widespread dependency on bottled water is an indicator of water injustice, both in the global North and in the South. It intensifies existing social inequalities, and its availability can enable governments to postpone or avoid making critical investments to repair or extend public tap water systems. Fourth, the diverse countermovements that are contesting bottled water consumption, pushing to expand access to public drinking water, and resisting commercial water extraction for bottling constitute an emerging and increasingly coherent movement for decommodification. Finally, while there are already partial linkages between bottled water opposition movements and the global water justice movement that opposes privatization and supports public and community water systems, these connections can be greatly strengthened, and these two sets of movements can more explicitly embrace each other's core concerns. They have the potential to unite around a shared critique of bottled water's distinctive threat to realizing the human right to water.

This book is based in large part on ethnographic field research that bridges multiple sites across national borders. Overall it is the product of over a decade of research, which took place between 2010 and 2021 in the United States, Canada, Mexico, and Brazil, including more than one hundred interviews with a wide range of participants involved in struggles over bottled water extraction and consumption, both in those countries and in other nations. They include local community residents on multiple sides of these conflicts; grassroots activists; staff and volunteers with local, national, and international organizations working on water, environmental, and consumer issues; representatives of bottled water firms; water researchers; staff of public water utilities and state agencies; local elected officials; and many others. The interviews were complemented by extensive observation at community meetings, conferences, international forums, protests, and other public events, and by analysis of documents, publications, and news media coverage.

Before exploring the larger concepts of commodification, privatization, accumulation, and decommodification and how they apply to water in its various forms, we first need to examine how drinking water was transformed from a market commodity into a public good, and then (partly) back again. The next chapter takes on both of these tasks, setting the stage for a deeper look at the phenomenon of bottled water and the movements that are complicating its continued growth.

CHAPTER I

A More Perfect Commodity

Since the turn of the twenty-first century, news media and popular culture have increasingly spread the message that the world faces a major crisis of water scarcity and that fresh water constitutes the new oil over which future wars will be fought.[1] *Newsweek,* for example, dramatically proclaimed that "the world is at war over water. Goldman Sachs describes it as 'the petroleum of the next century.' Disputes over water tend to start small and local. . . . But minor civil unrest can quickly mushroom, as the bonds of civilization snap."[2] Hyperbole aside, such coverage often leaves out a key insight: the "global water crisis" is socially produced, and it is exacerbated by human-caused climate change, ecological degradation, and overextraction by agribusiness and other industries.[3] Nevertheless, the lack of access to potable water around the world does indeed constitute a grave emergency. As of 2020, according to the United Nations, 771 million people worldwide lack access to any source of safe drinking water (although some argue the actual figure is far higher),[4] over 2 billion have no access to an "improved" or safely managed water source,[5] 2.3 billion lack basic sanitation services, and at least 3 billion are unable to wash their hands at home with soap and water.[6] Waterborne diseases including diarrhea and cholera account for approximately 5 percent of all deaths worldwide.[7] The increasing inequalities between "the water 'haves' and 'have nots'"[8] have been described as "one of the greatest crimes of the twenty-first century."[9] Lack of access to safe water is a key marker of social and environmental inequity on a global scale.

In many urban areas of the global South, only upper- and upper-middle-income residents are served by the public tap water networks that are common in the North, and even where such piped systems are available, supplies may be inconsistent, insufficiently treated, or contaminated.[10] Middle- and lower-income urban and peri-urban residents frequently rely on a mix of informal water sources, including refilling stations, water trucks, and vendors selling locally bottled water, all of which typically cost far more per liter than their wealthier neighbors pay for tap water.[11] Others turn to communal standpipes, household wells, or untreated surface water.[12]

Climate change is also making fresh water increasingly inaccessible in many parts of the world. A vicious cycle of drought, declining aquifer levels, and overpumping means that one-fifth of all groundwater wells "may be facing imminent failure, potentially depriving billions of people of fresh water" who are unable to afford or access alternative sources.[13] Droughts have increased 29 percent just since 2000, and without major action, three-fourths of the world's population will live in areas that are water-stressed at least part of the year by 2050, up from half today.[14]

Partly because of this growing crisis of quantity, quality, and inequality, drinking water has increasingly come to be viewed as a profitable commodity—a product to be sold to consumers at market rates. Yet when a good or a resource becomes a commodity, access to it depends on the ability to pay. Water is essential for life, and it is unsubstitutable. Allowing the market to set a price for water can push clean water out of the reach of people who lack what the World Bank calls "effective demand": the money to buy it.[15] This fundamental contradiction helps explain the intense opposition to the privatization of water that has emerged worldwide in recent decades.[16]

This chapter begins with a look at the history and ideology of the substantial push to privatize municipal (tap) water around the world, as well as the current status of water privatization. It briefly describes the opposition movements that have arisen to challenge these developments. This history lays the groundwork for the second section, which explores key conceptual approaches for understanding the privatization and commodification of nature, and of water in particular. Using these lenses, it identifies some crucial differences in the commodification of tap water versus bottled and packaged water, and explores what this distinction means for the future of access to safe, affordable drinking water for all.

PRIVATIZING PUBLIC WATER

Over the past four decades, there has been a major change in how water is governed and managed. Geographer Karen Bakker describes the shift from a "state hydraulic paradigm" that dominated from the postwar period through the 1970s, in which the provision of clean water was widely understood to be both a public good and a right of citizenship, to a neoliberal paradigm,[17] in which—to cite the influential Dublin Principles of 1992—water "has economic value in all its competing uses and should be recognized as an economic good."[18] Since the 1980s, a constellation of actors, led by the World Bank and large private water service firms, have argued that governments in the global South have failed to provide access to water to their growing and urbanizing populations and that only the private market can effectively expand water service.[19] Only by pricing water at its true value, the argument continues, can this scarce resource be conserved. To accomplish these goals, they insist, private firms should be responsible for providing drinking water, leaving states to merely regulate, and making water users into customers. To do so they would need "full-cost recovery"—guaranteed profit margins backed by higher water bills.[20] The result of this ideological shift has been the emergence of a $500 billion-plus global water utility industry,[21] roughly one-fourth of which is controlled by two French-based multinational corporations, Veolia and Suez,[22] which partially merged in 2022.

The start of the global push to privatize water dates back to the Pinochet dictatorship in Chile, which in 1980 implemented a new, radically promarket national water law, bearing the imprint of Milton Friedman and the Chicago school of economics.[23] But the trend truly gained momentum in the late 1980s and 1990s when international financial institutions, primarily the World Bank and the International Monetary Fund (IMF), mandated that most debtor governments offer up their public utilities to corporate bidders as a precondition for debt renegotiation under so-called structural adjustment programs, arguing that states had failed to provide "water to all" and that the private sector should fill this role.[24] Although many Southern governments have indeed been hard-pressed to extend water networks to keep pace with urbanization, that is often directly due to the legacies of colonialism and to the very same strictures of debt repayment under structural adjustment. In 1989, Margaret Thatcher augmented the global push for privatization with the sale of the ten public water authorities in England and Wales, along with their physical water and sewerage assets.[25]

Privatization has spread not only through pressure tactics applied by creditors but also through ideological intervention. Sociologist Michael Goldman describes how the World Bank since the early 1990s worked to manufacture a global "consensus" around private provision of water, creating "transnational policy networks" that integrated Southern bureaucrats, global water firms, some NGOs, and industry-heavy entities including the World Water Council. Goldman argues water privatization is emblematic of the World Bank's embrace of "green neoliberalism," a paradigm that combines sustainable development discourses with the expansion of market fundamentalism.[26] Bakker describes this approach as "market environmentalism."[27] By whatever name, this regime led to a dramatic increase in private water concession contracts and facilitated the growth of a global cartel of water service firms, the top five of which at their peak controlled over 70 percent of the private water market.[28]

In reality, the term *privatization* subsumes a wide variety of forms of market involvement, ranging from relatively uncommon outright asset sell-offs like those in the U.K., to long-term leases or concession contracts (outsourcing arrangements in which water systems remain publicly owned but private firms are responsible for operations, maintenance, and sometimes infrastructure investments), to "corporatized" water utilities that remain in public hands but are "run as though they were private firms,"[29] often raising rates and shutting off customers for nonpayment.[30] The large majority of the "privatizations" that have caused major social protest in the global South were multidecade concessions—often called "public-private partnerships" (PPPs) by their proponents—which typically guarantee the corporate operator a minimum profit margin and the freedom to set water rates.[31]

Water scholars disagree on the extent to which privatization and commodification overlap. Bakker argues that partly because of water's biophysical and sociocultural qualities, "private ownership and the introduction of markets do not necessarily entail commodification," since privatization has not always proven profitable[32]—in other words, that privatization can happen without true commodification. Others consider privatization a subset of commodification.[33] In this book, I use the term *water privatization* to refer to multiple arrangements that cede control over key functions of public or community water systems to private corporations, ranging from PPPs (concessions and leases) to outright water utility sales—but I do not employ it as a synonym for commodification. Rather, the privatization of water can be understood as one facet of the much broader dynamic of *commodification*—the absorption of formerly public, common-pool, or other nonmarket goods, resources, and services into the market and their transformation into commodities.[34] Although these two dynamics sometimes intersect, commodification can take place in the absence of privatization, a distinction that bottled water illustrates especially clearly. It is also important to recognize that academic debates over the boundaries between these phenomena are often of less concern to water justice activists, many of whom employ *privatization* as a broad term encompassing all forms of market control over water, which they view as antithetical to human rights and the public interest.[35]

What are the results of this wave of water utility privatization? The short answer is that it has not lived up to the promises of its promoters. A huge body of literature examines the outcomes of privatization in the global South—especially the dominant model of long-term concession contracts to large multinational water firms in cities like Buenos Aires, Johannesburg, or Manila. Some research has found increases in water connections to new households under private concessions in certain cities,[36] but the large majority of studies concur that market involvement has failed to substantially expand water service, let alone meet the stated goal of providing "water for all."[37] Even the World Bank itself has acknowledged that concessions have not generated significant numbers of new water connections.[38] The overall track record of municipal water privatization across the South includes large rate hikes, mass water shutoffs for nonpayment, deviations from promised levels of service, increased unsafe water events, sewage spills, and the spread of diseases including cholera.[39] In some cases, rate hikes have made tap water "so unaffordable that citizens are forced to drink water from contaminated sources."[40] When access to clean water is based solely on the ability to pay, some people will lose access. This, of course, can be deadly.

Privatization also has major gendered effects. Increasing water rates in pursuit of "full-cost recovery" can substantially alter not only the economies of poor households but also women's economic and physical security.[41] Political ecologist Rhodante Ahlers writes that "unequal gendered access to resources is perpetuated and legitimized by the introduction of market mechanisms in the water sector."[42] Given that the burden of collecting water in many parts of the world falls predominantly on women and girls, if households are disconnected from a tap water supply because of nonpayment and must travel to access water, they are the ones most likely to suffer in the form of added labor time, loss of school attendance, physical hardship or injury, and vulnerability to sexual assault.[43]

CONTESTING AND REVERSING PRIVATIZATION

Strong public opposition to the privatization of water has emerged across the globe. Substantial protests have erupted in Bolivia, Ecuador, Uruguay, Argentina, Nicaragua, El Salvador, Tanzania, South Africa, Nigeria, Indonesia, Philippines, and many other countries.[44] The landmark 2000 and 2005 "water wars" in the cities of Cochabamba and El Alto, Bolivia, have attracted particular academic and popular interest. In Cochabamba, after a World Bank imposed, single-bidder concession to a local subsidiary of the U.S.-based Bechtel Corporation led to water rate hikes of up to 200 percent, residents responded in April 2000 with massive protests that eventually forced the Bolivian government to back down and company officials to flee the country and cancel the contract.[45] A second water uprising in 2005 in the heavily Indigenous city of El Alto, against the privatization of the La Paz–El Alto water system by a subsidiary of Suez, also led to major demonstrations and a huge public strike, causing the contract's cancellation and contributing to the election of Evo Morales as president in 2005.[46] Several observers have characterized these antiprivatization outcomes as major early victories against the neoliberal model. Many grassroots activists in these antiprivatization uprisings have received support from an array of regional and international networks, unions, and organizations—including Red Vida in Latin America, the European Water Movement, and Public Services International, among many others—which collectively constitute the global water justice movement.[47]

Opponents of privatization argue that public utilities like water systems are structurally unsuited to market control, since the profit motive and the obligation to maximize shareholder return are incompatible with the need for long-term investments in water quality and system maintenance.[48] Because the rate hikes needed to achieve these profit margins cause the exclusion of those unable to pay, they insist that privatizing a life-sustaining resource is inimical to public well-being and that water must be managed on a nonprofit basis.

Concerns about privatization and lack of access to water among grassroots activists in the global South, as well as water justice–focused NGOs based in the North, coalesced into a major campaign to declare water a human right. The landmark 1945 UN Declaration of Human

Rights had explicitly included the rights to food, clothing, housing, medical care, and education, among others—but not to water.[49] "At that time," writes the prominent Canadian water activist and author Maude Barlow, "water was not perceived to have a human rights dimension and was also felt to be virtually limitless."[50] In 2010, after a decade-long push, the UN Assembly and UN Human Rights Council declared access to clean drinking water and sanitation to be a human right, a major victory for water justice advocates.[51] The resolution recognizes "the right to safe and clean drinking water and sanitation as a human right that is essential for the full enjoyment of life and all human rights."[52]

While this declaration is a valuable tool for citizens aiming to hold governments accountable, actually operationalizing the human right to water remains an enormous challenge, one that is further complicated by market control. Nor does it guarantee that this water will be provided by public entities: private water firms and even some UN officials insist that the right to water is compatible with for-profit provision.[53] This raises important questions about international efforts to attain the UN Sustainable Development Goals, one of which—SDG 6—aims to achieve "universal access to safe and affordable drinking water for all" by 2030.[54]

However, the pendulum has begun to swing away from privatization. Beginning in the early 2000s, many private water concession contracts were terminated by local or national governments because of corporations' noncompliance with contract terms, failure to make required investments in infrastructure or system extension, public opposition, and/or changing state political ideologies (especially in Latin America). Other contracts were ended by the private firms themselves in response to lack of profitability and public protests.[55] The German firm RWE, formerly the third-largest private water corporation, sold off much of its $10 billion in water holdings worldwide.[56]

The reasons for this deprivatization trend are particularly relevant to the contrast with bottled water that I draw later in this chapter. Providing universal tap water service to predominantly poor urban populations has proved not to be sufficiently profitable, for several reasons. First, the extensive costs involved in maintaining and expanding water treatment and delivery networks for the long term compete with the need to maximize shareholder return in the short term.[57] Second, delivering tap water is a low-margin business: it is hard for private firms to raise water rates high enough to deliver the profit levels they seek. Doing so has often led to mass disconnections of households for nonpayment, which can cause large-scale social conflict. Overall, it turns out that municipal water privatization is simply not lucrative enough in largely poor urban areas in the global South. Geographer Alex Loftus writes that "it is incredibly difficult to make the profits expected by private investors from the large, long-term needs of infrastructural development for poor people."[58] Finally, although antiprivatization protests have occurred in only a minority of cases, they played a key role in the cancellation of a number of privatization contracts because they increase investors' perception of risk and tarnish corporate brands.[59] As one large water services firm notes in an analysis of the private water industry, "It remains a challenging sector do business in. When 12% of all contracts awarded to date (120 contracts, covering 119 million people) in population terms have ended (for whatever reason) . . . [and] 8.9% of contracts have ended prematurely, that is a significant risk factor."[60]

Public opposition and the poor track record of privatization have led hundreds of cities to "remunicipalize" their water systems in the past two decades, terminating contracts with private firms and bringing water back into public management. The largest recent examples include the cities of Manila and Jakarta. A 2020 report by the Transnational Institute documents 303 cases of water remunicipalization in thirty-six countries, with the highest numbers occurring in France, the United States, Spain, and Germany.[61] Several nations have also incorporated outright bans on water privatization into their constitutions, including Bolivia, Uruguay, and Ecuador.[62] However, some remunicipalized water utilities in the South—such as SEMAPA, the Cochabamba water system whose privatization sparked the 2000 Bolivian water war—continue to struggle with low investment, indebtedness, and difficulty expanding water networks.[63] One model that water justice groups advocate to address such challenges is "public-*public* partnerships," in which water utilities partner with NGOs, labor unions, and/or utilities in other cities and nations to share expertise, pool resources, and increase efficiencies.[64]

PRIVATIZATION, AUSTERITY, AND AFFORDABILITY IN THE NORTH

The involvement of private capital in municipal tap water in the global North follows some of the same dynamics. The first drinking water systems in North America and much of Europe were largely developed by private firms in the eighteenth and nineteenth centuries, but they were plagued by disease, contamination, inadequate water pressure and supply, and other problems. "Across the country," write authors Alan Snitow, Deborah Kaufman, and Michael Fox, "the pattern was repeated: private water management often meant leaky pipes, pollution, and disease." As a result, by the early twentieth century, city governments had largely taken over the operation of water systems:

> The story was similar in cities across the United States and Canada. As populations grew, private water companies did not have the resources to meet the need. Citizens demanded and eventually won modern public water systems, financed through bonds, operated by reliable engineers and experts, and accountable to local governments. The nation built a dazzling system of community waterworks, which provide clean, reasonably priced water and sewer systems that still rank among the best in the world. Approximately 85 percent of Americans are presently served by the thousands of publicly owned and locally operated water systems. For several generations, water has been a public trust.[65]

The human impact of these developments was monumental. One study calculated that water quality improvements were responsible for half of the dramatic decline in U.S. urban mortality that occurred between 1900 and 1936, and three-fourths of the drop in infant mortality.[66]

Since the 1990s, however, pressure has emerged to privatize urban water systems.[67] Several large U.S. cities signed private concession contracts with Suez, Veolia, and other corporations to manage their drinking water and sewerage systems, including Atlanta, Indianapolis, New Orleans, Stockton, California, and more recently Baltimore. Yet largely because of their poor track record, most of these large private concessions have since been canceled, adding to the global remunicipalization trend.[68]

Nevertheless, public water systems in the United States are increasingly hampered by deteriorating public water and sewer infrastructure, resulting from decades of underinvestment and fiscal austerity. Major cuts in federal funding—a precipitous decline of 77 percent in real terms between 1977 and 2017—have left many local governments with increasingly dilapidated water and sewer systems and a maintenance backlog estimated at more than $1 trillion.[69] Canada faces similar problems.[70] This is emblematic of the retreat by governments from service provision under neoliberal austerity, which has especially come to the fore since the Great Recession.[71] While the great majority of tap water in the United States today is still quite safe to drink, this disinvestment has contributed to high-profile crises of unsafe tap water, which further increase consumer demand for bottled water.

The cost burden of maintaining this infrastructure has made "public-private partnerships" and outright system sales newly appealing to some cash-strapped towns and cities, especially in the growing number of states that have passed legislation incentivizing them, including Pennsylvania and New Jersey.[72] Private equity firms like the Carlyle Group have also increasingly sought to buy out both private and public U.S. water utilities in search of high short-term returns.[73] This indicates that water utilities in wealthy nations are a more attractive proposition for investors than their counterparts in much of the global South. Some observers describe these phenomena as cases of "disaster capitalism," emphasizing the way political elites and investors can seize upon crises, whether natural disasters or fiscal shocks, to advance the privatization or commodification of public resources.[74]

Despite these trends, the proportion of the U.S. population served by public drinking water systems has actually risen slightly, from 83 percent in 2007 to 87 percent in 2015. Studies have documented that customers of privately owned water utilities pay on average 59 percent more than those of public utilities, because of higher borrowing costs and the imperative to generate returns for investors.[75]

However, although tap water provided by the public sector is less expensive, the United States is experiencing a burgeoning crisis of water affordability. Municipalities forced to assume nearly the entire burden of water and sewer system maintenance have passed the costs on to users through rising bills. As of 2017, nearly 12 percent of U.S. households had unaffordable water and wastewater bills (defined as more than 4.5 percent of household income), a figure expected to rise to 35 percent by 2022.[76] An analysis of twelve U.S. cities in 2020 showed that water and sewer bills had risen an average of 80 percent in only eight years, making them unaffordable for up to 40 percent of residents.[77] And a nationwide study determined that low-income families paid an average of 12.4 percent of monthly income to cover water and sewer bills in 2020.[78] The epicenter of the affordability crisis is arguably Detroit, where more than one hundred thousand poor households have had their water service shut off for nonpayment of bills since 2014, but large-scale disconnections have occurred in many cities.[79] Water bills are the least affordable in cities with aging pipes, high levels of poverty, private water utilities, and state regulations favorable to privatization.[80]

In Europe too, most municipal tap water systems had long been in public hands, with the exception of France, home to Veolia and Suez. After the U.K. privatized its water systems under Thatcher, cities in

Spain and other countries began to follow suit. Since the global financial crisis of 2007–10, the "troika" of the IMF, the European Union, and the European Central Bank has imposed privatization mandates on Portugal, Ireland, Spain, and Greece.[81] Implementing such privatization and "full-cost recovery" policies has led to substantial water shutoffs in these countries too, generating major opposition movements, particularly in Greece and in Italy.[82]

COUNTERCURRENTS: THE PRESENT STATUS OF WATER PRIVATIZATION

Where does the push for private provision of water stand today? The overall panorama is mixed. As of 2015, private corporations supplied tap water to 646 million people, or just under 9 percent of the global population—up from only 50 million in 1990. Another 494 million people received sewerage service from private firms.[83] At the same time, there has been a substantial withdrawal by large multinational water firms from much of Latin America,[84] western Europe, and several large Southern cities, including Jakarta, Indonesia, and Accra, Ghana. Water researcher Gregory Pierce argues that this "strategic retreat" in the early and mid-2000s has been followed by a "shallow expansion" of privatization since the Great Recession, with multinational firms playing a lesser role, and greater involvement by domestic capital and private equity investors. This expansion of private water delivery is bimodal: on the one hand, there is substantial growth in middle-income nations, notably China, India, and more recently Brazil, where the far-right-wing Bolsonaro government implemented a major privatization push. On the other hand, privatization has been imposed from above by creditors in southern Europe,[85] offering a stark parallel to the earlier wave of structural adjustment in Southern debtor states. In fact, structural adjustment in the global South has been renamed but not rescinded: nations participating in the World Bank's Highly Indebted Poor Countries (HIPC) initiative still must implement utility privatization as a precondition for securing debt relief. Bakker writes that the current state of affairs is a "refinement, rather than a retrenchment, of the neoliberal project" in water.[86]

In sum, there are substantial countercurrents at work. A genuine remunicipalization and deprivatization trend in some regions coexists with a countertendency toward new privatizations, mainly in middle-income countries but also in parts of the North. The number of people who receive tap water from private corporations has increased, but more slowly than many boosters and critics had expected: 90 percent of households with water service worldwide are still served by public or community water systems.[87] However, as I discuss below, another mode of water commodification is advancing far more rapidly: bottled water.

PRIVATIZATION, COMMODIFICATION, AND ACCUMULATION

The commodification and privatization of nature have generated major social conflict, both around social justice issues of access to life- and livelihood-sustaining resources and around questions of sustainability and ecological impact. To better understand how water exemplifies these tensions, it is valuable to explore some influential conceptual approaches to commodification.

In his foundational work *The Great Transformation,* the Hungarian economic historian Karl Polanyi identified what he termed the "fictitious commodities" of land, labor, and money, highlighting how they are central to the destructive tendencies of a purportedly "self-regulating" market economy.[88] While genuine commodities are goods produced for sale in the market, Polanyi wrote, "Labor is only another name for a human activity which goes with life itself, which in its turn is not produced for sale. . . . Land is only another name for nature, which is not produced by man; actual money, finally, is merely a token of purchasing power. . . . None of them is produced for sale. The commodity description of labor, land, and money is entirely fictitious."[89]

Of course, nature is indeed commodified in numerous forms, yet the social and ecological consequences of treating nature as if it were a genuine commodity, produced for sale in the market, are dire. "To allow the market mechanism to be sole director of the fate of human beings and their natural environment," wrote Polanyi, " . . . would result in the demolition of society. . . . Nature would be reduced to its elements, neighborhoods and landscapes defiled, rivers polluted, military safety jeopardized, the power to produce food and raw materials destroyed."[90] These devastating effects of the industrial revolution eventually led to the rise of "movements of self-protection" against the tyranny of the market. The late nineteenth and early twentieth centuries witnessed what Polanyi called a "double movement," in which "the extension of the market organization in respect to genuine commodities was accompanied by its restriction in respect to fictitious ones,"[91] through increased government regulation of capital, the rise of labor movements, and the growth of welfare states. However, neoliberal globalization has once again removed or weakened many of those safeguards. It has also stretched the fictitious commodities even further into new areas, such as body parts and patented life forms, seeds, and genes.[92] Water, as an element of Polanyi's land, offers a compelling contemporary illustration of the dangers of this commodity fiction.

Other writers describe the conversion of water into a marketable commodity as a form of "primitive accumulation," drawing on Karl Marx's analysis that stressed the historical process through which capitalism has separated producers from their social means of subsistence and production.[93] Marx's prime example of this phenomenon was the English land enclosures of the sixteenth and seventeenth centuries, in which huge areas of common land used for pasture and food production were fenced for private use by large landowners, forcing peasants into cities and creating a wage-dependent proletariat. However, this process did not end with enclosure, nor was it limited to a "primitive" or original stage of capitalism. The economist and philosopher Rosa Luxemburg, for example, viewed this dynamic as a continuous process. She argued that capitalism inherently needs to expand into noncapitalist realms to access new sources of profit, including resources and (free or nearly free) labor, emphasizing the role of European colonialism and imperialism in forcing open new markets.[94]

More recently, the geographer David Harvey has extended these ideas further with his influential concept of "accumulation by dispossession," which stresses the ongoing nature of this dynamic in the present day. Accumulation by dispossession, Harvey argues, is a response by capitalists to a crisis of overaccumulation, "a condition where surpluses of capital . . . lie idle with no profitable outlets in sight."[95] In these conditions, which have prevailed since the late 1970s, capital must conquer new terrains in order to retain or return to profitability. "What accumulation by dispossession does," writes Harvey, "is to release a set of assets . . . at very low (and in some instances zero) cost. Overaccumulated capital can seize hold of such assets and immediately turn them to profitable use."[96] The essence of this process is commodification, the dynamic by which formerly common or publicly owned goods, assets, and services are brought into the market, converted into commodities. Harvey describes privatization as the "cutting edge" of accumulation by dispossession, stressing the role of the World Bank, the IMF, and the World Trade Organization (WTO) in imposing "the wave of privatization (of water and public utilities of all kinds) that has swept the world."[97] These modern-day "enclosures" combine commodification with exclusion, whether that exclusion is economic or physical. Harvey also describes the broad range of social movements that have arisen to oppose accumulation by dispossession worldwide, noting that they share an emphasis on "reclaiming the commons."[98]

Several observers have applied Harvey's framework directly to the commodification of water.[99] Geographer Erik Swyngedouw argues that "nature itself has long resisted commodification, but in recent years, nature and its waters have become an increasingly vital component in the relentless quest of capital for new sources of accumulation. . . . A local/global choreography is forged that is predicated upon mobilizing local H_2O, turning it into money, and inserting this within transnational flows of circulating capital."[100] In other words, the privatization of drinking water can be understood as one facet of a much broader, ongoing process of commodification of nature, linked to the strategies of global firms to ensure continued capital accumulation—that is, profit.

The overall tendency of capital, then, is a relentless drive to incorporate more and newer domains into the market, a trend characterized by more than one observer as "the commodification of everything."[101] Nonetheless, while many view commodification as an inexorable one-way street, it is also possible to move in the opposite direction. The concept of *decommodification* is increasingly used to describe a range of social movements and alternative practices that push back against the expansion of market control into new realms of nature and human society. Sociologist John Vail defines decommodification as "any political, social, or cultural process that reduces the scope and influence of the market in everyday life."[102] Canadian social scientists Gordon Laxer and Dennis Soron argue that the goal of decommodification is not necessarily the wholesale "rejection of commodities, consumption, and markets" but rather the imposition of limitations on the reach and power of the market.[103] As Polanyi wrote, the development of twentieth-century welfare states removed many vital goods, resources, and areas of social provision from the market economy, thus partially decommodifying society.

The constellation of water justice movements and activists who oppose water privatization and advocate the human right to water can be viewed as a Polanyian countermovement for decommodification, because of their aim to make the commodification of water unprofitable, difficult, unacceptable, and/or illegal and to reassert water provision as a public (or in some cases a community) function.[104] As I will discuss later in the book, the same can be said of the movements challenging bottled and packaged water.

BARRIERS TO ACCUMULATION

Before exploring how bottled water manages to overcome some of the major challenges involved in commodifying water, it is worth digressing slightly to look at a parallel example in a different realm: agriculture. For several decades, rural sociologists have debated a perplexing question: Why have some elements of farming proven so difficult to commodify, to industrialize profitably? Rural sociologist Jack Kloppenburg and others focus particularly on seeds.[105] They argue persuasively that seeds, because of their unique characteristics—they are self-reproducing and they can be saved—"have offered a particularly large stumbling block to capital accumulation."[106] The dual character of seeds, as both the means of production and also grain, "is antagonistic to the complete assimilation of seed (as opposed to grain) under the commodity form."[107] This work also examines the factors that have allowed capital to overcome such structural obstacles, particularly the role of genetic engineering and seed biotechnology. Genetic engineering (GE) has managed to surmount these barriers to commodification in several major crops, most notably corn, soy, sugar beet, and canola. This has contributed to a strikingly rapid concentration of the global seed industry through mergers, with the top three corporations now controlling over 55 percent of the market.[108] According to sociologists Gabriela Pechlaner and Gerardo Otero, GE represents "the key technology driving capital accumulation in the neoliberal food regime."[109] But while GE technology has been transformative in a few major crops, it has succeeded only because of the support of a broader political and institutional framework, including intellectual property rules in global trade and investment agreements like the WTO that offer legal recognition for seed patents and licenses.[110] The growth of genetically modified seeds and crops has also unleashed diverse and dynamic opposition movements, consisting of small-scale farmers, consumers, environmentalists, and others in both the North and the South who aim to resist or reverse seed commodification and defend food and seed sovereignty.[111]

Just as the technology of genetic modification has finally enabled capital to break down the structural barriers to accumulation that seeds have long posed, so too the growth of bottled water and its transformation into a global commodity have been aided by analogous tricks that allow it to surmount the hurdles water has posed to effective commodification.

What is it about water that makes it more difficult to profitably commodify than forests or minerals or fossil fuels? It is instructive to look at water's innate qualities. Snitow and colleagues write that because water is "heavy, slippery, and expensive to transport, it resists being made into a commodity."[112] That is, it does not cooperate sufficiently well with capital to ensure sustained high levels of profit. Bakker similarly argues that water's physical characteristics make it a poor fit for capital accumulation. She writes that water's geography, its sociocultural qualities, and its nature as a flow resource all hinder its profitability, rendering it an "uncooperative commodity."[113]

This valuable analysis helps explain why, despite a favorable political and legal environment, many private water utility concessions have been terminated by the contracting companies themselves. However, this framework does not capture the ways that distinct *forms* of water pose differential barriers to accumulation. And that, it turns out, is vital for understanding the rising fortunes of bottled water.

MAKING WATER MORE PERFECT

In contrast to the water that flows through the pipes of municipal tap systems, bottled water enjoys several structural advantages. Its lightweight plastic package and extreme portability, along with the political-economic shifts that have enabled its rapid growth, have allowed it to avoid many of the constraints described above.[114] Bottled water, regardless of its origin (whether from springs, wells, or municipal systems), does not present many of the impediments to capital accumulation that are posed by tap water networks. This renders it—in contrast with Bakker's description of water as an uncooperative commodity—a far more mobile and profitable commodity.

Let us explore this argument in greater depth. First, bottled water requires virtually none of the sunk (unrecoverable) fixed infrastructure costs and obligations that are inherent to municipal tap water systems. Private concession contracts for water utilities typically require companies to make major capital investments to maintain water quality and the physical water treatment and distribution network, as well as to meet strict public health and environmental standards, hire and train staff, manage billing, and handle other imponderables that can reduce profit margins or make returns unpredictable. Municipal tap water systems also constitute natural monopolies, because it is utterly impractical to build multiple competing sets of water pipes and treatment plants. They are inherently tied to a specific place.

In contrast, bottled water firms require a very limited set of investments, enjoying what political scientist Joshua Greene terms a "capital light modus operandi with an attractive risk-return profile."[115] They access the water itself either for free or at very low cost. A corporate report on the bottled water industry underscores the point: "Entry barriers are low, and decreasing by the day."[116] Author James Salzman concurs: "With the cost of bottling equipment around $100,000, this is an easy market to enter if you can find the right marketing angle."[117] This is especially true for brands that use already-treated municipal tap water, such as PepsiCo's Aquafina and Coca-Cola's Dasani, which bottle water at the same plants as other beverages and distribute them through their existing networks. Nor do they face many of the extensive public health and environmental regulations that private firms managing tap water systems must meet.[118] "From a regulatory point of view," writes journalist Ryan Felton, "companies that want to put vast quantities of public water into bottles for profit face few hurdles and minimal ancillary costs."[119] These factors all increase profitability and reduce risk. For this reason, the bottling of municipal tap water—which is treated by and moved through infrastructure paid for and maintained by a century or more of public investments in most places—represents an especially dramatic form of commodification.

Second, bottled water largely defies the locality of water. Bakker argues that because of its weight and high transport cost, water is typically "used and disposed of locally."[120] However—much as the technological package of genetically engineered seeds removes a crucial barrier to further commodification of the food supply—bottled water's protective plastic skin enables it to escape those fundamental constraints. The development of lightweight PET plastic bottles and their widespread adoption by the beverage industry in the 1990s were the key transformations enabling this shift. Over one-fourth of all bottled water crosses national boundaries, making it truly a global commodity,[121] and much more travels long distances domestically, moving far beyond local watersheds.

Third, bottled water is not merely more portable than tap water but also more profitable. This is true in affluent nations—where many brands cost thousands of times more than the same volume of tap water, and families spend hundreds of dollars or more per year buying it—but it also pertains to places in the global South where safe public water supplies are largely lacking. Private water utilities in the United States, regulated by state utility commissions, typically have net profit margins of 10 to 15 percent. The explosive water wars in Bolivia and other nations were sparked when private firms attempted to raise water bills enough to achieve contractually guaranteed returns of 15 to 17 percent. In contrast, profit margins for bottled water historically have been greater, frequently 20 to 35 percent and even higher in some market segments and nations.[122]

This higher profitability is augmented by water injustice. Although the uprisings in Cochabamba and many other cities across the South were sparked by water rate increases of 100 to 200 percent, the (often poorer) residents of the same countries who are not served by piped water networks, or cannot drink the water from their taps, already typically pay many times those rates for bottled or packaged drinking water.[123] The inability of the public sector to provide safe tap water to over two billion people—reinforced through debt and austerity—has paved the way for the rapid growth of packaged water, further exacerbating social inequality. Since water is essential for life, people who lack other options will often pay whatever the market demands. Where packaged water is the only safe option, its cost—and the resulting profits—can be quite high indeed.[124]

In contrast with municipal tap water, then, bottled water constitutes an ideal commodity for capital accumulation—because of its defiance of water's locality, its lower sunk costs and investment requirements, and its greater portability and profitability. This divergence is manifested in the rapid expansion of the bottled water industry worldwide, in contrast to the slower and more uneven growth in private water utilities, characterized by hesitant investment, cancelled contracts, and a remunicipalization trend in both South and North.[125] These disparate growth rates mean that within one to two decades the global bottled water industry will likely surpass the size of the private water utility industry. A more perfect commodity, and a more successful one.

FINAL OBSERVATIONS

According to Harvey, accumulation by dispossession is a continuous process, in which capital constantly seeks new realms into which to expand.[126] I have argued here that some forms of commodified water are more amenable to profit than others and that technological transformations have played a key role in enabling these shifts. Bottled and packaged water currently represents the cutting edge of water commodification, and its extraction and manufacture involve processes of accumulation by dispossession that are arguably farther-reaching, longer-lasting, and harder to reverse than those at work in the privatization of tap water.

To sum up, bottled water represents a more perfect commodity for capital accumulation because of several intrinsic and extrinsic characteristics. It differs in important ways from municipal tap water systems, and those differences enable bottled water to escape many of the obstacles to profitability that are posed by private operation of the massive piped water treatment and supply networks originally built by governments in the nineteenth and twentieth centuries.

However, without the political-economic and cultural developments that have enabled bottled water's rapid rise—including lifestyle shifts that increase demand for on-the-go convenience, neoliberal globalization and deregulation that have facilitated the growth of transnational corporations, and governments increasingly unwilling or unable to defend public goods—it would likely not be nearly as widespread as it is today. This suggests that we need to consider how the process of commodification unfolds differently in distinct contexts, as well as how social movements could most effectively respond to such developments. It also raises questions about the implications of a broader societal move toward such individualized, market-based approaches to meeting human needs.

In the next chapter, I trace the history of the bottled water industry and how it has managed to "turn water into water," taking a product long viewed as an unnecessary extravagance and building a global market for it so successful that billions of consumers now regard it as essential—not only in places where drinking water from the faucet is unsafe or unavailable for most people, but also where taxpayers fund and take for granted nearly universal safe tap water.

CHAPTER 2

Making a Market, Fearing the Tap, Building a Backlash

One reflection of a healthy beverage industry is the decline of tap water, and tap water consumption has been trending down in recent years.

—Gary Hemphill, Managing Director of Research, Beverage Marketing Corporation

How did the nascent bottled water industry make a market for a product that for much of the past century was largely viewed as an unnecessary or wasteful luxury good? This commodity has transformed in the space of only four decades into a ubiquitous consumer object that is now the primary, and sometimes sole, source of drinking water for billions of people globally. The story of bottled water's resurgence in places with abundant clean tap water revolves around the question of why people have increasingly come to distrust their tap water, and how the expanding bottled water industry has fueled and taken advantage of this phenomenon.

This chapter traces how bottled water emerged from a near-death experience to become a mass-market commodity and a huge global industry, focusing on the four corporations that have led this market. It shows that bottled water's rise in the global North was due not merely to aggressive advertising but to a confluence of political and cultural trends, technological developments, and opportunistic marketing that seized on emerging public concerns to make consuming bottled water into a normal, commonplace practice. It assesses the industry's claim that surging bottled water use has not come at the cost of lower tap water consumption. I address how the rapid spread of packaged water in the global South both mirrors and deviates from these patterns, examining the role it plays in settings where tap water is not widely accessible, reliable, or potable. I also explore tensions over packaged water's role in meeting international goals for improving clean water access, and examine the safety of bottled water itself.

The second half of the chapter examines the social and economic effects of bottled water consumption, particularly how the high cost of purchasing this commodity deepens existing socioeconomic inequalities. I then turn to the extensive environmental impact of packaged water and its role in the connected crises of plastic pollution and climate change. In the last few years these burgeoning impacts have generated a major public backlash, leaving the industry scrambling to respond. Its reactions raise intriguing questions about the future both of this commodity and of public drinking water.

. . .

Bottled waters have a very long history, reaching as far back as the Roman Empire.[1] In Europe, traditions developed among pilgrims and later the wealthy of "taking the waters" at natural springs for medicinal or spiritual reasons, and some of those waters were bottled and sold for consumption. Among these was Perrier, which received favored trade status in England in 1863 and later gained wide distribution across the British Empire. Bottled water played a key role in urban water supply on both sides of the Atlantic during the nineteenth century, when crowded conditions and poor sanitation contaminated local rivers and wells, making drinking water a perilous proposition. By 1900, the number of bottlers had proliferated further, delivering water from rural springs and wells in glass bottles of varying sizes. "Bottled water was very much an up and coming industry," writes hydrologist Francis Chappelle. "But then, disaster struck. This disaster . . . was the introduction of chlorinated municipal water in American cities. . . . By 1941, 4,590 of the 5,372 water treatment systems in the United States were using chlorination. Within a few short years, an important reason for using bottled water in the first place . . . [had] evaporated. . . . This just about killed the bottled water industry in America."[2]

For much of the twentieth century, bottled water faded into insignificance, never disappearing entirely but relegated to a handful of small companies attached to particular natural springs. Thus historically bottled water thrived when public drinking water supplies were either nonexistent or contaminated but became superfluous once those supplies were made safe and trustworthy. "Just like the buggy whip," writes

Chapelle, "bottled water became a commodity rendered unnecessary by the brave advance of civilization."[3]

The story of its resurgence in the past four decades in places with abundant clean tap water is a definitive break from that pattern. This paradox revolves around the question of why people in both the global North and South have increasingly come to distrust their tap water, and what role the bottled water industry has played in that shift. But bottled water's resurrection is also the result of a confluence of late twentieth-century cultural, technological, and commercial developments—some of which are linked to deeper conflicts over the relative power of governments and markets, the commercialization of the public sphere, and the rise of neoliberalism. This commodity has proven highly resilient, and its growth has appeared to be unstoppable. Yet the ubiquity of packaged water and its major social and environmental impacts have also sown the seeds of opposition, generating a counterreaction in recent years that is now threatening to disrupt the industry's momentum.

AN INDUSTRY APPEARS: THE BIG FOUR AND MORE

The effective start of the modern American bottled water industry occurred in 1978: the year Perrier launched its bottled water brand in the United States with a massive advertising campaign that aimed "to move from a small number of fancy restaurants to a mass market . . . to position its product as a healthy and chic drink," according to author James Salzman.[4] Many viewed this as akin to bringing the proverbial coals to Newcastle: Why would people spend good money for a bottle of imported spring water when they could get perfectly clean water straight from the tap? Unlike some of their European counterparts, Americans could also count on being served free tap water with restaurant meals. However, the critics underestimated the power of snob appeal. "Perrier," writes branding expert Douglas Holt, "pioneered the idea that drinking bottled water in small single-serve glass bottles was an affordable way to grasp a bit of European sophistication, which resonated among so-called yuppies."[5]

Perrier's success changed the game. At the start of the 1980s, the North American bottled water industry consisted almost entirely of local and regional spring water companies, many family-owned. But once consumption began growing by double digits, national and global brands started snapping up these bottlers. By 1988 only ten companies controlled over half of the U.S. market.[6] At the forefront of this buying binge was Perrier, which had long sold only its own imported French spring water until it purchased several major regional bottlers, including Poland Springs, Calistoga, Deer Park, Arrowhead, and Ozarka. Suddenly, Perrier controlled nearly a quarter of the U.S. market.

In 1992, Perrier itself was eaten by a much bigger fish: Swiss-based Nestlé, the world's largest food and beverage corporation. Nestlé, which had experimented with bottled water in its Vittel plant in France as early as 1969, began acquiring bottling firms in Europe and North America. By 2019, worldwide revenue from Nestlé's water business reached US$8.6 billion, out of total corporate revenues of $101 billion.[7]

Groupe Danone, a French conglomerate best known for Dannon yogurt, brought its Evian and Volvic waters to North America and acquired several Canadian bottlers. By the turn of the century, Danone, Nestlé, and the Japanese firm Suntory together accounted for half of U.S. bottled water sales.[8]

During the 1990s, this turbulent market underwent another upheaval: the entry of giant soda makers Coca-Cola (with its Dasani brand) and PepsiCo (with Aquafina). Holt writes that because these competitors "controlled the key distribution channels . . . they could easily wrestle control of the category simply by delivering the convenience that drinkers demanded. And that is what they did."[9]

FROM PRISTINE SPRINGS TO PURE LIFE

These soda giants not only had existing distribution networks and bottling plants. They also enjoyed ready access to a cheap, abundant supply of their prime ingredient: the same refiltered municipal tap water that they already used in their other beverages. This was an entirely different proposition from that of their competitors, whose bottles at the time contained groundwater from springs or wells. There was only one problem: most consumers of Dasani and Aquafina were completely unaware of this fact. Paying a dollar or two for a bottle of spring water was one thing, but spending that amount to buy the same water that came out of one's home faucet was another matter entirely.

In Europe—where the distinct tastes of mineral waters are considered a vital attribute—the vast majority of bottled water comes from natural springs or groundwater, as it also does in Canada.[10] However, this is no longer true in the United States. Initially, selling reprocessed or filtered tap water was controversial. In 2004, under pressure from the consumer group Corporate Accountability International, PepsiCo and Coca-Cola

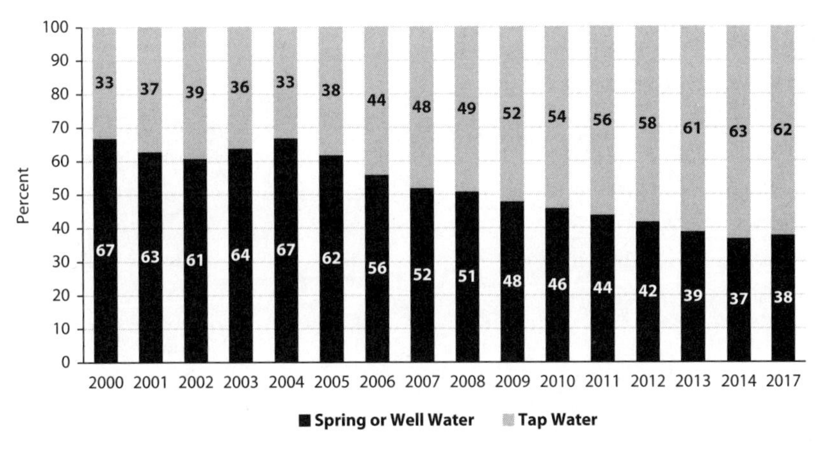

FIGURE 3. Share of U.S. bottled water, by source, 2000–2017. Sources: Food and Water Watch 2018a; Antea Group 2018.

were forced to admit publicly that Aquafina and Dasani contained tap water, albeit filtered and supplemented with minerals. Pepsi agreed in 2007 to acknowledge on its bottle labels that the water came from a "public water source," and Nestlé followed suit for its Pure Life brand. Coca-Cola, however, remained resistant. "We don't believe that consumers are confused about the source of Dasani water," a Coke spokesperson told the press. "The label clearly states that it is purified water."[11] The revelations generated a momentary scandal, but sales of Dasani, Aquafina, and Pure Life continued to soar. According to the Beverage Marketing Corporation (BMC), "The conventional wisdom in the bottled water industry is that the majority (but by no means all) of bottled water consumers do not recognize the distinction between spring water and drinking water packaged after it has been processed in municipal systems."[12]

Brian Ronholm of Consumer Reports argues that "these bottlers are essentially double-dipping—receiving low-cost water subsidized by taxpayers and then turning around and selling it back to the public at a significant markup."[13] For these and other reasons, the share of U.S. bottled water that comes from public tap water systems soared from one-third in 2000 to almost two-thirds in 2017, as figure 3 illustrates.

To sum up, the bottled water market in the global North—and to some degree worldwide—is bifurcated, led by two distinct groups of corporations: the European-based food conglomerates Danone and Nestlé (and Nestlé's successor BlueTriton in North America), selling mainly spring water; and the U.S.-based soft drink giants PepsiCo and

TABLE 1 MAJOR STILL AND SPARKLING BOTTLED WATER BRANDS AND PARENT COMPANIES, U.S. AND CANADA

Parent company	Nestlé Waters/ BlueTriton‡	Coca-Cola	PepsiCo	Danone	Other [Brand (Parent)]
Still water brands	Nestlé Pure Life	Dasani	Aquafina	Evian	Crystal Geyser (CG Roxane)
	Poland Spring*	Glaceau Smartwater	Lifewtr	Volvic	Fiji (Fiji Water)
	Ozarka*	Glaceau Vitaminwater	Propel		President's Choice** (Loblaw)
	Deer Park*				Naya (Naya)**
	Ice Mountain*				Eska (Eau Vives)**
	Arrowhead*				Compliments (Sobeys)**
	Zephyrhills*				
	Acqua Panna†				
Sparkling water brands	*Perrier*†	*Topo Chico*	*Bubly*		*LaCroix (National Beverage)*
	San Pellegrino†				*Polar (Polar)*
					Sparkling Ice (Talking Rain)
					Schweppes; Canada Dry (Keurig Dr. Pepper)

SOURCE: Euromonitor International 2021a, 2021b.

*Sold in U.S. only. **Sold in Canada only. †Imported.

‡ In 2021, Nestlé sold its North American bottling operations and seven still water brands to BlueTriton Brands; it retains the water brands Perrier, San Pellegrino, and Acqua Panna.

Coca-Cola, which sell almost exclusively retreated tap water in the United States, Canada, and other nations. Table 1 shows the Big Four companies and their North American bottled water brands, as well as other major brands and their parent firms.

EVERYONE'S PET BEVERAGE

So far, this discussion has omitted a critical element without which the bottled water industry would never have escaped its niche as a luxury good. One word: plastics. Although bottlers began using plastic bottles in the late 1970s for soft drinks, their early spread was slow. In 1989, manufacturers developed a cheaper, lightweight, and more durable plastic, PET (polyethylene terephthalate) to replace PVC (polyvinyl chloride) bottles, which according to Holt "allowed manufacturers to hit ever lower price points for water, and . . . [enabled] new consumer uses for bottled water."[14] The 1990s were thus the pivotal decade for bottled water, when this new packaging transformed it into a mass-market commodity—causing consumption to soar, plastic waste to proliferate, and, by the decade's end, the first real opposition to the industry to emerge. The conversion from glass to plastic continues. As of 2020, 45 percent of all beverage containers sold in North America were plastic, up from 36 percent in 2012.[15] According to BMC's Gary Hemphill, this increase was "due mostly to the success of the bottled water category."[16]

This dovetails with major social shifts in the global North since the 1970s, including changing gender roles, the rise of two-income households, longer workweeks, increased commuting, and greater mobility generally, all of which mean fewer meals consumed at home and increased consumption of single-serving food and beverage containers.[17] The newly available plastic water bottle fit this on-the-go, throwaway lifestyle perfectly.[18]

A critical factor in defusing public opposition to the resulting explosion of plastic was the advent of plastic recycling. The beverage and packaging industries strongly encouraged local governments in the global North to incorporate plastic into existing curbside recycling programs or to begin them if they did not exist.[19] Environmental critics counter that the existence of plastic recycling has provided "green" cover to the dramatic expansion of an unsustainable and polluting industry. According to Judith Enck, president of the organization Beyond Plastics, "The reason the public thinks recycling is the answer is that the plastic industry has spent 30 years on multimillion-dollar campaigns saying that. That was absolutely the wrong message. The message should have been: Don't use so much plastic."[20]

HOW THE INDUSTRY MADE A MARKET

A great deal of ink has been spilled on the topic of why consumers purchase bottled water, or how they have been persuaded to do so. Ultimately, academic analyses and market surveys show this boils down to four key factors: concerns about style, taste, health, and hydration—or the "Four F's": fashion, flavor, fitness, and frequent drinking. To these must be added a fifth F: fear, which I discuss later.

The simplest to address is fashion. It is undeniable that many consumers drink bottled water, or select particular brands, because of the perceived cool factor. While bottled water firms pay for celebrity endorsements and product placement, they also get plenty of free star validation anytime a celebrity is photographed with a bottle. President Obama appeared publicly with bottles of Fiji Water, a priceless promotional gift. More recently, "Dwayne Johnson and Gwyneth Paltrow, among others, have partnered with bottled water brands."[21]

Second, market research shows many consumers buy bottled water because they do not like the taste of their tap water. However, what some people perceive as poor tap water taste or smell in many cases is related to the very water treatment processes that make tap water safe to drink in the first place. "Chlorine's yucky taste is the reason a lot of otherwise sane people drink bottled water," writes author Elizabeth Royte. "But removing chlorine doesn't require much equipment: all you have to do is let your water stand a few hours in an uncovered pitcher or jug."[22] These perceptions of poor taste may also be a result of consumers' repeated exposure to bottled water—particularly the standardized, bland flavor of the re-treated tap water that constitutes most bottled water in the United States today. These brands use hyperfiltration, UV light, and reverse osmosis to strip out chlorine and minerals that impart taste to water. Then to make it palatable, they add proprietary mineral blends, known colloquially in the industry as "pixie dust." This trains people to expect a generic taste from drinking water, irrespective of location. And once trained, such preferences are very difficult to reverse.[23] However, in blind taste tests, the tap water in many cities frequently wins out over bottled water, suggesting the role that branding and packaging may play in perceptions of taste.

Third, the promotion of bottled water as a means of attaining (or regaining) health, fitness, vitality, and youth leverages a long and deep tradition in advertising and marketing. "To sell bottled water as a lifestyle choice," sociologist Andrew Szasz writes, "advertising associated it with images of desirable social identities, being young, fit, attractive, and active."[24] Consumer culture scholar Kane Race observes that this "aspirational attachment to health and fitness" is evident in bottled water advertising, where "the sporting body [is depicted] as the absolutely generic body of the consumer public."[25] Industry marketing also tapped into broader societal concerns about health in the 1990s and 2000s—

including the role of soft drinks in growing obesity rates. "Bottled water was given [another] cultural push," writes Holt, "by a media-generated moral panic against sugar, specifically high-fructose corn syrup."[26]

Fourth, beginning in the mid-1990s the bottled water industry began promoting its product as a solution to an ailment that had only recently been "discovered": a chronic lack of hydration, for which drinking eight glasses of water per day was alleged to be an essential remedy. Race writes that while "no reputable evidence justifying the 8 × 8 rule can be found" for sedentary human bodies,[27] this myth "is a conspicuous feature of efforts to market bottled water, allowing companies to appeal to scientifically framed principles and ideas of health . . . [including] new practices of water drinking and consumption in which the consumer appears to be always at risk of dehydration."[28] A culture of "frequent sipping," carrying bottled water at all times, was one of the practices to emerge from this hydration discourse.[29]

It is easy to chalk up the spread of bottled water consumption to a simple tale of manufactured need. However, its growth reveals a more complicated interplay between political and social trends, opportunistic marketing, and cultural dynamics. "Yes, the corporations waged a prolonged PR offensive to change attitudes to tap water," writes commentator Jeff Sparrow. "But . . . the marketing campaign worked because it connected with a basic human need for self-expression."[30] This yearning for individual choice is precisely the desire that neoliberalism has harnessed so effectively overall to gain public acceptance for policies that have increased inequality and dismantled protective welfare states.[31]

Branding further enables beverage firms to contrast their product with public tap water. As cultural studies scholar Gay Hawkins observes, "Branding has had the effect of implicitly undermining tap water by making it appear inferior or by generating doubt about where it has actually come from and whether it is safe or 'pure.'"[32]

The packaging itself also serves to undermine trust in public tap water. Szasz writes that "the most powerful signifier of all [is] the bottle itself. This water is *bottled*. . . . It *must* be cleaner, purer, better than water that has *not* been thus singled out, the mundane kind of water that just runs out of the kitchen faucet. It must have been cleaner in the first place; otherwise, why would they have thought it special enough to capture it, separate it, *give it a name?*" (emphasis in original).[33]

Some of the most interesting writing on bottled water explores the cultural shifts enabling the normalization of this commodity. Hawkins argues that "the [bottled water] consumer is not reducible to a gullible fool or passive victim of corporate hype and manipulation but [rather] a subject for whom bottled water has come to make sense as a rational drinking practice—the issue is how did this reasoning emerge and what forms did it take?"[34]

According to Holt, bottled water exemplifies a dynamic he terms "ideological lock-in," in which beliefs become part of everyday consumption practices that further reinforce those beliefs. "Americans drink bottled water," he writes, "because they believe that they are healthier for so doing, have developed routines around this belief, and circulate in a society that continually reinforces this ideology. . . . As long as these cultural mechanics are in place, the public's perceptions of tap water are not going to change."[35] These dynamics include the creation of new norms of hospitality, in which hosts may be expected to offer bottled water to guests and tap water is viewed as unacceptable. For critics, this indicates the necessity of challenging bottled water consumption on the terrain of culture, not merely politics.

Despite the ready availability of safe tap water for most people in the global North, the move to bottled water includes another cultural cost: hauling large volumes of heavy water has once again become normalized.[36] While the issue has not been well studied, it appears that this added labor is at least somewhat gendered, as is the labor of food shopping generally.[37] In contrast, for those who turn to bottled water because their tap water service has been shut off or is genuinely unsafe to drink, hauling bottled water is not a perplexing choice but an unjust imposition.

FADING FOUNTAINS

One final cultural change, which both is enabled by bottled water's availability and contributes to its spread, is the demise of the public drinking fountain. I first noticed this trend in one of my treks through the endless terminals of Chicago O'Hare Airport in the 1990s, changing planes on the way home to Wisconsin, looking for somewhere to top up my refillable bottle. Suddenly, it seemed all of the water fountains were either out of service or missing entirely, but three-dollar bottles of Aquafina and Dasani abounded at every newsstand, restaurant, and vending machine. I ended up awkwardly half-filling my bottle under the bathroom tap. Airline passengers might be the ultimate captive audience, but the results are similar just about everywhere.

As working water fountains disappear from public spaces, they are also vanishing from the official expectations of those public spaces. Several authors tell the story of the University of Central Florida's stadium, built without a single water fountain, which on a hot day in 2007 led sixty fans to be treated for heat stroke and sent eighteen to the hospital, after concession stands ran out of bottled water.[38] Some Canadian provinces do not require new public buildings to include any water fountains at all, and some universities—which have signed beverage exclusivity contracts with Coke or Pepsi—have built new facilities without them.[39] In 2015, a revision of the International Plumbing Code, which guides builders and architects in the United States and many other countries, cut in half the number of water fountains required.[40]

The result of these developments, writes Sparrow, is a vicious cycle: "As free drinking water becomes harder to find, bottled water becomes more and more popular, thus reducing the pressure on authorities to care for fountains and the like, until the dire state of the facilities makes buying water almost mandatory. An ideological shift takes place as well, with fountains perceived as unsanitary and gross, and bottled water a signifier of health and vitality."[41]

This is highly ironic, given the history of the public drinking fountain. "The modern era's first free public water fountain was unveiled in London in 1859," writes journalist Kendra Pierre-Louis. ". . . The poor were drinking water bottled from the sewage-infested Thames. Water-borne diseases such as cholera and typhoid were rampant. . . . By 1879, London had 800 fountains. American cities followed suit. . . . By 1920, most municipalities were providing free, chlorinated water. The public health benefits were obvious."[42] Public water fountains represent the one truly shared point in a municipal drinking water network. Although our home faucets are connected to that network, they are individualized. The drinking fountain also symbolizes a societal commitment to protecting public health. The loss of this shared public resource is more profound than might initially appear. It not only obligates people to seek privatized sources of water but also works to erase our collective memory of what existed only a few years earlier, limiting our ability to envision alternatives.

MAKING A MARKET IN THE GLOBAL SOUTH

These dynamics, however, are substantially different in places where household tap water coverage is not widespread, where the public water supply is not reliably safe to drink or does not flow consistently, or where drinking fountains may never have existed. Here the problems are more complex, as there may not be a readily available alternative to packaged water for many people. The interaction between public water supplies and packaged water in these settings highlights stark inequalities between the water haves and have-nots, raising profound issues of social injustice.

The bottled water industry promotes packaged water as a solution not only to the unavailability of potable tap water in so-called developing nations but also to poor tap water quality. BMC director John Rodwan Jr. writes that "bottled water serves as at least a partial solution to the problem of often-unsafe water found in many economically developing countries."[43] While that might be accurate as a description of the current state of affairs in many places, critics view this status quo as unacceptable. Canadian water activist Richard Girard contends that the industry is profiting "from human suffering by exploiting the inability of municipalities, governments and institutions to find the correct and most sustainable way of delivering water services that is safe and managed publicly."[44]

Especially in large cities and peri-urban areas in the global South, rapid urbanization has often outpaced the expansion of water infrastructure. Many fiscally strapped governments have been unable or unwilling to dedicate sufficient resources to water system maintenance and expansion. The packaged water industry is happy to fill this gap, and sales have grown at a rapid clip in the past two decades, especially in Asia but also in parts of the Middle East, Latin America, and Africa.[45] The Big Four have expanded into the South, buying up local bottlers,[46] but domestic firms still hold a significant share of the market in most places. In several countries, between half and 80 percent of the population now relies on packaged water as their primary source of drinking water.[47]

Packaged water consumption across the South is highly classed. Wealthy and upper-middle-income households typically consume costly branded water, often owned by one of the Big Four. Much of this water comes in large plastic jugs, frequently delivered to homes, but consumption of single-serving bottles has also grown rapidly. In contrast, most lower-middle- and low-income families turn to the "refill water" or "microtreatment" segment of the market: hundreds of thousands of small-scale businesses that draw water either from the public supply or from local rivers, springs, or wells and filter it using reverse osmosis, UV light, or other technologies. Consumers bring in empty plastic jugs and watch while the containers are sterilized and refilled, or get delivery by bicycle, motorbike, or truck (figures 4–5). This water typically sells for one-quarter to one-half the cost of branded water.[48] These businesses are usually informal, often unregulated and unregistered, and the safety

FIGURE 4. Microtreatment business (*purificadora*), San Cristóbal de las Casas, Chiapas, Mexico. Photo: Author.

and quality of the water they sell are highly variable, yet many families still consider it safer than other affordable options. In West Africa and some other regions, disposable plastic bags of water called sachets dominate this lower-income niche instead.

However, there is substantial variation between and within nations. It is instructive to examine two countries with very high packaged water consumption: Indonesia and Mexico, which sit in fourth and third place respectively in total consumption volume, behind only China and the United States (see table 2 later in the chapter). The commonalities and differences between these two settings illustrate the interaction between colonial legacies, neoliberal austerity, state (in)capacity, social inequality, access to and trust in tap water, and the growth of packaged water.

Indonesia has the tenth-highest bottled water consumption on a per-person basis, but sales are growing so fast that it will soon be in the top five. Here, write Teddy Prasetiawan and coauthors, "drinking tap water has never been accepted as the norm" because of the country's colonial history, in which "piped water . . . networks [were not] planned, designed, and built for the majority" but rather for "the white European population

FIGURE 5. Customer at *purificadora,* San Cristóbal de las Casas, Chiapas, Mexico. Photo: Author.

. . . and local elites, while people of lower status relied on traditional water sources such as shallow wells and surface water for domestic purposes."[49]

Expansion of the water system since independence has been extremely slow, with only 10.2 percent of households served by piped water in 2015. In the capital, Jakarta, between one-fourth and one-third of people are connected to the grid.[50] Yet even for those fortunate enough to be connected, the water, while originally treated to a safe standard, is often contaminated as it passes through the system or while in home storage. An added complication is inconsistent flow resulting from rationing of piped water by municipal authorities.

Packaged water increasingly fills this gap. Almost 90 percent of households buy refillable nineteen-liter jugs (called *galon*), and 80 percent buy single-serving disposable bottled water.[51] After experiencing massive growth following the Asian financial crisis of 1997,[52] refill water now constitutes more than half of Indonesia's packaged water consumption.[53] In contrast, affluent young urban residents are the target market for branded bottled water such as Danone's Aqua and VIT brands, which constitute one-third of all sales.[54]

In Jakarta, tap water (costing on average only US$0.38 per cubic meter) is far cheaper than refill water (US$18.33 per cubic meter) or branded bottled water (US$56.16). However, according to political ecologist Carolin Tina Walter and colleagues, residents view packaged water as the cheapest drinking water source because of the cost of the gas needed to boil tap water to make it safe to drink. They write that although most of the population spends under 5 percent of household income to buy packaged water, one-fourth of the poorest Jakarta residents still cannot afford even low-cost refill water.[55]

Mexico illustrates many of these same dynamics but diverges in important ways. Although its per capita bottled water consumption has been the world's highest for more than a decade (over 74 gallons or 281 liters in 2020), access to tap water is widespread. Official statistics state that 94 percent of the population has access to an improved water source, and in-home connections are ubiquitous, although coverage is lower in poor and informal neighborhoods.[56] These networks were largely built and extended through massive government investment, especially between the 1940s and 1970s. In most Mexican cities, drinking from the tap was commonplace as recently as the 1980s, and in a few northern cities it is still the norm.[57]

How, then, did Mexico become the global epicenter of packaged water consumption? Joshua Greene identifies four causes of the appearance and dramatic growth of bottled water: the severe economic crisis beginning in 1982, in which Mexico was forced to submit to World Bank–imposed austerity and cut its social and infrastructure spending drastically; the 1985 Mexico City earthquake, which destroyed much of the city's water system, causing contamination and leading the state to recommend boiling all drinking water; the nationwide cholera epidemic of the 1990s, during which authorities advised drinking bottled water; and the actions and growing political influence of the bottled water industry.[58]

The Mexican state implemented a strongly neoliberal national water law in 1994 that promoted privatization, eased industries' access to vast quantities of free groundwater, and shunted responsibility for funding and maintaining drinking water systems to municipal governments, but without giving them the fiscal resources to do so.[59] President Vicente Fox (2000–2006), the former CEO of Coca-Cola FEMSA,[60] further promoted the role of private capital in the water sector and presided over the period of fastest growth in the bottled water market. The budget for the federal water agency Conagua has been cut significantly, falling to just over US$1 billion in 2020, a figure dwarfed by Mexicans' annual spending on packaged water at US$14.8 billion.[61] According to journalist J. J. Lemus, 91 percent of Mexican water utilities are in a state of budget deficit, unable to invest in even basic system maintenance.[62]

As a result, in many parts of the country water service is inconsistent and water pressure often low. In Mexico City, although the water is largely treated to be safe to drink, contamination is frequently introduced during storage at home, while in many smaller cities the treatment systems cannot fully remove bacteria and other pathogens.[63] The biggest problem, however, is rationed water service or *tandeo,* which is the norm across the country. According to a report by Leo Heller, UN Special Rapporteur on the Right to Water, 70 percent of Mexico City residents receive water service less than twelve hours per day, while in poorer neighborhoods and smaller cities service can be far less frequent or consistent, often just a few hours per day or a few days each week.[64] Some communities report going entirely without water service for weeks or months at a time.[65] Those who can afford to do so install rooftop storage tanks or cisterns.

This perfect storm of disinvestment, poor water quality, and inconsistent service has resulted in widespread loss of public trust in tap water, which according to Greene is "so entrenched . . . that even when improvements are made, utilities are unable to convince consumers that the water is safe."[66] The bottled water industry has stepped in with abundant "information" to inspire trust in its product.[67] Danone's Bonafont and Levité and Coke's Ciel are Mexico's top-selling brands, backed by massive advertising campaigns. Today, only 11 percent of Mexico City residents report drinking any water from the tap, while an estimated 90 percent (and 80 percent of all Mexican households) regularly consume packaged water, two-thirds of it in the form of twenty-liter refillable plastic or glass *garrafones*.[68]

The economic impact of this dependency on packaged water can be severe, with "households that pay water rates for water from the public utility pay[ing] several times that amount more each month for bottled water or tanker delivery."[69] Water scholar Delia Montero found that in Mexico City the poorest fifth of households spend an average of 15 percent of household income on packaged water; adding on the tap water bill means that over 25 percent of their total income is spent on acquiring water. Two-fifths of families in the city dedicate at least 13 percent of their household income to pay for water.[70] Inconsistent supply aggravates this injustice: "In cases where water shortages leave families utilizing bottled water for all household water needs, including bathing, laundry, mopping, and cooking, the costs quickly become exorbitant and reproduce existing inequalities."[71]

Unlike Indonesia, then, Mexico represents a situation in which tap water *does* reach the vast majority of the population, yet residents depend almost entirely on packaged water to drink. Although the initial causes for bottled water's appearance and growth were a natural disaster and an epidemic, those events translated into long-term degradation of water systems and distrust in public tap water because Mexico was subjected to severe fiscal austerity, unable to restore the state's role in guaranteeing clean water provision. In this context, government and the public have come to accept or at least tolerate packaged water as a de facto permanent replacement for formerly potable public water supplies.[72] Yet even more than in Indonesia, buying packaged water constitutes a heavy financial burden for a major swath of the population.

However, a major nationwide civil society movement, under the coalition Agua Para Todos (Water for All), has been pushing for over a decade to restore governmental capacity to maintain the water system, replace the national water law, reverse privatization, and remedy the crises of drinking water access and injustice.[73] "The water bottlers have political power," Elena Burns, the coalition's cofounder, tells me in late 2019. "That's where we have to convert the needs of everyday people into an equivalent or greater political power so that we can get this [new water] law passed . . . in order to guarantee equal access to quality water. And that includes water fountains and other public sources of potable water."[74]

So, does packaged and bottled water have any legitimate role to play in addressing the lack of widespread access to safe drinking water in the global South? This question has made its way into debates over how to reach international targets for universal drinking water access, embedded within the UN's Sustainable Development Goals (SDGs). In 2017, an agency of the United Nations for the first time included packaged water as an "improved water source" for the purposes of achieving SDG 6.1, "Universal and equitable access to safe and affordable drinking water for all."[75] This controversial change means that states can now count the availability of packaged water as progress toward their obligations to meet this goal.

The World Health Organization (WHO) and UNICEF assert that this redefinition barely changes existing statistics on countries' progress toward the SDG.[76] It could be argued that the change merely acknowledges the existing reality on the ground in many places, where packaged water is already providing the only potable water. However, critics have assailed the decision as granting governments an escape clause. Greene argues that "this new reclassification may prove key in allowing [Mexico] to achieve the Target 6.1, without investing public funds in upgrading, maintaining, or expanding water infrastructure." He adds that "the model is advantageous to policymakers and corporate actors and investors as it proves to be a highly profitable way to provide for a vital human need while eliminating the role of the state."[77] From within the UN itself, the Special Rapporteur's report on Mexico insists that "dependence on bottled water is definitely not a way to meet the country's obligations relating to the right to water, as it undermines the necessary accessibility and affordability of water provision."[78]

A MOMENT OF CHANGE

Before looking closer at the relationship between bottled water and tap water, it is important to get a handle on where the commodity of packaged water stands today. In broad strokes, there is great variation between nations and world regions. Packaged water consumption is growing at lightning speed in parts of Asia, and rapidly in the Middle East and elsewhere in the global South, but it has expanded more slowly in the North and stagnated in western Europe.

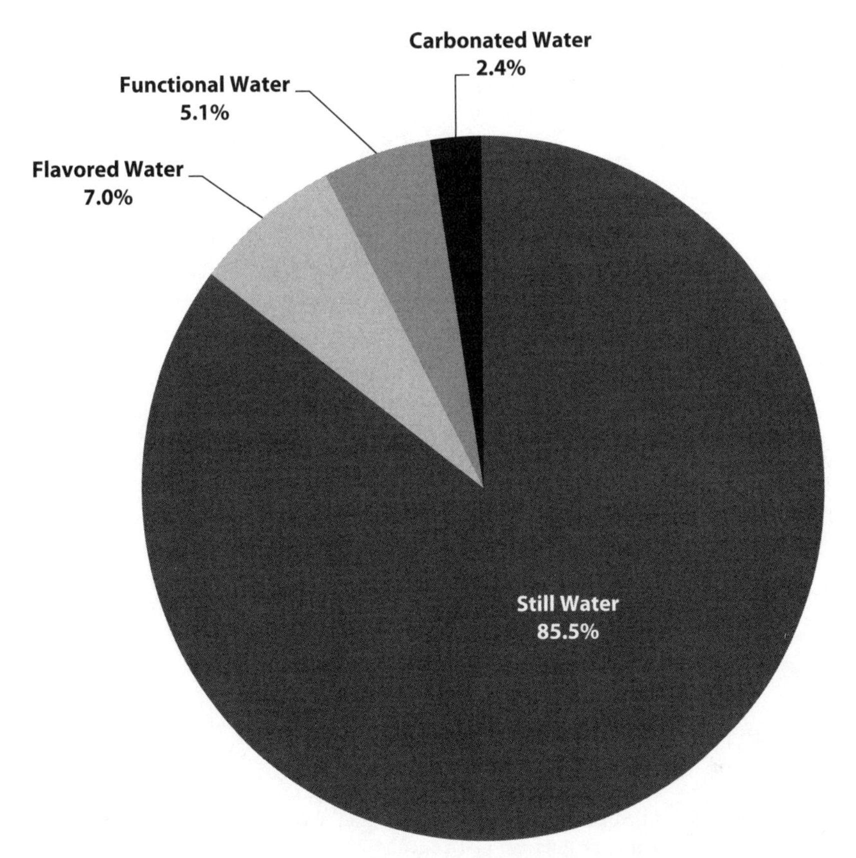

FIGURE 6. U.S. bottled water market: category share by volume, 2022. Source: Adapted from Euromonitor International 2022a.

In the United States, demand for bottled water continues to increase, with sales surpassing $40 billion in 2021.[79] Even before the Covid pandemic, however, this growth was starting to slow. After rising an average of 6.8 percent annually from 2014 to 2017, bottled water consumption grew by only 3.5 percent per year on average from 2017 to 2022.[80]

These aggregate numbers mask big divergences between different types of water. The main trends can be summed up this way: U.S. consumers are slowly shifting from drinking plain or "still" bottled water toward sparkling water, flavored water, and so-called functional waters containing additives like caffeine, electrolytes, protein, or probiotics that make claims of improved mental acuity, memory, mood, and more. However, "still" bottled water remains dominant, representing over 85 percent of the market by volume, as figure 6 shows, and 67 percent by

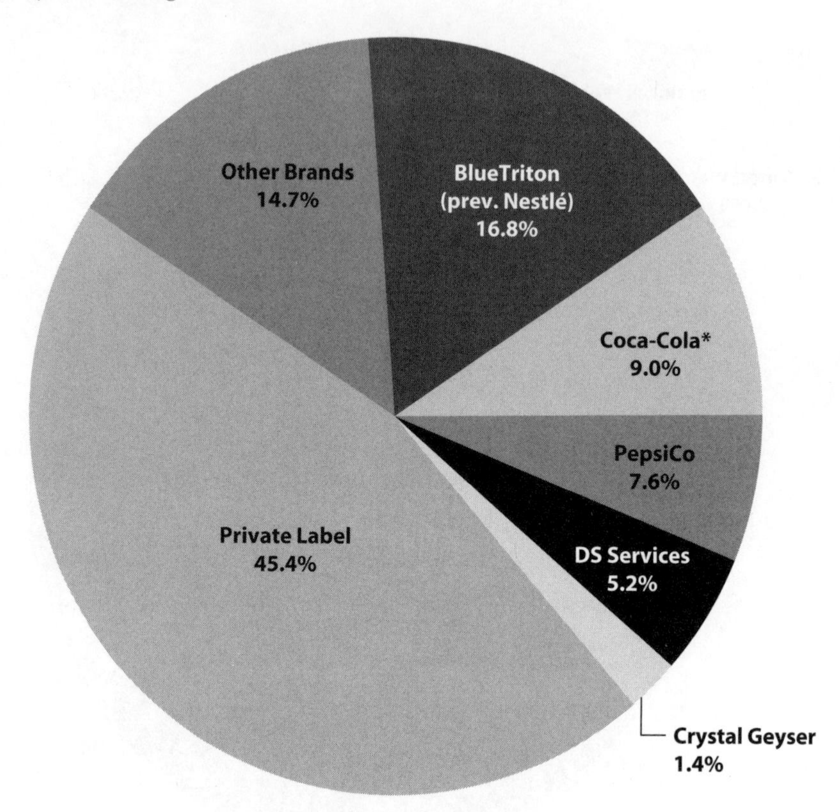

FIGURE 7. U.S. bottled water market: company share by volume, 2022. Source: Adapted from Euromonitor International 2022a. *Coca-Cola distributes Danone's water brands in North America. Includes subsidiary Energy Brands.

value.[81] Cheaper private-label (store brand) waters are growing fastest, stealing market share from the Big Four. The Big Four brands containing filtered tap water, and the spring water labels owned by Nestlé until 2021, have largely stagnated.[82] Figure 7 shows the U.S. market shares of the major bottled water makers in 2022.

The type of packaging the water comes in has also shifted. Over two decades there has been a move away from large returnable water-cooler style jugs (now less than 10 percent of sales) toward single-serving, single-use PET plastic bottles, which soared from only 24 percent of the market in 1999 to over 70 percent today.[83]

Canada's bottled water market has some notable differences from its southern neighbor. The average Canadian per-person packaged water

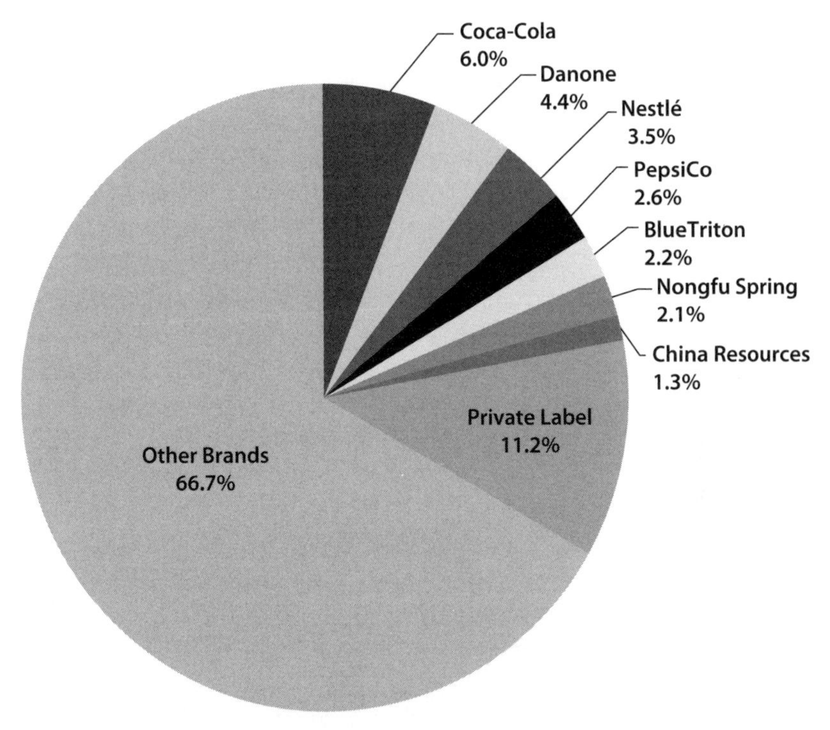

FIGURE 8. Global bottled water market: company share by volume, 2022. Source: Adapted from Euromonitor International 2022b.

consumption in 2020 was less than half that of the United States, at only 19.2 gallons (72.7 liters), and this figure has been stagnant for several years while U.S. per capita consumption continues to rise.[84] The Canadian market is dominated not by the Big Four but rather by the "Big One" plus several national brands, with Nestlé (now BlueTriton) representing nearly 31 percent of all sales.

Globally the packaged water industry is far less concentrated than in North America, but the Big Four still control the largest share, with Coca-Cola topping the list in 2022, as figure 8 illustrates. Two Chinese firms, Nongfu Springs and China Resources, round out the top seven. Table 2 shows the top ten consuming countries in terms of total volume and average growth rates. India and China stand out with annual growth over 6 percent, followed by the United States, Mexico, and Indonesia. When it comes to per capita consumption of packaged water, however, China does not even break into the top twenty. As table 3

TABLE 2 TOP TEN NATIONS IN TOTAL BOTTLED WATER CONSUMPTION, 2020, AND GROWTH RATE, 2015–20

	Total volume, 2020 (billions of gallons)	Compound annual growth rate, 2015–20
China	27.78	6.3%
U.S.	14.96	5.4%
Mexico	9.96	4.3%
Indonesia	8.51	3.7%
Brazil	6.46	3.8%
India	6.42	6.9%
Thailand	3.96	1.8%
Italy	3.48	1.0%
Germany	2.75	–1.5%
France	2.24	1.5%

SOURCE: Adapted from Rodwan 2021.

TABLE 3 TOP NATIONS IN PER CAPITA BOTTLED WATER CONSUMPTION, 2012 AND 2020, AND CONSUMPTION GROWTH

	Gallons per capita		Compound annual growth rate, 2012–20
	2012	2020	
Mexico	62.2	74.4	2.3%
Italy	47.7	58.8	2.7%
Thailand	46.9	57.0	2.5%
U.S.	30.9	45.2	4.9%
U.A.E.	25.3	35.4	4.3%
Spain	30.9	35.3	1.7%
France	35.8	34.1	- 0.6%
Belgium/Luxembourg	34.6	34.0	- 0.2%
Germany	36.6	33.3	- 1.2%
El Salvador	(N.A.)	33.1	—
Saudi Arabia	26.4	31.6	2.3%
Indonesia	20.1	31.3	5.7%

SOURCE: Adapted from Rodwan 2018, 2021.

illustrates, Mexico remains the biggest per capita consumer, followed by Italy, Thailand, and the United States. Western Europe, with a longer history of imbibing spring waters, still consumes large amounts per person, but its total consumption is barely growing, and in Germany, France, and Belgium it is actually falling.

FEARING THE TAP

In *Merchants of Doubt,* the book (and film) by Naomi Oreskes and Erik Conway, the authors trace the tactics used by the fossil fuel industry to dispute the scientific consensus on human-caused climate change back to strategies originally developed by the tobacco industry. They quote a 1969 internal memo from an executive at the leading tobacco firm Brown & Williamson: "Doubt is our product since it is the best means of competing with the 'body of fact' that exists in the mind of the general public. It is also the means of establishing a controversy."[85]

While the cigarette makers protected their sales and fended off lawsuits for decades by sowing doubt about the health harms of their product, doubt can be deployed in many ways. The bottled water industry has both subtly and overtly spread doubt about the safety of public tap water, positioning its product as the safer alternative. In what some critics have called a "war on tap water," individual bottled water firms and industry groups have impugned the quality of public drinking water.[86] They have been aided enormously in spreading doubt by media coverage of infrequent but high-profile local drinking water emergencies. In the process, critics claim, we have been (and are being) taught to fear the tap. This is the fifth "F." "Fear is an effective tool," writes water expert Peter Gleick in the book *Bottled and Sold.* "If we can be made to fear our tap water, the market for bottled water skyrockets."[87] Many people, of course, do not need to find incontrovertible evidence of health risk in order to lose confidence in their drinking water supply; mere doubt about the trustworthiness of tap water is sufficient.

In one of the most widely reported examples of this strategy, Fiji Water ran a full-page advertisement in national magazines in 2006, with the caption "The Label Says Fiji Because It's Not Bottled in Cleveland." Subsequent tests, however, showed that Fiji Water contained volatile plastic compounds, high levels of bacteria, and arsenic, while Cleveland's public water supply was far cleaner. Royal Spring Water ran an advertisement that proclaimed, "Tap water is poison."[88] And a current online advertisement for Primo Water depicts a corroded pipe and warns consumers, "Your tap water could be full of harmful bacteria and contaminants."[89]

Anthropologist Andy Opel describes this approach as "the two-pronged strategy of the bottled water industry: establish the purity of their sources while raising the fears of contaminated public drinking waters."[90] A bottled water firm representative I interviewed illustrated this rhetorical stance:

> If I stopped off in one of these truck stops where everybody stops to go in and get some candy or food or something . . . and my only choice is to go into the men's room and get water out of the tap there, I'm not going to do that. I have no idea who cleans that, I have no idea how sanitary it is. . . . And that's the only source of tap water that's available. So I would want to have that alternative to buy a bottled water that I know. The product that they are paying for is not just for the water, but the sanitary conditions that the water is packaged in and everything that goes into the protection of that product.[91]

Industry group representatives have tended to couch their criticisms of tap water in somewhat more subtle terms. International Bottled Water Association (IBWA) spokesperson Stephen Kay stated in 2001 that "the difference between bottled water and tap water is that bottled water's quality is consistent,"[92] even though in the United States and elsewhere it is far less strictly regulated than municipal water, as I discuss below. In 2019, the BMC's Rodwan wrote, "Even where tap water *may be safe,* people may prefer bottled water, which they deem as better tasting" (emphasis added), a statement that conflates the very different issues of taste and safety.[93] Similarly, a 2021 industry market report advises that "positioning bottled water as the safest water option [is] necessary for keeping consumers engaged."[94]

Do these efforts cast doubt on the quality of tap water work? In a word, yes. "Bottled water's success is partly the result of a decades-long campaign against tap water," writes Jennifer Kaplan.[95] Since the 1970s, opinion polls have shown a major decline in public trust of tap water quality in the United States. The proportion of Americans reporting they were extremely concerned about drinking water pollution rose from 32 percent in 1973 to 66 percent in 1988,[96] and it has since risen as high as 80 percent.[97] Ironically, this soaring concern about tap water safety is partly a product of the modern environmental movement and the policy gains it achieved. According to Pierre-Louis, "In response to activist pressure, the government drafted measures like 1974's Safe Drinking Water Act. The legislation made [tap] water much safer . . . But it had an unintended consequence: because municipalities had to notify residents of contamination immediately, Americans who had grown up trusting tap water were now getting bombarded with warnings of possible risks."[98] Holt, in contrast, attributes growing fear of tap water to a broader loss of trust in government, fed by news coverage of corruption or regulatory failure.[99]

Whatever the causes, studies find a strong correlation between people's perceptions of poor tap water quality and higher bottled water consumption.[100] Whether these perceptions correspond to the actual safety of the tap water is another question. According to water researchers Ariana Javidi and Gregory Pierce, "Perception of tap water as unsafe has at least three potential sources: health-related contamination which should incur a violation of the primary standards of the Safe Drinking Water Act; non-health-related but often more perceivable contamination due to the contaminants' sensory qualities . . . [;] or pure misperception."[101]

Regardless of the reasons, people in North America and around the world are increasingly learning to fear their tap water. And fear, of course, is the antithesis of trust. But why has the campaign against tap water proven so effective? In the book *Shopping Our Way to Safety*, Szasz describes bottled water as a prime example of a phenomenon he terms "inverted quarantine": the pursuit of individual protection from perceived environmental risk through the consumption of products believed to be safer. This response involves efforts to construct a protective "personal commodity bubble," as opposed to pursuing policy or structural change. "Rather than do something politically or collectively to improve the public water supply," he writes, ". . . [people] try, individually, to assure themselves a supply of water that they think is safer to drink."[102] Szasz describes inverted quarantine as the antithesis of a social movement, which pursues collective solutions to societal problems. This response diminishes the intensity of public demands for policy change because it produces a "false sense of security, undercutting political support for reform."[103] The growth of such consumption-based strategies can be seen as an artifact of neoliberal ideology that emphasizes individual and market-based, rather than collective, approaches to social problems.[104] In a vicious cycle, argues Szasz, the embrace of bottled water reduces pressure on government to maintain public water infrastructure, which can imperil tap water quality, further weakening trust and accelerating the shift to the private solution of bottled water.[105]

WHO'S THE COMPETITION?

The packaged water industry vociferously opposes the idea that consumers' choices between tap water and bottled water have larger political implications, or that they represent a zero-sum game. On the contrary, it insists that demand for its product has grown entirely because consumers are shifting away from other less-healthy drinks. The BMC's director writes that "although it has at times been characterized as competing with tap water, bottled water achieved its position atop the beverage rankings by enticing consumers away from other packaged beverages."[106]

This portrayal of bottled water as merely one of an interchangeable set of beverages is also strongly promoted by individual beverage firms. For example, the bottled water firm representative told me that "if bottled water disappeared today, people would not be turning to tap water. People would be turning to all the other packaged beverages that we're in competition with right now. Those with sugar, alcohol, caffeine, and artificial flavors and sweeteners. So we firmly believe that there is a valuable place in society for our product, and we essentially want to be treated equally with all other packaged beverages."

This argument—that bottled water is in competition with other beverages but not with public tap water—is one worth assessing seriously. Let's review the available evidence. To begin with, consumers in the global North are drinking far fewer sugary soft drinks and more bottled water, as figure 1 in the Introduction shows. From 2010 to 2020, the volume of soda consumed in the United States shrank by over 2 percent per year, while bottled water volume grew at a 5.7 percent average annual clip.[107] On a per-person basis, soft drink consumption fell by more than nine gallons per year in the same period, while bottled water consumption rose by over seventeen gallons.[108]

But are the changing fortunes of soda and bottled water directly linked? When a 2019 market survey asked consumers who said they were drinking more bottled water than in the past why they did so, 54 percent responded that they were "drinking less of other beverages."[109] It appears many consumers are indeed switching from buying soda to buying water in single-use bottles.

However, that does not address whether these shifts are *also* hurting tap water consumption. In a press release, the bottled water industry trade association, the IBWA, touts a U.S. Harris poll it commissioned as proof that its product does not compete with tap water:

> If plain bottled water is not available, 74 percent of people *who identify bottled water as among their most preferred beverages* said they would choose another packaged drink: soda (19 percent), coffee (9 percent), sparkling bottled water (7 percent), tea (7 percent), juice/fruit drinks (7 percent), sports drink (6 percent), flavored or sweetened sparkling or still bottled water (5 percent), functional water (5 percent). . . . Among the remaining 26 percent,

> 1 percent said they would stay thirsty, half (12 percent) would drink filtered tap water, 7 percent would drink from a public water fountain, while 5 percent would drink unfiltered tap water.[110](emphasis added)

Because nearly 25 percent of even this bottled water-loving group of respondents stated that if bottled water were not available they would instead drink tap water in some form, it is disingenuous to claim that bottled water does not compete with tap water for what the industry terms "share of stomach."

It is true that from the beverage industry's perspective, bottled water really is "just another beverage" and is packaged and sold as such. Dasani and Aquafina, for example, are distributed by Coke and Pepsi via the same distribution networks as their sodas, teas, and energy drinks—often on the same trucks—and they share space in these firms' branded store coolers alongside their calorie- and additive-heavy brethren. To these multinational firms, bottled water has no special status; it is merely the most successful of a wide range of beverage commodities jostling for market share. In other words, it is just another flavor.

What about data on changes in how much tap water we actually drink? Gleick writes that between 1980 to 2006, as annual bottled water consumption in the United States rose by twenty-five gallons per capita, consumption of tap water fell by a whopping thirty-five gallons per person.[111] More recently, at a 2018 packaging conference, the BMC's managing director of research gave a presentation—quoted in the epigraph to this chapter—with a slide showing that U.S. tap water consumption fell every year between 2001 and 2017, with the exception of five years during the Great Recession when it partly rebounded (meanwhile, bottled water sales grew every year except for 2008 and 2009). Over that sixteen-year period, per capita tap water consumption fell by nearly 11 percent, on top of the previous losses.[112] Figure 9 depicts the annual changes in bottled water and tap water consumption in the United States.

Statements from top beverage industry executives add further context. Robert Morrison, the chairman of PepsiCo's beverage division, told an interviewer in 2000, "The biggest enemy is tap water. We're not against [tap] water—it just has its place. We think it's good for irrigation and cooking." Nor is this the only such declaration. Shortly before her firm was acquired by PepsiCo, Susan Wellington, the president of Quaker Oats' U.S. beverage division (maker of Gatorade), told industry analysts, "When we're done, tap water will be relegated to showers and washing dishes."[113]

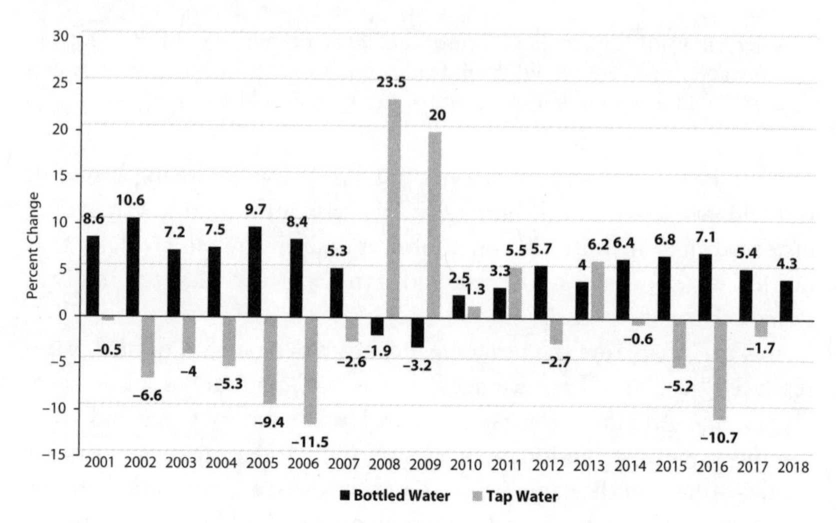

FIGURE 9. Percent change in U.S. per capita consumption of bottled water and tap water, 2001–2018. Sources: Hemphill 2018; Rodwan 2019, 2011.

Recent market reports illustrate just how successful the industry believes it has been at achieving those aims. According to a 2019 survey by Mintel, 85 percent of U.S. adults had consumed packaged water at some point in the past three months, while only 62 percent had drunk *any* tap water, down from 67 percent in 2017.[114] The IBWA claims that bottled water accounted for fully 25 percent of "volume share of stomach" in the United States in 2021 (up from 23.4 percent two years earlier), compared to only 9.4 percent for "tap [water] and others" (down from 11.2 percent).[115] While these IBWA figures do not square with the peer-reviewed studies described throughout this chapter, they do suggest that rivalry with tap water is central to the industry's paradigm.

Let's attempt to settle the dispute over whether bottled water competes directly with tap water or mainly with soda and other beverages. To some extent this *is* a zero-sum game, as the industry claims: bottled water sales have clearly benefited from the dramatic reversal of soda's fortunes. Many consumers have indeed switched directly from buying plastic bottles of sugary soft drinks to plastic bottles of water.

However, bottled water *also* unquestionably competes against tap water, whose consumption has fallen every year except during the Great Recession. The "war on tap water" is real—both as a marketing strategy and as a description of the actual effect of packaged water's growth on our consumption of tap water. As the statistics above show, people are drinking far less public water from taps and water fountains than two decades ago. Among some demographic groups in the United States, bottled water already constitutes more than half of total drinking water volume.[116]

The bottled water industry's argument to the public—that its dramatic growth has not come at the expense of the tap—quite simply does not hold water. This claim is belied not only by the data on declining tap water consumption but also by the industry's decades-long campaign against tap water, illustrated in advertising, public statements, and its own internal discourse. When industry representatives are speaking to one another, as in the quote in the epigraph to this chapter, they are remarkably candid: falling tap water consumption is essential to their profitability. Packaged water's gain is not only soda's loss; it has also weaned hundreds of millions of consumers off the habit of drinking from the tap, in part by teaching us to distrust our tap water. If tap water is the enemy, the war against it has been proceeding quite well indeed.

AN UNFAIR FIGHT

What does this war on the tap mean for our public tap water systems? To answer this, we need to consider how bottled water interrupts the relationship between residents and their local, overwhelmingly public, water utilities. There is a fundamental distinction between beverage firms and the public sector in providing drinking water, which involves the undifferentiated, bulk nature of tap water supply versus the wide (albeit cosmetic) variety of bottled water brands and packaging.[117] The bottle itself further weakens public water provision, argues Hawkins, "by differentiating the water it contains from the tap. In this way, responsibility for safe drinking water is transferred from urban authorities to the bottle, the individual and their choices."[118] However, as the Flint water disaster and the Covid pandemic make painfully clear, the idea that we no longer need public water infrastructure is ludicrous, even if many people do avoid drinking from the tap.

Another factor is how the packaged water industry relates to drinking water crises. When cash-strapped cities with aging pipes experience lead contamination or other tap water emergencies, this is a tragedy for the often low-income residents who cannot drink their tap water. However, "Beverage companies see a state failure as a market opportunity," opines Hawkins.[119] In a context where distrust of the tap has already been sown by advertising and the normalization of bottled water, the crises in places like Flint land in a receptive medium; they work to confirm what people may already feel or think about the safety of tap water. In this way, the war on tap water is effectively a war on public water utilities.

Faced with such a threat, why don't those public utilities fight back? Some have—for instance, Cleveland's water department pushed back against the Fiji Water campaign that impugned its tap water. Yet in this competition, most public water utilities are vastly outmatched. Metaphorically, they find themselves facing down a super soaker armed with a small glass of water. "For the most part," write Snitow and colleagues, "public water agencies . . . don't have significant advertising budgets, and they are required to stay out of politics. Private companies, however, are aggressive in shaping public opinion and influencing elected officials. They employ teams of publicists, lawyers, and political-campaign consultants as part of a concerted effort to expand and maintain their control of the water market."[120]

Most water utilities in the global North (including those that reach 87 percent of U.S. households and over 90 percent of Canadians) are nonprofit, public agencies under local governments, staffed by highly trained civil servants, biologists, chemists, engineers, and administrators, among others. Their imperatives are to protect public health, maintain ultra-high-quality water, ensure consistent delivery, and keep rates reasonable. Their job descriptions do not include profit maximization, and they are not typically trained to actively market their usually excellent "product" or assertively defend it in the media.

There is also the prospect of legal action. Gleick writes that after Miami–Dade County, Florida, ran radio advertisements in 2008 promoting the county's tap water as cleaner, cheaper, and safer than bottled water, the nation's top bottled water firm threatened a lawsuit: "A law firm representing Nestlé sent a letter to the county demanding they pull the ads, and the company also sent a complaint to the Florida attorney general. The *Miami Herald* newspaper reported that the International Bottled Water Association threatened 'similar action,' though the issue seems to have been dropped when the ads ended."[121] The mere threat of such costly lawsuits may have a chilling effect on similar efforts by other cities to promote tap water.

Moreover, water utilities, typically on slim budgets, will never be able to compete with the massive marketing campaigns of the Big Four. Nestlé's spending to promote its Pure Life brand in 2009 alone was $9.7 million.[122] Coca-Cola's slick ad campaign for Dasani in 2003 cost $20 million.[123] Beyond this is advertising implying that bottled water is safer than tap water, as discussed above. The IBWA points to its Advertising Code of Standards, which states that its members "should not allege that public water supplies that comply with federal and state safety standards are dangerous to health, unhealthy, or will cause disease or illness" and also "should not attempt to alarm consumers who may be concerned about the safety of their drinking water."[124] However, these standards are entirely voluntary, and many water bottlers—including Coca-Cola and PepsiCo—are not IBWA members to begin with. According to Gregory Pierce and Silvia Gonzalez, "Improving [public] perception may require that proponents of tap water employ more adversarial, fact-based tactics in debunking the myths about alternative sources of water supply. Given its potentially confrontational nature, this is likely to be a role which non-profits rather than public agencies are best suited to play."[125]

Thus there is a structural mismatch in which public water utilities find themselves at a great disadvantage against bottled water firms' legal departments and marketing. Of course, public agencies should not have to compete against a well-heeled global industry, but the dramatic shift toward bottled water and the ensuing decline in tap water consumption threaten a core government service. For this reason among others, water utilities will need to enter the battle over public perception and communicate to residents that tap water is not only far more affordable and environmentally sustainable than bottled water but also in most cases as safe or safer.

HOW SAFE IS *BOTTLED* WATER?

All this discussion about bottled water and fear of the tap has so far avoided a basic question: How safe is bottled water itself? This is an important issue, given that the industry's marketing is premised on the implicit or explicit claim that its product is safer and purer than public tap water.

In the United States, the story of bottled water safety is one of an industry governed by different and largely weaker regulations than those that apply to tap water. Municipal tap water in the United States is regulated by the Environmental Protection Agency under the Safe Drinking Water Act. Bottled water, on the other hand, is deemed a foodstuff and thus is regulated by the Food and Drug Administration. The difference between EPA and FDA regulation is effectively the difference between night and day. Salzman sums up the distinction: "If contaminants are found in tap water, which is tested daily, the water utility must quickly inform the public. If contaminants are found in bottled water . . . manufacturers must remove or reduce the contamination but there is no similar requirement to notify the public. Perhaps most important, FDA regulations only apply to goods in interstate commerce, i.e., traded across state lines. Yet anywhere from 60 to 70 percent of bottled water never enters into interstate commerce. As a result, *two-thirds or more of bottled water is effectively exempt from federal regulation*" (emphasis added).[126] Most large public water utilities actually test water safety far more frequently than daily. Washington, D.C., for example, tests its drinking water thirty thousand times per year.[127]

The issue is not that there are higher tolerance levels for contaminants in bottled water. "The FDA allows in bottled water the same complement of disinfection by-products, pesticides, heavy metals, and radioactive materials that the EPA allows in tap," writes Royte. The main difference "is that public water utilities are required in their annual reports to let you know, while the bottled-water industry has spent millions to make sure you don't, lobbying hard to keep such information off its labels."[128] FDA regulators inspect bottling plants but do not test the actual water themselves, leaving that to manufacturers. Consumer Reports found that the number of FDA bottled water inspections fell by one-third between 2008 and 2018 and that "the FDA might not become aware of contamination until long after it happens." Incredibly, even when water is found to be contaminated, "in most cases manufacturers don't have to stop bottling or alert the public."[129]

The two-thirds of bottled water that doesn't cross state lines, escaping FDA scrutiny, is still subject to state regulation, but ten states don't regulate bottled water at all, and only fourteen states require bottlers to notify regulators immediately if excessive contamination is found.[130] The Natural Resources Defense Council (NRDC) found that forty-three states did not have even one staff person assigned to bottled water regulation, and the FDA had fewer than two staff regulating bottled water for the entire nation.[131]

In other words, if you don't actively look for contamination, or don't have the staff to do so, you probably won't find it, and consumers won't know about it. Bottled water firms are not required to publish the results of their tests, although some do. For this reason, most of the data available on bottled water quality comes from testing by nonprofit consumer organizations or academic researchers. A 1999 study of over one hundred bottled water brands by NRDC found that 33 percent either violated a state standard or exceeded purity guidelines for cancer-causing chemicals or arsenic.[132] A 2008 analysis of ten major bottled water brands in the United States uncovered thirty-eight pollutants, including arsenic, bromates, and trihalomethane.[133]

More recent reports document a troubling pattern of ultra-light-touch regulation. A series of investigative articles by Consumer Reports, some copublished with *The Guardian,* found high levels of arsenic, above the FDA's limit of ten parts per billion (ppb), in Keurig Dr. Pepper's Peñafiel water brand. Consumer Reports later found that the FDA had been aware of the high arsenic levels in Peñafiel since 2013 but did not act. The reporting led the company to issue a voluntary recall.[134] Whole Foods Markets' Starkey bottled water brand was recalled for exceeding allowable arsenic levels in 2017, but the reporters again found arsenic in Starkey water in 2019 at just under the 10 ppb threshold.[135] Jill Culora, the IBWA's vice president for communications, disputes these reports, claiming that "Consumer Reports and *The Guardian* are unnecessarily scaring consumers about the safety of bottled water." She points to the IBWA's Bottled Water Code of Practice, which requires IBWA bottlers to "undergo a mandatory annual plant inspection conducted by an independent, third-party organization."[136] However, these are the very same guidelines that the IBWA refers to on its own website as a "set of self-regulating industry standards" and a "valuable marketing tool," and which Royte says "aren't legally binding or enforceable."[137]

In sum, public water utilities in the United States are operating on an unfair playing field against the bottled water industry, which escapes most of the scrutiny to which they are subjected. The industry is largely allowed to police itself and is in effect shielded from most negative public revelations. Thus consumers have little chance of learning when bottled water is found to be contaminated or is recalled. "Assuming bottled water is safer than tap water may make us feel better," writes Salzman, "but there is little reason to think this is necessarily so."[138] The deck is stacked against public tap water. This is, quite literally, a double standard.

What about the safety of the plastic bottles themselves? Substantial concern has emerged among consumers in the past two decades about the safety of plastic food packaging, including plastic bottles, food containers, and baby toys, particularly around estrogenic or hormone-mimicking chemicals like Bisphenol A (BPA) and phthalates. Today virtually all single-serving water bottles and some larger sizes as well are made of PET, which many observers consider to be quite stable.[139]

However, that is not the whole story. A 2008 study found that average phthalate levels in PET plastic bottled water were over twelve times higher than water sold in glass bottles.[140] Other research has found that when subjected to heat at levels commonly occurring in cars and trucks, or when stored more than six to eight months, PET plastic does sometimes leach phthalates and antimony chloride—both estrogenic chemicals—into the water. Bottles stored at seventy degrees Centigrade for twelve days reached the EPA's antimony contaminant threshold of 6 ppb, and seven days at eighty degrees raised the level to 14.4 ppb.[141] Overall, "The evidence suggests that PET bottles may yield endocrine disruptors under conditions of common use, particularly with prolonged storage and elevated temperature."[142]

While PET plastic gets contradictory reviews in terms of leaching, the hard polycarbonate plastic that constitutes most larger, clear water containers—including the common five-gallon water cooler jugs—is another story. Polycarbonate (PC) is known to leach the notorious endocrine-disrupting toxic chemical BPA, which has been banned or restricted in Europe, Canada, and other countries.[143] Some newer replacement plastics are being touted as BPA-free, but these turn out to contain *other* compounds with estrogenic activity (EA). According to neuroscientist George Bittner and coauthors, "BPA-free did not mean EA-free."[144]

Finally, there is a major recent controversy about what we consume when we drink bottled water. This involves ingestion of microscopic pieces of plastic—microplastics. Chemists from SUNY Fredonia tested over 250 bottles of eleven global bottled water brands purchased in nine countries and found that 93 percent of the bottles were contaminated with microplastic polymer fragments, with an average of 325 particles per liter of water.[145] Bottles of Nestlé Pure Life water from the United States had the highest level of microplastics in the study, averaging 2,277 particles per liter, with one sample containing an astonishing 10,390 particles per liter.

Microplastics are now being found everywhere—in oceans and waterways, in the air, in seafood, crops, and processed foods, in wildlife, and in human tissues. While the levels of microplastic in the environment and our bodies raise public alarm, their human health impacts are contested. A report by the WHO concludes that the evidence does not yet exist to show any health harms,[146] but other researchers contend that microplastics are correlated with illnesses such as diabetes, heart disease, cancer, and reproductive and neurological problems. While 90 percent of microplastics are likely excreted, the tiniest particles may pass directly into the lymphatic system, bloodstream, and internal organs, and can be passed from mother to fetus.[147]

The same researchers who found microplastics in bottled water also tested tap water. Crucially, they discovered that tap water contained far fewer microplastic particles per liter than bottled water.[148] Another study found that on average, a person drinking only bottled water consumes twenty-two times more microplastics than someone consuming only tap water—ninety thousand particles per year versus only four thousand.[149] Although bottled water firms contest these findings, they represent the starkest head-to-head comparison so far between microplastics in bottled and tap water, and a serious challenge to the industry's claims of greater purity. According to plastic pollution researcher Sherri Mason, "You're going to be drinking significantly less plastic from tap water out of a glass than if you go and buy bottled water."[150]

THE REAL COST OF PACKAGED WATER

The price of bottled water has frequently been the focus of both ridicule and concern. While it is often priced competitively with soda and other beverages, this commodity is vastly more expensive than the tap water it often replaces for consumers. A 2018 report by Food and Water Watch contrasted the cost to treat and deliver one gallon of tap water—one-half of one cent per gallon[151]—to the price of gasoline at $2.35 per gallon and bottled water at $9.47 per gallon.[152]

In an attempt to get a better handle on these cost differences, I compared the prices of several major bottled water brands and a store brand in different volumes, both individual bottles and multipacks, at my nearest large grocery store, Fred Meyer (owned by retail giant Kroger). The results, in table 4, show the retail price in dollars per gallon and the cost of satisfying the average annual U.S. consumption of forty-seven gallons of bottled water per person in 2022. Depending on the brand and package, the total cost of the average yearly bottled water consumption for a family of four ranged from a low of $254 to a high of $2,690 per year, while the equivalent volume of tap water cost less than *one dollar* for four people—a mere 23.5 cents per person.

Of course, the issue is not just bottled water's astronomically higher cost, but the way this translates into families' expenditures on bottled water and how that affects household income. It has long been argued that bottled water consumption is positively correlated with income—that "the higher the household's income, the more bottled water [is]

TABLE 4 RETAIL PRICE OF SELECTED BOTTLED WATER BRANDS, COST FOR AVERAGE ANNUAL U.S. BOTTLED WATER CONSUMPTION, AND COST OF EQUIVALENT TAP WATER VOLUME, 2022

Bottled water brand and size	Retail price	Volume in gallons	Retail price per gallon	Cost for 47 gallons (avg. annual US consumption in 2021)	Cost for family of four (avg. annual consumption)	Cost of 47 gallons of tap water (1 person)**	Cost of 47 gallons tap water (4 people)**
Aquafina* (Pepsi) 20-oz. bottle	$2.29	0.16	$14.31	$672.57	$2,690.28	$0.235	$0.94
Arrowhead (Nestlé/ BlueTriton) 12-pack of 12-oz. bottles	$3.99	1.125	$3.53	$165.91	$663.64	$0.235	$0.94
Fiji Water 6-pack of 1-liter bottles (on sale)	$14.99	1.59	$9.43	$443.21	$1,772.84	$0.235	$0.94
Dasani* (Coca-Cola) 24-pack of 16.9-oz bottles	$5.99	3.17	$1.89	$88.83	$355.32	$0.235	$0.94
Fred Meyer Brand 24-pack of 16.9-oz. bottles	$4.29	3.17	$1.35	$63.45	$253.80	$0.235	$0.94

* Contains refiltered municipal (tap) water.

** At $0.005 (one-half cent) per gallon, the average consumer cost of tap water in the U.S.

SOURCE: FredMeyer.com, online search for Portland, OR, store.

consumed."[153] According to one market report, "Households with higher disposable income purchase significantly more bottled water and ice products than lower-income households."[154] The report adds that the reverse is also true: "Due to the plentiful supply and less expensive nature of [the] substitute, tap water, demand for bottled water is highly elastic. When per capita disposable income declines, customers become more price-sensitive and seek to reduce their purchases of bottled water."[155] This is what happened during the recession of 2008 and 2009.

However, recent studies paint a very different picture about the link between income and bottled water consumption. A 2019 survey by Consumer Reports reached precisely the opposite conclusion: poorest families spend the *most*. Households with annual incomes under $25,000 spent an average of $15 per month on bottled water, those earning between $25,000 and $49,000 spent $12 per month, and the wealthiest (over $50,000 annual income) spent only $10 per month. The survey also found that Black households spent an average of $19 per month on packaged water and Latino/a households spent $18, while White households spent only $9 per month.[156]

Other research has found similar patterns. In a survey of over six hundred U.S. families, African American and Latino/a households (with a median annual income of $18,000) spent an average of $20 per month on bottled water, compared to $12 for non-Latino/a Whites (whose median income was $42,000). Bottled water spending accounted for a median of 1.0 percent of total household income for the Black and Latino/a households, but only 0.4 percent for the White households. However, some Latino/a and Black families in the survey spent as much as *12 to 16.7 percent* of their entire household income on bottled water. Fourteen percent of the Latino/a families, 12 percent of African Americans, and 6 percent of White families said that they had to give up other needed items in order to purchase bottled water.[157] Health scholar Asher Rosinger and colleagues found that as respondents' incomes increased, they drank *less* bottled water and more tap water. They also found bottled water consumption was much higher among Blacks and Latinos/as, people born outside the United States, and people with less than a high school degree. In fact, the *majority* of plain water consumed by Black and Latino/a adults was bottled water.[158] And in a 2020 paper examining recent U.S. data, nutrition researcher Florent Vieux and coauthors confirmed these trends: "Contrary to expectations, tap water consumption was higher at *higher* incomes, whereas the consumption of bottled water was higher at *lower* incomes" (emphasis added).[159] In other words, bottled water use and its impact on household income map onto—and exacerbate—existing social inequalities along the lines of race, ethnicity, and class.

These patterns intersect with people's perceptions of tap water safety. Javidi and Pierce found that only 5 percent of White respondents believed the tap water in their home was unsafe to drink, compared to 8 percent of Black respondents and 16 percent of Latinos/as. Of those who felt their tap water was unsafe, 79 percent of African Americans and 73 percent of Latinos/as chose bottled water as their primary alternative water source, versus only 55 percent of Whites.[160]

Researchers studying the shift away from soft drink consumption—a positive outcome of campaigns targeting the harmful health effects of Big Soda—have also found that race, ethnicity, and income are strong predictors of whether U.S. consumers who give up soft drinks switch to drinking tap water or bottled water. Public health scholar Colin Rehm and colleagues found that the people with lowest incomes, as well as African Americans and Latinos/as, were the most likely to become bottled water consumers instead of (re)turning to the tap.[161]

Taken all together, these conclusions point to a fairly recent phenomenon, in which demand for bottled water is now actually higher—and less elastic—among the communities that are on the losing end of the deterioration of U.S. public water infrastructure, and whose tap water is perceived (and/or documented) to be unsafe to drink: low-income people, Blacks, and Latinos/as. This indicates that the market for bottled water is now a bimodal one, in which "higher income adults drink bottled water for convenience, whereas lower-income adults may drink bottled water because of tap water access issues."[162]

In fact, market researchers are explicit in recommending that bottled water firms should target their advertising campaigns at African Americans and Latinos/as: "Black and Hispanic consumers are important consumer groups for packaged water brands. Both groups believe bottled water tastes better than tap water. Black consumers are significantly less likely than total consumers to use refillable water bottles and drink tap water, possibly because they prefer the taste of bottled water; marketing messages around taste/flavor may resonate strongly with Black consumers. Water brands can steal Black and Hispanic consumers from the sports drink market."[163] The leading packaged water firms have faced substantial criticism for focusing their marketing on these groups, as well as immigrants from nations where tap water may not be safe or reliable. According to Food and Water Watch, "Bottled water companies have honed their marketing to target lower-income groups, people of color and immigrant communities in the United States—especially Latina mothers, children and women generally. Industry marketing strategies are designed to hype the healthfulness and safety of bottled water to people who historically have lacked access to safe tap water (especially recent immigrants)."[164]Thus the social groups who on average can least afford to pay for a constant supply of bottled water are precisely those who tend to trust their tap water the least, who are targeted by the industry's advertising, and who spend the highest percentage of household income to buy packaged water.

Of course, families who shun their tap water because of fears about its quality or safety do not stop paying utility bills for tap water, even if they choose not to use it (or cannot use it) for drinking. They are paying double for their drinking water. In places where the tap water is genuinely unsafe to drink as a result of industrial contamination or water system failure, such as parts of the Appalachian coalfields,[165] regions with fracking-related groundwater pollution or widespread septic system leaks, agricultural areas with nitrate contamination, or cities with lead leaching from aging pipes, this becomes an even more serious problem of linked economic and environmental injustice. The cost of replacing tap water with bottled water for a household's entire drinking *and* cooking needs—roughly four gallons (fifteen liters) per person per day for an average-sized family—was between $983 and $4,757 per year in 2015, according to Javidi and Pierce. In fact, "The minimum annual expenditure by all U.S. households who perceive the tap as unsafe and buy bottled water to meet necessary consumption standards [was] $5.65 billion" in 2015—just under 40 percent of total U.S. bottled water sales for that year.[166] Adding these costs to water bills that now average more than 12 percent of income for poor families represents a significant economic burden.[167]

Such injustice is rendered even more egregious if, like up to fifteen million other Americans, people are obliged to turn to bottled water because their home tap water service has been shut off for nonpayment or if, like another two million people in the United States, they lack access to running water or indoor plumbing altogether.[168] In Detroit, writes Ryan Felton, poor and predominantly Black families whose water supply was disconnected for past-due water bills of as little as $150 are meeting their drinking and cooking needs with bottled Dasani water that comes from the Detroit municipal water supply maintained by their tax dollars, thus paying hundreds of times more for the very same public water they were cut off from.[169] Adding insult to injury, these families must take on the extra labor required to constantly haul cases of bottled water to their homes. In these ways, bottled and packaged water intersects with, highlights, and exacerbates social and racial injustice.

LEAVING A HUGE FOOTPRINT

It is the numerous environmental impacts of bottled water, however, that have generated the most controversy and played a major role in stoking public opposition to this commodity. First, there is its much higher energy and carbon footprint. According to a study by Gleick and Heather Cooley, bottled water consumes between 1,000 and 2,000 times more energy per unit volume compared to local tap water. They found that U.S. bottled water production and consumption required the energy equivalent of between thirty-two and fifty-four million barrels of oil, or one-third of 1 percent of total U.S. primary energy consumption, based on 2006 figures. Since that date, U.S. bottled water volume has grown by a further 90 percent.[170] A newer study calculated bottled water's overall environmental impact as 1,400 to 3,500 times higher than that of tap water.[171]

Bottled water also requires significant water use—beyond what is contained in the bottle itself—in plastic manufacturing, cleaning, and water wasted at all stages of production. Estimates of the water footprint of bottled water vary wildly. A study commissioned by the IBWA claims that only about 1.4 gallons of water are used for each gallon of water sold.[172] However, Royte contends that "manufacturing and filling plastic water bottles consumes twice as much water as the bottle will ultimately contain," adding that "plants that use reverse osmosis to purify tap water lose between three and nine gallons of water . . . for every filtered gallon that ends up on the shelf."[173]

When the source is springs or groundwater, there is also the impact of the water extraction itself on local aquifers, springs and rivers, ecosystems, agriculture, residential wells, and other nearby water users. Although the volumes extracted for bottling are often tiny relative to total groundwater usage—a point strongly emphasized by the industry to respond to its environmental critics—this water mining can have significant and even irreversible hydrological and ecological effects in specific watersheds and localities, including aquifer depletion, loss of irrigation and well water, and drinking water contamination.[174]

A TSUNAMI OF PLASTIC

Yet there is one externality that dwarfs all the others: the massive ecological crisis of plastic waste that has engulfed the planet, largely in the past three decades. I recall clearly the first time I came face to face with the rising tide of single-use plastic pollution. It was in 2003, on a small island in southern Thailand. All of the island's western-facing beaches were inundated by a multicolored belt of floating plastic garbage, in some places extending fifty feet out to sea, in others a few hundred, making swimming not just unappealing but impossible. I had seen marine garbage before, but nothing remotely like this. The one item that indisputably dominated the flotsam? Plastic water bottles discarded by tourists and local residents.

The evidence is unequivocal: we are facing a veritable tsunami of plastic waste. An article in the journal *Science Advances* concluded that fully half of all the plastics ever produced worldwide (8.3 billion metric tons) had been manufactured in just the previous thirteen years. Only 9 percent of this plastic had been recycled, 12 percent was incinerated, and the rest ended up in landfills or as litter dumped on land, or in rivers, lakes, or the ocean.[175] Approximately eleven million metric tons of plastic garbage enter the marine environment every year, including between twenty-one billion and thirty-four billion plastic beverage bottles.[176]

This tidal wave of plastic trash, and the microplastic fragments it breaks down into, has been spread via winds and ocean currents to reach the most remote places on the planet. Plastic has been found in the bodies of over half of all seabirds and 100 percent of sea turtles. The five enormous subtropical ocean gyres sweep much of this waste into several massive garbage patches, a mix of large objects and tiny microplastic particles. A frequently cited estimate is that by 2050 the oceans will contain more plastic by weight than fish.[177] Plastic bottles are central to this picture: one report found that bottles and their caps together were the number-one marine garbage item, and another concluded that they account for fully half of all marine waste.[178]

In 2006, three hundred billion single-use plastic beverage bottles were produced worldwide. A decade later that number had jumped to 480 billion. In 2021, the figure was estimated to be 583 billion—more than 1.1 million bottles per minute.[179] Who is responsible for producing these bottles? A large-scale study of garbage collected on beaches and shores in 42 countries found that Coca-Cola (which uses three million tons of plastic packaging annually) was the top source of plastic garbage entering the oceans, followed by PepsiCo, Nestlé, and Danone.[180] While the Big Four generate the largest share of the problem, they bottle a wide range of beverages in plastic, not just water. However, according to Rosemary Downey, a leading expert in plastic bottle production, the majority of plastic bottles used across the globe are for drinking water.[181]

What about recycling? While the industry promotes and labels its bottles as recycl*able,* the actual story of plastic bottle recycl*ing* is one of high hopes, big promises, and so far, dashed expectations. Single-serving PET bottles now constitute over 70 percent of all bottled water sales in the United States.[182] Yet the U.S. recycling rate for PET plastic bottles, never very high, fell to only 26.6 percent in 2021.[183] Moreover, only 7 percent of the bottles that do go into recycling programs are made into new beverage bottles; the rest are "downcycled" into lower-value products such as carpeting, which cannot be recycled again. And bottles, it turns out, are the most frequently recycled plastic product of all. The overall U.S. plastics recycling rate was a paltry 5 percent in 2021, down sharply from 9 percent in 2018—compared to 25 percent in China and nearly 30 percent in the EU.[184]

The beverage industry publicly encourages more recycling. Its promotion of municipal plastic recycling programs has been key to gaining public acceptance of the ongoing shift to plastic packaging.[185] However, the industry also uses its lobbying muscle to kill or weaken public policy measures that would *actually* result in more recycling, including new or expanded container deposit laws or "bottle bills." As *National Geographic*'s Laura Parker writes, "Beverage companies have long strongly promoted recycling, and vigorously opposed bottle deposit legislation, arguing bottle bills cost them too much money" and would raise the price of their products.[186] The industry's lobbying defeated a proposed U.S. national bottle bill in 2009. As a result, only ten U.S. states and Guam have bottle bill laws, and only six of those include all bottled water containers.[187] Yet these laws work: bottle recycling rates in those six states far exceed the national average, ranging from 44 percent in Connecticut to 86 percent in Oregon.[188] Germany's nationwide bottle bill has yielded recycling rates over 98 percent.[189] The beverage industry also opposes new fees or taxes to cover the costs to local government of dealing with the plastic waste.[190] In sum, the industry has fought vigorously to defend the profitable status quo: voluntary recycling, paid for entirely by taxpayers and ratepayers.

Beverage firms have also dragged their feet over increasing the levels of *recycled* plastic content—called rPET—in their bottles. Under pressure from environmental groups, the Big Four have repeatedly pledged to increase the percentage of rPET that they use, yet time after time, those pledges have failed to materialize. "Danone, like much of the industry, has made promises about using recycled material before only to break them," writes Saabira Chaudhuri in the *Wall Street Journal*. "A decade ago it pledged to use 50% recycled plastic in its water bottles by 2009. The very next year it slashed that target to between 20%-30% by 2011. Today, just 14% of the plastic in the bottles across its brands is recycled material."[191] For other companies the figures are worse: Coca-Cola uses a total of 7 percent recycled plastic in its bottles, and Nestlé Waters 6 percent. The average for the industry as a whole is 6.6 percent.[192]

But environmental critics of the industry have their sights on far more than mandating more recycled content in bottles and raising recycling rates. Many reject the entire premise of plastics recycling, insisting on a conversion to returnable, refillable beverage containers like those that vanished by the 1970s in most of the global North. According to the group Oceana, a wholesale switch to refillable plastic bottles would reduce raw material use by 40 percent and greenhouse gas emissions by 50 percent.[193]

Others go further, condemning the production of all plastics as inextricably linked with the fossil fuel industry, and thus the climate crisis. The production and incineration of plastics already account for 4.5 percent of all global greenhouse gas emissions.[194] According to journalist Zoe Carpenter, a dramatic expansion of plastics manufacturing is now under way, as a hedge against falling oil demand. Plastic, she writes, "is becoming an increasingly important source of profit for Big Oil, providing yet another reason to drill in the face of climate change."[195] With over $200 billion invested since 2010 in building 333 new petrochemical facilities in the United States alone, global plastic production is slated to double by 2030 and quadruple by 2050, when it is projected that one-fifth of all fossil fuels will be used to make plastic.[196] These emissions will consume "10 [to] 13 percent of the entire remaining [global] carbon budget."[197] To call the current production and use of plastic unsustainable would be a profound understatement.

For the bottled water industry, however, the climate crisis represents a market opportunity. According to a 2022 market report, "Consumers will adopt a 'prepper' mentality following the increase in frequency and severity of extreme climate events; more consumers will keep packaged water stocked 'just in case,' benefiting the regular bottled water and jugged water categories."[198]

A DOUBLE-EDGED SWORD

Yet none of this fully explains why the issue of single-use plastics has exploded into public consciousness in the past few years. Why is it suddenly being treated as an urgent global crisis by many residents of wealthy nations? The prime reason is China's momentous decision to stop accepting imports of most recyclables from the rest of the world as of 2018. This policy, called "National Sword," has almost completely upended the world of recycling and waste collection. It bans twenty-four types of waste, including many plastics. The reasons for China's policy shift include persistent contamination of recycling shipments, water and air pollution, and a desire to stop being the dumping ground for the rest of the world.[199]

Before National Sword, the United States was exporting about 80 percent of its mixed plastics and one-third of all recyclables, the vast majority of it to China, which in turn imported 45 percent of the world's total plastic waste.[200] Without China, local governments in the rich world are now swamped with mountains of plastic, backing up at recycling plants with few markets to be found. Many communities are now sending the recyclable plastics they collect to the dump instead, leaving those dedicated to diligently separating their household recyclables confused and disillusioned. The closing of the escape valve of China's market vividly illustrates the notion that there is no "away" when it comes to waste.

The plastic waste is not only piling up in the rich world. After National Sword, governments and businesses in the North scrambled to find new buyers, and many quickly shifted their waste exports to Southeast Asia. Vietnam, Indonesia, and particularly Malaysia virtually overnight found themselves swamped by inconceivable quantities of dirty plastic. According to journalist Dominique Mosbergen, "Activists say these Chinese recyclers have been setting up factories, often illegally, across Malaysia and have been processing and disposing of waste without regulatory oversight. Whatever they manage to recycle is allegedly flowing back to China, where it's used for manufacturing. . . . Without any environmental regulations to worry about, the recyclers can leave contaminated water untreated."[201]

As illustrated in the documentary film *The Story of Plastic,* these unregulated waste imports have created nightmarish landscapes in rural and urban areas—thick blankets of dirty plastic dumped in backyards and farm fields, blowing across the landscape, choking local waterways. The workers employed in the hazardous manual sorting and melting of these plastics for recycling come from the poorest and most marginalized social groups, and women are disproportionately represented, constituting a little-recognized environmental justice crisis, particularly in Asia.[202]

Burning the 70 percent of the plastic that is not recycled—either in open waste fields or in industrial incinerators—generates toxic fumes and ash that are responsible for four hundred thousand to one million deaths per year worldwide, on top of the climate impacts, according to the NGO Tearfund.[203] The implications of a fourfold increase on these figures by 2050 are virtually unimaginable.

AN UNPRECEDENTED BACKLASH

It is hard to overstate the speed and intensity of the reaction against disposable plastic. The years 2018 and especially 2019 appear to have been a genuine tipping point, due to a convergence of factors: media coverage of the marine plastic pollution crisis, the impact of National Sword on wealthy nations, and soaring international awareness of the climate crisis, catalyzed by youth-led activism—all amplified via social media. The connection to the global climate justice movement is especially important. When 7.6 million people took to the streets in 185 countries in September 2019 in a global protest against government inaction on climate change, many inspired by the young Swedish activist Greta Thunberg, it was a pivotal movement. Thunberg has also called attention to marine plastic pollution, and others are quick to make the linkage as well. Annie Leonard, director of Greenpeace USA, writes that "the plastics crisis and the climate crisis are two fronts in the same battle. We cannot end the era of single-use plastics without stopping the fossil fuel industry, and we cannot stop the fossil fuel industry without ending single use plastics."[204]

The growing movements against plastic pollution, and the mounting costs to governments of dealing with no-longer-exportable waste, have generated an avalanche of policy change. The ever-growing list of governments banning, restricting, or taxing single-use plastic products reflects a synergy of fiscal self-interest, environmental concern, and activist pressure. Under the European Union's Single Use Plastics Directive, which took effect in 2021, all member states must entirely ban disposable straws, utensils, stirring sticks, polystyrene containers, and so-called degradable plastics—but not plastic bottles, as many campaigners demanded. Instead the policy requires a minimum of 25 percent recycled plastic content in bottles by 2025 and mandates that 90 percent of all plastic bottles be collected for recycling by 2029.[205] California has mandated at least 50 percent recycled plastic content in single-use beverage bottles by 2030, but has taken no steps to limit their sale.[206] Meanwhile, China now prohibits plastic straws, and a phased-in ban on plastic bags and utensils will take full effect in 2025. China's policy has even been flagged by the International Energy Agency as a threat to global demand for petroleum, but it too does not include bottles.[207]

The convergence of all these developments has rapidly shifted public attitudes toward waste, consumerism, and environmental degradation. And plastic beverage bottles—which remain largely unrestricted—have now become the next target. As Parker writes, "The backlash against the glut of discarded bottles clogging waterways, polluting the oceans and littering the interior has been swift. Suddenly, carrying plastic bottles of water around is uncool."[208]

The beverage industry has begun to respond to this backlash, at least rhetorically. The big beverage corporations claim (yet again) they are working to increase the percentage of rPET in their bottles, with Coca-Cola, Nestlé, and PepsiCo promising at least 50 percent recycled material in their packaging worldwide by 2030. These firms also say they continue to make their water bottles lighter (using less plastic), and some are including so-called bio-based—but not fully biodegradable—substances to replace part of the plastic.[209] However, they have firmly rejected calls to abandon single-use plastic altogether. A top Coca-Cola executive told an audience at the World Economic Forum in Davos that "the firm would not ditch plastic outright, as some campaigners wanted, saying this could alienate customers and hit sales."[210]

Industry critics counter that this stance is wholly incompatible with the urgency of the linked plastic waste and climate crises. "You can only do so much by lightweighting bottles [to] reduce the amount of plastic required," Gleick tells me. "But if you have a single-use plastic bottle, which is what the industry is built on, that is and will continue to be a disaster."[211]

PLASTIC PUSHBACK, PRODUCER PANIC

Far more than the plastic bag, in the current zeitgeist it is the plastic water bottle that has become a newly politicized object of rejection, the epitome of an unsustainable culture of overconsumption. This backlash is both dramatic and generational. According to market analysts, Generation Z—young people born between 1997 and 2012—are much more hostile toward single-use plastic bottles than either their parents or Millennials. Look in the backpack or bag of anyone under thirty (and many older folks too), and you'll likely find a metal refillable water bottle. They are a cultural signifier—a trendy consumer item that paradoxically signals a rejection of disposable consumerism. "Buying a refillable bottle is opposite to the hyperindividualism of buying a private phone or musical headset," opines Royte. "Refillables announce a commitment to public water, a heartening step away from . . . inverted quarantine."[212]

Whatever the motivations, all these changes amount to a major, rapid shift in attitudes and consumer behavior. A 2019 survey found that 53 percent of U.S. consumers say they carry a refillable bottle with them when they leave home.[213] Another in 2020 reported that 59 percent of all bottled water buyers, and two-thirds of Generation Z buyers, believe companies should stop using plastic bottles, even if it makes the product more expensive.[214]

The industry recognizes these trends and has become concerned. Reading market research reports is normally a very dry affair, but some recent reports on the bottled water market are remarkable for their barely contained urgency. They speak in insistent terms about young people's rejection of plastics and bottled water, urging companies to respond meaningfully to the backlash. "Though the number of Americans who have reduced their single-use bottled water purchases is small (7% of bottled water buyers)," states one report, "the bottled water attrition rate will grow as concerns around plastic bottles also grows. There is a strong possibility that young consumers have an extremely negative view of bottled water and some may *never* become bottled water purchasers. It's entirely possible Gen Z and future generations will completely turn against bottled water, with their perception of drinking bottled water akin to that of smoking" (emphasis in original).[215]

The backlash was especially pronounced for Nestlé and Coca-Cola, who saw their U.S. bottled water sales fall by 1.8 percent and 0.2 percent respectively in 2019, the first such drop since the Great Recession. Brands containing tap water were hit hardest: sales of Nestlé Pure Life, which according to Reuters "has been suffering from the tap water revival,"[216] dropped nearly 6 percent and Dasani fell 1.8 percent. On a global level, Nestlé's total water sales shrank by almost 2 percent that year.[217] Facing these trends, in March 2021 Nestlé announced that it had sold its U.S. and Canadian bottling operations and water brands to the private equity consortium BlueTriton.[218]

Meanwhile, the public reaction against single-use plastic bottles marches on, and marketers' worry over lost sales and lost consumers is palpable. "The bottled water market is at risk of seeing slower growth, or indeed decline, due to people opting for tap water instead, as plastic packaging waste has hit the headlines," writes one analyst.[219]

In response, beyond renewed promises to increase recycled plastic content, water bottlers are hedging their bets and dabbling with alternatives to plastic. Some companies are trialing water in cardboard cartons, aseptic box packaging, and aluminum cans in an attempt to shed the reputational stain of plastic bottles. Many flavored and sparkling waters are now sold in cans, but plastic continues to dominate the huge market for still water.[220] According to one analyst, "Still bottled water is likely to remain loyal to plastic packaging and the furthest shift within the subcategory will be towards recycled PET."[221] However, another warns that "the innovations won't bring back environmentally conscious consumers who have already switched to reusable water bottles."[222]

Given these trends, how are bottled water companies to hold onto their customer base? The report argues that they "need to consider developing packaged water products that appeal to young consumers . . . [or] products that allow the company to capitalize on the sustainability trend."[223] That is, they are being advised to monetize the backlash against single-use bottles, and they are taking this advice to heart. In 2018, PepsiCo purchased SodaStream, the maker of home water carbonation devices, for over $3 billion. Coca-Cola has gone further, debuting point-of-sale refilling stations:

> Coca-Cola's Dasani PureFill station was conceived in 2016. . . . Coke has piloted the machine on college campuses for the past two years and this fall will expand it to more schools as well as zoos and aquariums. So far, the most popular option has been the free one: chilled, filtered water. . . . Dasani's brand director for North America [said,] "Sometimes these units will be placed right beside a traditional water fountain. And you'll actually see people line up to get water from these machines." But the company also has found that many students are willing to pay a fee of 5 cents per ounce to add flavors or bubbles or both. Consumers can pay directly through the app, use Apple Pay or . . . swipe a credit card. In some cases, a 20-ounce pour comes out to roughly the same price as a 20-ounce bottle of Dasani.[224]

Nestlé and PepsiCo have also introduced similar products.

How else should the industry respond to this anticommodification trend? One market report recommends that bottling firms "need to develop new solutions to commercialize Gen Z's movement away from packaged water, which could include . . . investing in refillable water bottle companies."[225] And this they are doing: Keurig Dr. Pepper has invested in Life Fuels, a "smart" battery-powered refillable water bottle with an RFID chip and insertable plastic flavor pods, linked to a phone app that tracks the user's hydration and nutrition data, with a price tag of $179.

Are these nonplastic, refillable, and noncontainerized trends an improvement? Has the explosion of environmental concern among Generation Z finally forced the industry to reckon with the vast ecological damage caused by the production, transport, and disposal of more than half a trillion single-use plastic bottles annually? Although machines dispensing water into users' own bottles do generate less plastic waste, these efforts can be understood as an attempt to subvert the youth backlash against single-use bottles, to stanch the bleeding among one consumer segment—not a fundamental rethink. The evidence suggests the beverage firms do not plan to chuck plastic water bottles any time soon. They are reluctantly being forced to increase recycled content in a few places, like the EU and California, but single-use plastic remains their dominant model, unless policy changes or social movement pressure make this untenable.

More fundamentally, water from a Pepsi- or Coke-branded water refilling station, like Dasani or Aquafina, is still commodified public tap water—treated, distributed, and maintained by public tax dollars but sold to consumers via a new channel. This is the advent of the branded corporate drinking fountain, hooked directly into the public water supply. Even if some of these models dispense unflavored, uncarbonated tap water free of charge for the moment—and the free introductory sample is a time-tested marketing tactic—they serve to reinforce the industry's claims of tap water's impurity and the need for their private technological intervention to render it fit to drink. And they train us to view this as an acceptable, normal way to access drinking water.

A BOTTLED WATER WORLD

This chapter has traced how bottled and packaged water rose from obscurity to become one of the most successful marketing stories of all time. The emerging bottled water industry in the global North took advantage of a confluence of societal trends, including greater mobility; concerns with diet, health, fitness, and hydration; the disappearance of drinking fountains; and growing public distrust of tap water. That distrust was intensified by longer-term deterioration of public water infrastructure, and rare but visible tap water emergencies. Bottled water firms and their industry organizations supplemented these trends with aggressive marketing that often included subtle or overt disparagement of public tap water.

The result has been the normalization of bottled water across society in only a few decades. This is a dramatic shift, in which consumers in the global North are increasingly likely to get their drinking water in plastic containers from grocery stores or vending machines rather than from public drinking fountains or the overwhelmingly safe faucets located in their kitchens, bathrooms, and workplaces. The development of lightweight PET plastic bottles and their widespread adoption in the 1990s changed the game by greatly increasing portability and profits, making packaged water ubiquitous, and enabling beverage corporations to position themselves—rather than public water utilities—as the most trustworthy source of hydration.

The wholesale shift from tap to bottled water also has substantial social and economic effects. People's perceptions of poor tap water quality and their expenditures on bottled water differ along lines of race, ethnicity, and income, meaning that this commodity contributes to widening already huge racial and class inequalities. Buying packaged water imposes non-negligible economic costs on families, which at least in the United States are highest for communities of color and low-income people. The industry recognizes these patterns and targets its marketing at these communities. Yet bottled water in the United States is less regulated than tap water and no safer overall; it has been shown to contain higher levels of microplastics and sometimes other harmful substances; and consumers are far less likely to find out when those contaminants are detected.

Three decades after the marketing of water in plastic bottles began in earnest, oceans, waterways, communities, and rural and urban dwellers worldwide are figuratively choking on a rising tide of plastic waste, and many are literally choking on the fumes from burning plastic, not to mention the climate impacts. This immense socio-ecological crisis has recently spawned a counterreaction by consumers and governments against single-use plastics, including packaged water—one that may still be in its infancy but is already triggering major policy change around the world. There is now a genuine, unprecedented backlash against plastic bottled water, particularly among young people, making the industry increasingly nervous.

Over those three decades, the packaged water industry has taught us to fear the tap—although such fear is not usually justified by the evidence—with very profitable results. But what about places within wealthy nations where, as in many parts of the global South, the tap water may not be safe to drink? What are the root causes of these crises? How does bottled water appear in these settings? How do both residents and the bottled water industry respond to such emergencies? And what are the implications of the ongoing war on tap water for the public utilities that actually supply our water, and for the future of public drinking water more generally? The following chapter examines these critical questions, beginning with the environmental justice disaster of Flint.

CHAPTER 3

Flint: Corroding Pipes, Eroding Trust

We intend to capture new customers as we capitalize on favorable consumer trends . . . [including] concerns about deteriorating municipal water quality.

—Primo Water Corp. (formerly Cott Corp.), maker of several bottled water brands, in report to investors

"I've learned, when you're a victim of something, it gives you that extra oomph to fight back. And that's exactly what happened to me [in] 2014, which was the beginning of the Flint water crisis. They switched [the water source] from Detroit to [the] Flint [River] on April 25th, and I'll never forget that day, because it's my daughter's birthday and she was turning five."[1]

Gina Luster speaks clearly and intently, in the manner of someone who's had plenty of experience talking to the media. She wears a black T-shirt emblazoned with "Flint Lives Matter" in big white letters under a blue down vest and a scarf. Her face is serious, yet she seems to be constantly on the verge of breaking into a smile. She talks about her personal experience of the Flint water disaster.

> I started noticing changes between April and July. I lost a massive amount of weight. I think I lost about forty pounds from April to July. I mean it was like every morning I would wake up and I lost like two pounds in my sleep, kind of thing. And I had zero appetite. And when I would try to force myself to eat, I couldn't eat with anything metal. I could only eat with plastic. . . . I was getting some type of reaction or something.
>
> I used to be a store manager for a retail chain. So it's Fourth of July weekend. No one really wants to work. So I tell my managers, you guys have the day off. I work, I go to open up the store, and we have four registers that we have to count. I'm in the store, basically alone at this time. I just remember counting the second register and then [I] collapsed at work.

She recalls how her worsening illness confounded the doctors:

> Next thing I know, I'm in the hospital and they can't figure out what's wrong. I mean, they checked me for HIV, lupus, thyroid . . . the checklist went on and on. Never checked for lead. Never on those records do you see where they checked my blood for lead. Because it's 2014. . . . Why would we be needing to be tested for lead? So for me it was like a two-year journey of just trying to figure out what the heck was wrong with me. . . .
>
> My employer was really nice. They put me in the Family Medical Leave Act and they extended it, because it was like a big mystery. "What's wrong with Gina? You know, she's not herself. She doesn't look like herself." My skin—I'm about two shades darker than I used to be. The weight—they saw the weight literally falling off of me when I would come to work. They're like, "Man, those pants fit you last week. This week you can't even wear them." So there were telltale signs, but I knew nothing about lead. . . .
>
> My daughter was sick. I was sick. . . . And I've had like four or five surgeries. They thought I had breast cancer. They went in and it was just pockets of bacteria. And then that's when I started really getting into the science of what the Flint water had in it. And we found out [there were] way worse things in it than lead. Some of the bacterias, Legionnaires' being one of the main ones, can kill you instantly.

The case of Flint is a dramatic illustration of the unequal distribution of risks and harms from decaying public water infrastructure and fiscal austerity.[2] It raises complex questions about the relationship between threats to tap water quality, falling public trust in that tap water, and a growing crisis of water affordability. A human-caused disaster like Flint also suggests different sorts of questions about the role bottled water should play in society.[3] For people exposed to unsafe tap water, is bottled water an important lifeline? A regrettable necessity? Or is it an "opportunistic commodity"[4] that exploits drinking water crises like Flint to advance a larger war against public tap water? How should residents, activists, and public officials respond when tap water is proven unsafe to drink—even if only for a short period of time? Does bottled water have a legitimate role to play in emergencies or disasters? Are there alternative modes of providing safe water, even in these situations? And how can future crises like Flint be avoided? This chapter addresses these questions and also explores the grassroots response to the crisis. While the Flint struggle has been widely publicized, less well known is the way community activists have forged a statewide coalition that has connected the city's toxic tap water crisis to both the human rights emergency of mass water service shutoffs in Flint and Detroit and the corporate extraction of groundwater in nearby rural communities.

That groundwater fills the very same bottles being consumed by people whose water has been rendered unsafe to drink, shut off for nonpayment, or both.

A WATER CRISIS UNFOLDS

Flint, whose ninety-five thousand residents are 54 percent African American and 39 percent White, with a poverty rate of 39 percent,[5] was governed by a series of four different unelected emergency managers (EMs) between 2011 and 2015. It was one of several Michigan cities that Republican governor Rick Snyder and his predecessor John Engler had placed under emergency management, arguing that they required fiscal discipline to resolve their municipal debt crises. "The chief objective of these appointees," writes Benjamin Pauli, associate professor at Flint's Kettering University, "is to re-establish fiscal solvency, which they accomplish by restructuring or eliminating departments, firing public employees, renegotiating or terminating union contracts, selling off public assets, and various other cost-cutting measures."[6] Critics charge that this practice violates local sovereignty and constitutes a form of "disaster capitalism" by removing authority from elected local officials and prioritizing the needs of capital over those of labor and residents.[7] They also point to the racial dynamics of emergency management, in which EMs have at some point ruled over 50 percent of Michigan's Black population but only a tiny fraction of its White population.[8]

In March 2013, under pressure to reduce costs, Flint's elected city council approved switching the city's water source from the Detroit water system, which it had used for nearly fifty years, to a new pipeline under construction that would draw water from Lake Huron. Later that year, two consecutive EMs, Ed Kurtz and Michael Brown, signed unpublicized contracts to switch Flint's water supply to the highly polluted Flint River for two years while the pipeline was being completed.[9] In April 2014, with the pipeline still unfinished and the Detroit water system canceling its contract with Flint, new EM Darnell Earley approved the switch to the Flint River.[10] Because Flint's water utility had not needed to treat its own water supply (having used pretreated Detroit water), its inexperience led to widely fluctuating levels of chlorine and other treatment chemicals in the weeks and months after the switch, compounding the harmful effects of shifting to the Flint River.[11]

In addition, the Michigan Department of Environmental Quality (MDEQ) made the fateful choice not to require adding inexpensive anticorrosion agents to the water (which were added to Detroit's water).[12] This decision meant that the river water rapidly leached lead, other heavy metals, and bacteria-laden biofilm from the city's aging pipes directly into its water supply, resulting in widespread contamination and causing Flint residents to drink toxic tap water for at least eighteen months. Researchers detected levels of lead in some tap water samples that were thousands of times higher than the EPA's allowable limit, some even exceeding the agency's definition of hazardous waste.[13]

Residents almost immediately began complaining about foul-tasting, discolored water that caused rashes, digestive problems, hair loss, and other mysterious symptoms, but public officials consistently denied there was a problem.[14] According to the *New York Times*, "Through months of complaints from residents of this city about the peculiar colors and odors they said were coming from their faucets, the overriding message from the authorities here was that the water would be just fine."[15] However, community members started to share information, to organize grassroots groups such as Water You Fighting For?, the Flint Democracy Defense League's Water Task Force, and the Coalition for Clean Water, and to protest. Luster continues:

> So, [I] lost my job. I'm at home sick watching TV, and I see hundreds of people on the lawn of our Flint city hall protesting, and mind you, they're up there in snow up to their knees. And it was like the Oprah aha moment. I was sick. I couldn't walk. I was walking with a walking cane. And I told my daughter, "We're going to go out there with those folks and protest," because they were protesting to get rid of the emergency manager. . . . They were out there literally with water bottles of water from their taps. And I always say it looked like apple cider almost—it wasn't water. It didn't resemble water at all.
>
> And so that day, because of that protest in front of city hall, the emergency manager resigned. . . . This is now 2015. So it's saying that that motivated me and showed me the power of the people. It didn't have to be ten thousand of us, like the Million Man March [or] the Women's March. It was just some regular citizens who are sick and tired of being lied to and watching their family members be so sick where they couldn't work, they couldn't function.

Luster identifies this moment as the beginning of her own activism. "From that day on," she continues, "I started inserting myself into city council meetings, county commissioner meetings. You know, in all of my thirty-eight, thirty-nine years, I had never been to a Flint city council meeting. I had been to council meetings in Detroit and Houston, but never Flint. Because when I left Flint, it was like the number four place to live. It was fine. Everyone had a job. And I came back, and it was a water crisis. So that really just motivated me."

After the switch to the Flint River, problems emerged almost immediately. Flint's General Motors plant demanded to switch back to the Detroit system (and was allowed to do so) because the acidic, chlorine-heavy river water was corroding its auto parts. Residents were ordered to boil their tap water twice during 2014 after fecal bacteria were discovered. Yet it was not until early 2015 that firm evidence of serious lead contamination emerged publicly, because of leaks by an EPA whistleblower and studies by scientists from Virginia Tech. By March 2015, the city council was concerned enough that it voted to return the city to the Detroit water system, but it was overruled by the new EM Jerry Ambrose, causing Flint residents to be exposed to toxic water for several more months.[16]

Further studies found a doubling of the number of Flint children with elevated blood lead levels, a figure that likely understates the actual level of harm, according to a key expert.[17] Later it was revealed that both city water department and MDEQ officials had had clear evidence of the extreme lead levels but chose to discard samples that were high in lead and adopted sampling procedures that minimized evidence of contamination, "thereby avoiding inconvenient remedial action and public outcry."[18] Lead is a potent neurotoxin that is particularly dangerous to children, and long-term exposure is associated with often irreversible developmental and learning disabilities, hyperactivity, attention deficits, and lower birth weight. Health experts widely agree there is no safe level of lead in the human body.[19]

Popular organizing among Flint residents grew throughout 2015, some of it focused on the same legally disempowered city officials. Luster recalls,

> I was at a city council meeting and I had my daughter and my niece . . . and we went up to the podium doing public speaking, and we all spoke. And we got a standing ovation. Yeah, and we said what we had to say and we walked out. We didn't sit back down. We left. So we're about to get on the elevator and I see this lady running towards me. Her name was Nayyirah Shariff. . . . She's like, "Have you ever thought about being an activist?" And I looked at her, and I was like, "What is an activist?" I come from a business and a pharmaceutical background, not an activist, and she's like, "You know, what you just did up there, with . . ."—she thought they were all my kids—". . . with your kids was activism."

This connection soon developed into collaboration: "She called me, we talked, and she was already being very active. Nayyirah was known as being an activist way before the Flint water crisis. And I asked her, 'Can I shadow you?' And she said sure. And that's basically where I got my feet wet with being an organizer and an activist. . . . And so we just partnered up with Melissa Mays, and Nayyirah and I . . . and that's how we founded Flint Rising."

By the time Governor Snyder finally signed the order to switch Flint's drinking water back to the Detroit system in October 2015, enormous damage had been done. According to scientists and health officials, the toxic water caused elevated blood lead levels in many of Flint's children, along with at least twelve deaths and dozens of hospitalizations from Legionnaires' disease.[20] The eighteen months of corrosion had also heavily damaged much of Flint's water system, and the cost of replacing all the corroded pipes was estimated at $1.5 billion. The strong racial and class element in the Flint case has led many critics to charge that this represents an extreme case of environmental injustice.[21]

FIGHTING FOR THE RIGHT TO (BOTTLED) WATER

Bottled water has been an inescapable presence in Flint since the earliest days of the crisis. Whether purchased by residents, donated by individuals, charities, foundations, or beverage corporations, or distributed by the state of Michigan, these hundreds of millions of bottles raise important questions about the place of this commodity in addressing genuine emergencies, versus its role as a longer-term fix for failing public trust in tap water. Beginning in the early months of the crisis, when information was scarce—and particularly after the data on lead levels were finally revealed in 2015—it is indisputable that the many nonprofit-run and later also state-operated water distribution sites, or "pods," spread across the city were a lifeline for residents.[22] Luster was the subject of a national TV news segment on the Flint crisis:

> That was during the time of the Democratic debate of Bernie and Hillary, and they held the debate in Flint. . . . So that day, CNN, full production, came into my little two-bedroom townhouse and shadowed me. And they counted every single water bottle that I and my daughter and niece had to use that day. And so, the debate is running and they're showing a commercial, 'cause they're doing a headline story on the Flint water crisis. And here I am there in my house. I'm on CNN. Wow. And I would say from that, it just really, people start[ed] following our Facebook pages. . . . A group of students from China flew in, you know, Germany, Italy. It was just amazing. You know, I was getting invited to speak at different places. I spoke at Yale, you name it. It was a whirlwind of a time. And for me it was a learning curve.

CNN reported that Luster, her daughter, and her niece went through 151 single-use (16.9-ounce) plastic water bottles in one day: 36 bottles for cooking, 36 for washing hair, 27 for drinking, 24 for washing dishes, and the remainder for tooth brushing and other purposes.[23]

Flint residents speak about their dependence on bottled water with a mix of resignation, resentment, and anger. "It's not like we want bottled water," declared Flint activist Melissa Mays in a video produced by the Story of Stuff Project, while using bottled water to cook rice in her kitchen: "Nobody wants to live like this. I've been mid-recipe before, and run out of bottled water. I'm not going to put tap water in there. . . . We have to make sure we have enough water, and if we run out, we have between noon and six [p.m.] to rush over to one of the water pods, to get in line to pick up our rations—our cases of [bottled] water to use for every day, everything."[24]

Yet confronted by poisonous tap water, for many Flint residents access to bottled water was more than a regrettable necessity—it became a political demand. As Pauli, author of the book *Flint Fights Back,* tells me,

> The demand for bottled water has been a really prominent demand within the activist community, but also just within the community more broadly. We have so many people, and at this point of so many different descriptions, relying on bottled water. I live [in] one of the two most affluent, Whitest neighborhoods in the city. My next door neighbors will never drink tap water again. . . .
>
> So we have a lot of people here who feel like they have basically no choice but to use bottled water. Their trust in public water is so broken. But they also feel entitled to it as a matter of right. And so that's where it gets translated into a demand, from the activists' perspective. Their attitude was—and this was directed at the state, specifically, the state of Michigan—"You guys are the ones who fucked up and poisoned us, and you have an obligation to provide us with the safest water available, at the very least until you've replaced all of our dangerous service lines."[25]

Of course, what residents were actually insisting on was not bottled water per se—it was safe water. A month after Karen Weaver was elected Flint's mayor in November 2015, she declared a citywide state of emergency, one of the catalysts that turned the crisis into a national story. Within another month, both Governor Snyder and President Obama had approved emergency declarations, freeing up state and federal aid to Flint. The Michigan National Guard was mobilized to distribute bottled water in the city, with media images that—if it were not winter in the Midwest—could have been interchangeable with the aftermath of Hurricane Katrina or another natural disaster.

However, there was a crucial distinction. "In Flint they [residents] originally were hoping that it would be declared a disaster," explains Peggy Case, president of Michigan Citizens for Water Conservation (MCWC). "Not an emergency, but a disaster, so they could get the big water buffaloes [bulk water tankers] and stuff from FEMA, and get treated like a true disaster. And they'd get a whole lot more out of that. But they never were able to get that declaration."[26] Pauli says many Flint activists felt that "if you could get the disaster declaration, then it would mobilize the Army Corps of Engineers, and they would come in and they would replace the pipes ASAP, but they would also bring in water buffaloes. . . . You drive the truck in or whatever and people come, they have a reusable jug that they bring and they fill up . . . because there is a general feeling that having to be so reliant upon bottled water is a bad thing."

This refusal by the state and federal governments to provide Flint residents with bulk water distribution was pivotal. It meant that bottled water would effectively become *the* sole mode of drinking water delivery for the duration of the crisis—and for many in Flint, it still is. Moreover, millions of images of Flint residents stocking up on cases of bottled water have traveled around the world on TV, internet, and social media, reinforcing the conscious or subconscious impression that bottled water is a natural or acceptable replacement for problematic tap water. It is difficult to imagine more valuable publicity for the packaged water industry.

Struggles over access to bottled water have marked the Flint crisis throughout. In response to a lawsuit brought by Flint residents and several Michigan-based and national organizations, a federal judge in 2016 required the city and state to distribute bottled water directly to the doors of all Flint residents who lacked properly installed lead-removing faucet filters. In April 2018, when the state announced it would end the free bottled water distribution sites that had been open for two years, residents rushed to stock up on as many bottles as possible. Mayor Weaver vowed to sue the state to continue the program. The next month, she announced that Nestlé would be donating one hundred thousand bottles of water per week to Flint residents for several more months, donations that still continued as of 2022.[27] In December 2018, Democratic governor-elect Gretchen Whitmer also vowed to restore state-run bottled water distribution, although that promise was not fulfilled.

By that point, the state's preferred interim solution to the crisis had changed: no longer bottled water but water filters, paid for with emergency funds. But these on-faucet filters were contested by many Flint residents. According to Pauli, "Showing that filters had been provided to all residents of the city—became the state's main strategy for ending its provision of free bottled water. Refusing the filters—and, as a corollary, demanding that the state continue to provide free bottled water—became an archetypal expression of political resistance in Flint."[28] Many residents argued the filters were difficult to install correctly, didn't actually remove lead, and were breeding grounds for bacteria.

After a fraught, multiyear process complicated by the pandemic, Flint's city government announced in 2021 that the enormous project to replace its corroded lead and steel water service lines—the result of a settlement in the above-mentioned lawsuit—was nearly complete.[29] Using nearly $100 million in state and federal funds, the city had excavated over twenty-six thousand lines and replaced nearly ten thousand with new copper pipe.[30] It reported that tests showed lead levels in the water system had fallen well below the EPA's action threshold of 15 ppb.[31] However, Flint water activists challenged the idea that the water had been rendered safe. "As of right now," charged Mays in January 2021, "the damaged city's distribution mains still remain in the ground and our corroded interior plumbing, fixtures and appliances have not been replaced. As long as our water flows through these pipes up into our taps, it will continue to be contaminated."[32] I ask Luster whether the people she knows have returned to drinking water from their taps: "I don't know any. I'm sure there are some. I don't know any. And the ones that are drinking from the tap, they're doing it because they have a filter on their kitchen sink faucet that was provided by the state in the beginning, and so they think that it's safe, but we know better because we brought in scientists to test this water that [is] supposed to be filtered."

In short, in Flint trust has been severely damaged. Whether the damage is reparable is an open question. Many Flint residents feel that one vital step in restoring faith would be to hold those responsible for the poisoning and the cover-up accountable, but developments on that score have been a discouraging roller-coaster. In June 2019, Michigan's attorney general, Dana Nessel, infuriated many residents when she announced that all pending criminal charges against state officials in the Flint crisis would be dropped because of flaws in the case.[33] In January 2021, new charges were filed against nine former state and local officials, including ex-governor Snyder, but in 2022 almost all of those charges were once again dismissed.[34]

In August 2020, Nessel announced a $641 million civil settlement between the state and attorneys for Flint residents, most of which will pay for the future costs of health treatment for all of Flint's children. Residents are divided on the settlement, some deeming it insufficient because it does not compensate poisoned adults without documented proof of harm.[35]

The City of Flint signed a new thirty-year supply contract with the Detroit water system in 2017. Although the last emergency manager left in 2015, state receivership of Flint's city government was not fully terminated until 2018. As mentioned above, the great majority of Flint's lead and galvanized steel service lines were removed by 2021, although some remain. As of late 2022, three nonprofit-run bottled water distribution sites remained open, and thousands of residents continued to use them.[36]

When I ask Luster what her organization wants to see now in Flint, she responds,

> Our three demands are simple. We put it on all of our literature and our business cards have it, and it says, "Flint Rising Demands: We don't pay for poison. Fix what you broke." Meaning our infrastructure, our pipes. "Make us whole." People lost hot water tanks, hot water heaters, refrigerators—anything that needs water to operate, people in Flint have lost that. The last thing is "Our families deserve to be healthy." And when we say that, we basically are saying Medicaid for all Flint residents. We should not have to worry about, can we afford to go see a doctor? We know we're all sick.

Of those three demands, only part of one—the replacement of pipes—has so far been achieved.

SHUTTING OFF A LIFELINE

Among the less well known aspects of the Flint story is the parallel crisis of water shutoffs. Even before the switch to the Flint River, under emergency management Flint had disconnected water service to thousands of households because of unpaid water bills—essentially a punishment for poverty.[37] According to Luster and Pauli, Flint's early water activism actually centered on opposing these shutoffs. This fostered collaboration with community activists in Detroit, where the city government made international headlines beginning in 2014 for its unprecedented mass water shutoffs, which by 2020 totaled over 127,500 (a figure including households with multiple disconnections).[38] This policy generated mass protests, widespread public condemnation, and a declaration from United Nations officials that the shutoffs constituted "a violation of the human right to water and other international human rights."[39] Lauren DeRusha, campaign director with the group Corporate Accountability, tells me that some Detroit families "have had their children removed from their homes after water has been shut off . . . their lives ripped apart because they couldn't pay their water bill."[40] Lack of running water in the home can be considered evidence of child neglect in twenty-one states.[41]

In a court hearing challenging the shutoff policy, one of Detroit's lawyers argued that the availability of other water sources—including bottled water—meant that disconnected households were not being denied the right to water. "Just because a person is out of water doesn't mean they can't get water," the lawyer stated.[42] These residents are only a small share of the roughly fifteen million people in the United States who experience water service shutoffs each year—an alarming 5 percent of the national population.[43] Water rates rose an average of 80 percent in the United States between 2010 and 2018, and 40 percent of residents in some cities cannot afford their water and sewer bills, a figure likely to increase further.[44] Nevertheless, the cost of substituting bottled water for a household's entire drinking, cooking, and washing needs is far greater.

Even after the lead poisoning in Flint was revealed, its water utility continued shutting off households for unpaid bills throughout the crisis, threatening to place property liens on their homes. The new mayor Sheldon Neeley, elected in 2019, implemented a citywide moratorium on water shutoffs,[45] and the state also suspended the practice for several months during the Covid pandemic. Nonetheless, Pauli tells me, water shutoffs remain "a threat that's constantly hovering in the background. . . . To me it's just a total scandal that it's ever considered to be acceptable to completely shut off people's water." The Flint residents who were not shut off have had to keep paying their water bills, aside from twenty-two months when the state retroactively subsidized a portion.[46] "I know people who have a $10,000 water bill," says Luster:

> I have relatives who didn't have running water for three or four years because they could not afford the water bill. We in Michigan have the highest water rates in the nation! We pay more than Beverly Hills. . . . The average home in Flint is paying anywhere from $200 to $300 a month just for a water bill . . . [but] Flint is 43 percent at, or below, poverty [line]. So imagine that if you're on a fixed income, like one of my neighbors who only gets about $790 a month, and you get a $300 water bill—where does that leave you?

> Do you eat? Do you have water? Or do you provide a roof over your head? These are the decisions that residents of Flint have to make every day.[47]

In other words, Flint residents have been forced to choose between continuing to pay for expensive tap water they couldn't use for drinking, cooking, bathing, or sometimes even washing clothes—on top of the cost of any purchased bottled water—or having their household water supply shut off entirely.

CONNECTING TAPS, BOTTLES, AND SPRINGS

In addition to fighting for safe water and opposing water shutoffs, activists across Michigan have begun drawing connections between the environmental and social justice crises in Flint and Detroit and the issue of groundwater extraction by the bottled water industry in western Michigan and beyond. They are using the physical movement of bottled water across the state, and the contradiction between the ease (and low cost) of access to clean water for the bottled water industry on one hand, and expensive water and shutoffs for residents on the other, as leverage to forge an unprecedented coalition around water injustice and the human right to water.

"When we started digging deep into why there was a Flint water crisis," Luster tells me,

> all roads led back to Detroit. . . . And when we started sitting down with the folks in Detroit, it was a no-brainer what happened to us. It was about money. It was about a pipeline. It was about privatization. It was about emergency managers. . . . It was a good thing that those groups had already been put in place in Detroit, that had been fighting [against] the shutoffs, and had been fighting against Nestlé for years. Some people since the '70s. That we realized, wow, we really have to join forces because this is a . . . mountain that we're going to have to climb, and we need folks who have already been in these spaces, know who the enemies are—as I always say, know all the characters to the play or the opera. And they were really a good catalyst for us Flint folks.

Perhaps it was inevitable that Nestlé Waters, the nation's top bottled water producer, would figure prominently in water activism in Michigan. The majority of the bottled water being donated and distributed in Flint was Nestlé water—primarily its Ice Mountain spring water brand, extracted from several sites in western Michigan, and Pure Life, which is refiltered tap water taken from municipal water supplies, including the Detroit water system. This geographic proximity alone made connecting that water to its source a logical next step for Flint activists and their allies. An ongoing legal and advocacy battle over Nestlé's spring water extraction in rural Osceola and Mecosta counties, spearheaded by MCWC, had been in the news since 2001.[48] This struggle centered on the impact of the company's pumping on nearby streams and rivers, the issue of how much it should be allowed to extract, and the negligible fees paid by Nestlé to access large volumes of water.

Connections between MCWC and groups in Flint and Detroit had been made in the early days of the crisis, but by 2016 they began taking more tangible form. After Nestlé applied for a state permit to nearly double its groundwater pumping at the headwaters of two trout streams near the town of Evarts to 210 million gallons (800 million liters) per year, MCWC forced the state to hold a public hearing in the town of Big Springs. According to Case,

> We had a bus from Detroit and one from Flint. And so those folks came up and they testified, and it was really good 'cause they really pointed out the injustice of all this—you know, [Nestlé pays] $200 a year for a filing permit and then you get 210 million gallons of water for that . . . And then you do the math on how much the people of Flint would pay for their water if they were paying the same price Nestlé's paying. It's like seven cents a year or something like that—they wouldn't have water bills! . . . So we have been keeping this, the more statewide view of things going for some time, and connecting the bottled water piece to the human right to water.

These urban-rural alliances have proven mutually beneficial and drawn media attention. The tactics involved have occasionally gone beyond testimony at public hearings, press conferences, and protests to direct action. Luster recounts,

> [Filmmaker] Michael Moore is from Flint, and he reached out and basically said, "What can I do to help?" . . . So he provided us a way to get to the Nestlé plant [in Stanwood, Michigan]. . . . [We had] a busload, so, thirty [people]. And we have . . . the film crew. . . . Next thing we know we're locked inside of this lobby. We're demanding that they come out and speak to us . . . They disappeared, and you could see them scrambling like rats, running away from us.
>
> Next thing we know, here's every kind of police you could think of, saying, we're taking you guys to jail. Wait a minute! You're pumping our water for $200 a year. You're a part of the problem. You're charging us to buy this bottle of water that we don't want. We never had to buy bottled water [in the past]. . . . Now we're forced to buy bottled water. . . .
>
> But instead, they didn't want to answer our questions. . . . We had every federal agency outside, even like the national wildlife people kind of pulled

> up, park rangers. . . . I guess they didn't have a paddy wagon, so they just called all surrounding agencies to come take the Flint folk to jail. . . . Michael Moore, with his producers, maybe thirty, forty-five minutes of talking these people down, and eventually they let us go get back on our buses. . . . It was one of those whew! moments. But that was our Nestlé story. Haven't been back since.

National organizations supporting grassroots activists in Flint and Detroit have made similar linkages between tap water crises and the packaged water industry. Mary Grant, campaign director at Food and Water Watch, framed the connection succinctly: "What is the relationship between bottled water [and tap water]? Flint relied on bottled water when they didn't have access to safe [tap water]. . . . A lot of the groups we work with in Detroit, when people experience water shutoffs, they have to rely on bottled water. But at the same time you have companies like Nestlé that are taking public water resources for essentially free and then selling them for a major profit to communities that don't have safe water, [or] that don't have water that they can afford to pay."[49]

Interestingly, Nestlé too linked its bottled water production directly to the Flint disaster. In a thirty-second TV advertisement showing the company's Stanwood bottling plant and a bottled water distribution site in Flint, a member of a local church says in a voiceover, "Some people had forgotten about Flint, but Nestlé Waters never did. Three years later, they're still providing spring water. . . . In Flint we haven't given up, because we have friends. Like Nestlé Waters."[50]

Activists also invoke the connections between Nestlé and Flint by highlighting the company's relationship to former governor Rick Snyder's administration. Snyder's chief of staff during the Flint crisis, Dennis Muchmore, who was involved in many key decisions regarding Flint, was married to Deborah Muchmore, a lobbyist and public relations consultant who served as Nestlé Waters' Michigan spokesperson.[51] Luster says that "at the time of the Flint water crisis, our governor, Governor Rick Snyder, had ties to Nestlé. . . . How ironic that a friend of yours [works for] Nestlé and this is the water that's being sent to Flint?"

Concern about the role of bottled water amid the crises in Flint and Detroit has even become the subject of a congressional investigation. In March 2020, the U.S. House Oversight and Reform Committee's Environment Subcommittee held public hearings on this issue. Michigan representative Rashida Tlaib, the subcommittee vice-chair, who represents the Detroit area, told Michigan Live that "where my constituents are having their water shut off due to exorbitant bills, we have Nestlé up the road profiting millions off the water my community is being denied."[52] Tlaib and California representative Harley Rouda sent a letter to the CEO of Nestlé Waters North America, demanding information about the company's bottled water sales, revenues, advertising, and extraction practices around the country. "The subcommittee is concerned," they wrote,

> that Nestlé is taking a critical public resource from communities in need without equitably reinvesting in those communities. . . . Between 2005 and 2015, Nestlé extracted 3.4 billion gallons of groundwater in Michigan. In 2016 alone, Nestlé made more than $343 million in sales of bottled water coming from water sourced in Michigan. When Flint, Michigan was in the midst of a lead contamination crisis, Nestlé continued to extract spring water from Michigan communities like Evart for the purpose of selling the water outside of the state, though this town was located just two hours from Flint.[53]

Nestlé promised to cooperate with the investigation, adding that the company recognizes "the significant responsibility we have as a bottled water company to operate responsibly and sustainably."[54]

The collaborations between grassroots Flint and Detroit activists and water protection groups like MCWC expanded further still into a regional effort. In September 2017, all of these actors, along with local groups opposing Nestlé's bottled water extraction in neighboring Ontario, Indigenous communities on both sides of the border, and several U.S. and Canadian water justice organizations, met in Flint for a summit titled "Water Is Life: Strengthening the Great Lakes Commons." Out of that gathering emerged the Water Is Life Alliance, which described itself as "a community-based effort of Michigan, Ontario, and Indigenous residents opposing commodification and privatization of water."[55] Although this particular initiative did not survive the pandemic, the work to link bottled water movements to justice issues of water shutoffs and affordability on a statewide level continues under the umbrella of the People's Water Board Coalition, composed of more than thirty faith, labor, environmental, and social justice organizations in Detroit, Flint, and Benton Harbor, along with MCWC and For Love of Water (FLOW) in western Michigan, and Food and Water Watch.[56]

FLINT: THE BIGGER PICTURE

It is worth stepping back to assess the significance of Flint and Michigan in the broader context of struggles over access to safe water, both domestically and internationally. The role of bottled water in Flint poses some genuine paradoxes. Flint activists demanded access to, and distribution of, free bottled water as the main alternative source of potable water in the midst of the toxic water crisis. When free state bottled water distribution ended, their insistence that it continue was echoed by the winning candidate for governor. How can this be squared with these same activists joining forces with groups opposing the extraction of that very same bottled water? Of course, demands for a federal disaster declaration in Flint, which would have allowed access to bulk supplies of clean water, were denied. The residents of Flint, Detroit, and other similar communities around the United States and Canada who have experienced shutoffs or unsafe drinking water had few choices other than to consume this far more costly product as a stopgap.

Yet what is significant about Flint is the fact that advocates did not stop at merely demanding a return to safe water in their city. For one thing, these activists have aimed to situate the local crisis within a larger context of infrastructure deterioration, austerity, threats to tap water safety, and the national crisis of water affordability. Certainly this framing, and the collaborations it has fostered, are one important outcome. "You realize it's not just you, and it's bigger than you," Luster tells me.

> . . . In the beginning of the water crisis, we didn't realize that this had been happening in other [communities]. . . . It had happened in D.C. It had happened in Pittsburgh. And to go and be able to sit down and talk with those people who had been through it in D.C. and Pittsburgh, it really gave us a tool to use and [ideas] of what to do and what not to do. And coming into these spaces, it broadens my scope. . . . It makes me a better organizer and activist to sit down next to someone who has accomplished what we're trying to accomplish.

These issues also came to be linked with activism around water shutoffs in Detroit and groundwater extraction for bottling in western Michigan. The latter concern is geographically removed from Flint and might not appear to be a high priority for grassroots activists dealing with life-or-death water safety and affordability issues. Yet they have connected the dots between these phenomena as one axis of a diverse urban-rural alliance. Those connections were facilitated by the physical presence of Nestlé's bottled water—the primary source of drinking water during the Flint emergency—and by the radically disparate standards and costs involved in accessing clean water for Flint or Detroit residents versus Nestlé. In this context, the commodification involved in water bottling—despite occurring over a hundred miles away in a rural region with very different socioeconomic and demographic traits—became viscerally relevant to urban activists who were confronting crises of safe and affordable water access.

These alliances turn a disturbing tale of government malfeasance and environmental injustice into a deeper and more complex story involving issues of corporate power, neoliberal austerity, privatization and commodification, and threats to local democracy. This final theme—that the Flint disaster was first and foremost a crisis of democracy—emerges repeatedly in the statements of local residents and activists. As we end our interview, Luster tells me: "What happened to us in Flint really made me more active for other causes. Because when you sit down and you get in a space and you hear about other communities, it all kind of leads back to one thing. The failure of government, the failure of democracy, the failure of dignity. If you can have dignity, and our democracy back, which means get rid of the emergency manager law across the board. It should not be allowed in any city, state or town. And give us our dignity."

TOXIC INEQUALITY

Much of the reporting on the Flint crisis observes that Flint is just the tip of the iceberg—that it is emblematic of much larger dynamics of deteriorating infrastructure and the unequal distribution of unsafe tap water in North America and beyond. The cost of the needed repairs and maintenance for water and sewer systems, which in the United States has been estimated at up to $3.2 trillion over the next twenty years,[57] has been pushed by a stingy federal government onto states and municipalities, which despite dedicating their own revenues to the problem are often unable to keep up with the needed upgrades. And this deterioration is breeding distrust in tap water. "In some places, water infrastructure appears to be at a breaking point," writes Ryan Felton. ". . . The lack of investment is widespread: Detroit public schools shut off all drinking water last year because of high copper and lead levels. A town in West Virginia has been on a boil-water advisory since 2002. . . . 34 percent of Americans—or 110 million people—say they regularly avoid drinking tap water at home because of safety concerns. . . . About one-sixth say they don't drink their home tap water at all."[58]

In fact, the Flint disaster has been nearly replicated across the state in Benton Harbor, Michigan, whose nearly ten thousand residents are 85 percent Black. Local clergy and residents complained for three years about high lead levels in tap water, while city and state officials delayed and fought over anticorrosion treatments. In late 2021 several advocacy groups, including Flint Rising, filed an emergency petition asking the U.S. EPA to take action. The governor then urged residents to use bottled water and directed state and federal funds to replace the lead pipes. Every home in Benton Harbor has received bottled water deliveries, provided first by churches and later by the state.[59]

The largely Black residents of Jackson, Mississippi, the state capital, have suffered for decades from problems with the city's fragile and underfunded water system, including frequent boil water orders, high lead levels, and loss of water pressure. The situation culminated in near-total breakdowns after extreme weather in 2021 and 2022, with the entire population forced to rely on bottled water. Critics charge that largely White state officials have long starved the Jackson system of resources for direly needed repairs while adequately funding other communities, and they link the crisis to a history of racial segregation and environmental racism.[60]

Recent reporting shows that tap water problems are more widely distributed than was previously believed. A report by the Natural Resources Defense Council documented that systems providing drinking water to approximately sixteen million people in the United States were in violation of the Lead and Copper Rule in the Safe Drinking Water Act.[61] These systems, mostly in smaller communities, represent less than 5 percent of the U.S. population, and many of the violations are remedied quickly. Recent media coverage of contamination by pesticides, nitrates, and PFAS or "forever chemicals" has also raised alarm for some communities.[62] Peter Gleick tells me that for all kinds of contaminants, the solution is straightforward: stricter regulation. "If the EPA and Congress fail, as they have failed for decades, to really expand protections of drinking water regulations, that will drive more people to bottled water and it will not drive municipal agencies to upgrade their system. If the standards were stricter and more broad, there would be a requirement that municipal water agencies improve water treatment, which we need to do, and they're not doing so."[63]

Can advocates for public water speak honestly about the threats to tap water safety that are posed by decaying century-old infrastructure and underfunding of maintenance (and contaminants from industry and agriculture) without scaring even more people away from tap water that in most cases is still quite safe to drink, or weakening public pressure to fix these problems? Water justice advocates whom I spoke with acknowledged the difficulties of promoting tap water use and opposing bottled water amid a steady drumbeat of revelations of new urban and rural water problems. Maude Barlow tells me, "There is a real connection between underfunding of infrastructure and the loss of faith in public water. Here in Canada, we recently had a major report that there are serious levels of lead in many of our aging systems. This set our campaign back, I can assure you. The problem is often local, such as a school where the pipe from the main water source into the school is lead lined, so it might not even be the municipality that is at fault. Nevertheless, it's a huge issue."[64]

Some go further along these lines. "The thing is, we have to be honest about the limitations of public water provision, at the very same time that we're fighting to defend it," says Pauli. "That's really the challenge for activists who want to work on this issue moving forward. Because every time you have a situation where officials are exaggerating what is known about the safety of tap water and then people find out that there's actually more to the story, it just puts another dent in people's willingness to trust these big public systems."

Of course, this infrastructure deterioration—and the resulting distrust of the tap—is not evenly distributed. A 2017 U.S. Gallup poll reported that 80 percent of non-White respondents were worried "a great deal" about drinking water pollution, compared to 56 percent of Whites. Seventy-five percent of those earning less than $30,000 per year worried a great deal, versus only 56 percent of those earning more than $75,000.[65] A study of U.S. drinking water system violations by *The Guardian* found that water systems in poor and rural communities had the most serious and chronic problems, and that Latino/a residents were the most affected by contaminants in their drinking water.[66] Another study found that Black and Latino/a populations are the strongest predictor of violations of the Safe Drinking Water Act in low-income communities.[67] And an analysis by the NRDC concluded that "drinking water systems that constantly violated the law for years were 40 percent more likely to occur in places with higher percentages of residents who were people of color," as a result of "residential segregation and other forms of discrimination . . . [that have] led to aging, underdeveloped, and underfunded water infrastructure."[68]

These patterns have direct implications for people's consumption of bottled water. Economist W. Kip Viscusi and coauthors found that "consumers are more likely to believe that bottled water is safer or tastes better if they have had adverse experiences with tap water or live in states with more prevalent violations of EPA water quality standards."[69] Pauli concurs, adding further context: "Research suggests that people of color, particularly African Americans and Hispanics, tend to be less trusting of tap water in general. With African Americans, I think that one reason for that . . . is that they tend to live disproportionately in urban environments where you have issues like lead and aging infrastructure [and] other water quality issues that stem from particular social contexts. . . . And the alternative of choice for most folks is not to go down to Home Depot and get some sort of point of [use] filter—it's instead to purchase bottled water."

Media coverage of drinking water crises can also harm people's trust in tap water well beyond the immediately affected communities. Coverage of the Flint disaster, write Gregory Pierce and Silvia Gonzalez, "has drawn national attention and increased suspicion of tap supplies," with studies showing a dramatic drop in confidence in tap water across Michigan.[70]

However, many observers caution there is also a danger of *over*emphasizing the scale of current tap water safety problems. They stress that the vast majority of tap water in the United States and other Northern countries remains safe to drink. "Since 2013, the percentage of the U.S. population serviced by community water systems with at least one reportable health-based quality violation has stayed below 10 percent," notes Felton.[71] Water justice advocates are insistent on keeping the scale of water safety problems in perspective. "More than 90 percent of water systems are fully in compliance with all federal regulations," Grant tells me. "Water quality violations disproportionately impact . . . low-income, rural [people and] communities of color. . . . These are where the health violations are occurring disproportionately. So . . . we need to dedicate funding toward communities that need the help the most."

AUSTERITY, RACE, CLASS, AND WATER

The imposition of emergency management on both Flint and Detroit, in which unelected EMs overrode local democracy to make ostensibly cost-saving decisions such as the switch to the Flint River, privatized public assets, and more, is a quintessential example of what geographer Jamie Peck terms "austerity urbanism." Peck describes how austerity, which has long been an element of the neoliberal toolbox, has particularly come to the fore since the Great Recession in both the United States and Europe. Austerity, he writes, "is something that Washington does to the states, the states do to cities, and cities do to low-income neighborhoods. . . . [It] is the means by which the costs of macroeconomic mismanagement, financial speculation, and corporate profiteering are shifted onto the dispossessed, the disenfranchised, and the disempowered." These draconian measures are justified under the logic "that the imposition of strict fiscal discipline and government spending cuts is the (only) way to restore budgetary integrity."[72]

Such policies are imposed disproportionally on low-income communities, and especially on people of color. Geographer Katie Meehan and coauthors argue that "water provision to largely Black and brown communities has been devalued and subordinated to the goals of fiscal solvency in ways that exacerbate social inequalities and threaten lives."[73] The sociologist Laura Pulido adds that the residents of Flint have been abandoned by both capital and the state: "This constitutes *racial capitalism* because this devaluation is based on both their blackness and their surplus status, with the two being mutually constituted" (emphasis added).[74]

This type of urban austerity blurs the lines between the so-called developed and developing worlds. The situations in Flint and some other predominantly non-White cities, Indigenous communities without safe tap water, towns in California's Central Valley with contaminated groundwater,[75] and rural communities in many resource extraction zones constitute a figurative "South within the North." In these and other settings, fiscally starved (or structurally adjusted) local governments cannot afford to remove some contaminants or upgrade decaying pipes and treatment plants, so (without greatly increased federal aid) there is a higher chance their tap water quality will deteriorate. Thus residents can lose trust in their tap water irrevocably—as have many people in Flint—and they turn to the market for an answer, in the form of bottled and packaged water. In such cases, this "inverted quarantine" response is not a reflex of privilege but rather an additional cost and labor burden imposed on poor and working-class people.[76] When these residents' tap water supply is not safe to drink, or if it has been shut off, or if they lack running water in their home entirely, their structural situation parallels that of the residents of many urban and peri-urban areas in the global South, who are obligated to purchase packaged water (if they can afford it) in order to access safe water. Pulido writes that "this is not new. This is what happened in the Global South in the 1980s. . . . One Flint resident actually said that the community felt like a 'Third World Country.' . . . Flint's water disaster signals the inability to provide a hallmark of the 'first world' to its surplus residents: clean, tap water."[77]

In all of these settings, widespread *dependence* on bottled water is a marker of water injustice. It serves as a red flag indicating that the human right to water—the ability to access sufficient, reliable supplies of affordable, safe water for drinking (and cooking), in or very near the home—is being denied.

Of course, bottled water consumption is not only a marker of water injustice. In fact, as we saw in the previous chapter, it has historically been linked with higher incomes and discretionary spending. But newer data—both opinion polls and academic research—show that in the United States, bottled water consumption is now increasingly correlated with *lower* incomes and is growing most rapidly among low-income people and communities of color.[78] Thus there is now a bimodal market for bottled water. The first group is economically advantaged people whose tap water is generally fine but who can afford to opt out, whether in response to media coverage of tap water emergencies, industry advertising, or shifting norms around fitness and hydration. The second group consists primarily of non-White and low-income people who are more likely to feel, often justifiably, that their tap water is not entirely safe to drink.[79] They buy bottled water despite the high cost burden, either because their tap supplies have documented problems or because they perceive a risk. This latter group is now the fastest-growing segment of the bottled water market.

A MARKET OPPORTUNITY; AN EASY WAY OUT?

The bottled water industry has been quick to seize on widespread awareness of the Flint crisis and other problems with deteriorating tap water quality. "Concerns in places like Flint do bring bottled water to people's attention as a safe and sealed source of drinking water," stated Nestlé Waters spokesperson Jane Lazgin.[80] Journalist Jennifer Kaplan writes that the industry's "expectation that sales will rise has to do with . . . crumbling infrastructure,"[81] as the epigraph to this chapter suggests. Indeed, the anticipation of business opportunities from deteriorating infrastructure dates back further than the Flint crisis. In 2009, the then-CEO of Nestlé Waters North America, Kim Jeffery, delivered a PowerPoint presentation titled "The Future of Bottled Water."[82] The text of one key slide reads:

What Is The Future of Bottled Water?

1. Water in total is a winner.
2. We believe tap infrastructure in the U.S. will continue to decline.
3. People will turn to filtration and bottled water for pure water needs.
4. We are bullish on water over the next 10 years.

The fact that Nestlé has understood for well over a decade that the trend of water infrastructure deterioration presents a major market opportunity is noteworthy. It offers further evidence that the company viewed tap water as its competition but also suggests that bottled water's use in contexts like Flint is, in the industry's vision, not merely an emergency fix for a short-term problem, but rather a longer-term solution. "Bottled water," writes Gay Hawkins, "has to be understood as an opportunistic commodity, something that makes sense in the face of threats to or the absence of other forms of safe water provision."[83]

However, it is not only beverage firms who see positives in this product substituting for stressed tap water infrastructure. Bottled water can also serve as a crutch for some public officials. It is a far cheaper and politically easier solution, at least in the short term, than raising taxes or water rates—especially when residents, not government budgets, pick up the tab for purchasing the water. Put differently, bottled water's availability can offer local officials a way to avoid the immediate obligation to fix water system problems. Mary Grant tells me that she sees this dynamic

> most sharply with the situation of lead piping in schools. For example, a lot of schools have relied on bottled water for more than a decade now. . . . Baltimore has been on bottled water since 2007 in the [public] school system. . . . Instead of providing that state funding to fix and remove lead plumbing within schools, they're relying on bottled water, because it's . . . year-to-year budget making. So . . . in the long term, even though [these investments] pay off, they just don't have the resources and capital right now to make those improvements that they need to make in order to get off bottled water.

Other cities have experienced similar problems: Atlanta, Detroit, Chicago, Newark, and Portland, Oregon, to name just a few, have found lead in their schools' water pipes, responding to the problem by turning off water fountains and sinks and providing bottled water to students instead, often for years, because of the high cost of pipe replacement.[84]

However, not everyone agrees that bottled water is an ideal cop-out for cash-strapped public officials. "It costs the state a lot to provide bottled water the way it was provided here in Flint," Pauli tells me. "The state never wanted to be doing that again . . . just because of the immense costs to that. Yes, replacing infrastructure is costly too, but I don't think the state wants to be in the bottled water distribution business."

This observation is an important one. Yet it is also worth acknowledging that aside from two years of state-funded bottled water distribution in Flint, across the United States it is overwhelmingly individual households and sometimes private donors—not city or state governments—who end up paying the cost of bottled water to substitute for tap water that people know or believe to be unsafe. And those residents must also keep paying their water bills. Thus local officials may indeed, consciously or unconsciously, find bottled water a useful release valve to lower public pressure for immediate fixes to unsafe tap water. In such a context, there is a risk that decision makers and even residents may come to view bottled water as a tolerable medium-term—or even long-term—"solution" for deteriorating tap water quality.

WHAT'S AN EMERGENCY?

Even some of the most vocal critics of water commodification concede that bottled water has a legitimate role to play in true disasters or emergencies. For example, Gleick views it as acceptable as a "short-term solution for emergencies when other safe alternatives are not available."[85] Yet such statements only generate further questions. First, what constitutes a genuine emergency? Does the socially produced lead and bacterial poisoning of Flint residents—a direct outcome of imposing an unelected manager on a predominantly poor and Black city—qualify? Is the terminology of "emergency" or "disaster" even appropriate here, or should we instead be using the language of criminality and human rights violations? Even a "natural" disaster like Hurricane Katrina, and the responses to it, can be understood only by recognizing how it is layered on top of deep existing class and racial inequalities. Given all that, is the provision of bottled water really an acceptable solution for the poisoning of an entire urban population via imposed austerity, and its prolongation by delay and cover-up?

Second and equally important, we need to ask when the disaster ends. At what point does the imperative to repair the underlying causes of the damage kick in? What is the moral statute of limitations for neglect? If bottled water's availability reduces the political impetus for officials to actually repair decaying water infrastructure, allowing them to extend such states of "emergency" indefinitely, there is a systemic problem. Mays, in a 2018 video, alleges that "if we didn't have bottled water as a safety option or a backup—as a wet Band-Aid for Flint—our pipes would've been fixed a long time ago."[86] Similarly, political scientist Raúl Pacheco-Vega warns that the presence of donated bottled water in disaster-stricken communities could "end up promoting it as a permanent solution to a public policy problem instead of a temporary relief measure, thus leading to a dereliction of duty on the part of governments, who will not see it as their duty to ensure that all citizens are able to have universal water access."[87]

The crucial point is that bottled water's ready availability diminishes the perceived urgency of fixing and maintaining public water infrastructure. DeRusha tells me this weakening of political pressure has real effects, particularly at the federal level, which would be the source of any sizable increase in infrastructure funding:

> We need the federal government to step back in and to provide funding. And the industry paints it as a fairy tale that the federal government could invest in our water systems again, but it's not a fairy tale. The federal government used to invest in it—and it is simply a question of political will. The money exists. It's just being spent elsewhere. And if everybody just says, "I don't trust my tap water anymore and therefore I'm only ever going to buy bottled water," people are voting with their dollars to rely on Nestlé, maybe for the rest of their lives, or Pepsi or Coca-Cola, instead of advocating for their public water systems and providing the backing that public officials really need.

The clearest path to addressing this corrosion of both pipes and trust, then, is to dramatically increase federal spending to repair, maintain, and improve dilapidated water and sewer systems. Rebuilding this public infrastructure is the only long-term solution commensurate with the scale of the problem. For several years, the Water Affordability, Transparency, Equity, and Reliability (WATER) Act, introduced by Vermont senator Bernie Sanders, California representative Ro Khanna, and Michigan representative Brenda Lawrence, which would greatly increase ongoing funding to restore water and sewer systems across the United States, has languished in Congress. In May 2021, Congress approved a modest down payment on this vision with a one-time $35 billion water and wastewater infrastructure funding bill, 40 percent of it earmarked for rural, tribal, and underserved communities.[88] The bipartisan Infrastructure Investment and Jobs Act followed in November 2021, appropriating $55 billion for restoring water and sewer systems, including $15 billion for removing lead pipes, half of it in the form of loans. While a historic investment, this funding is still just one-fourth of the cost of resolving the problem of lead in water pipes nationwide, only 7 percent of the amount needed for all utilities to meet EPA water quality standards, and wholly insufficient to address the root causes of disasters like Flint.[89]

DRINKING WATER IN A PANDEMIC

All of these dynamics were thrown into sharp relief by the Covid-19 pandemic. While early on utilities had to combat (inaccurate) public fears that tap water might carry the virus, the pandemic made it painfully clear that *not* having access to reliable in-home supplies of tap water for handwashing *does* increase transmission of the virus. In this context, water utility shutoffs are not just a violation of the human right to water—they can literally be deadly.

In the first months of the pandemic, these stark realities became apparent to public officials. Many cities and twenty states enacted bans on utility disconnections during 2020, and a few, including Michigan, also mandated restoring water service to previously disconnected households. However, most of these shutoff moratoria have since expired, once again leaving households at risk of losing their water service if they fall behind on bill payments.[90]

Several studies document that these shutoff bans did indeed reduce infections and save lives. A paper by Cornell University researchers argues that many more deaths could have been avoided: "A nationwide water shutoff moratorium might have saved more than 9,000 lives and prevented nearly half a million people from being infected with COVID-19" during 2020 alone.[91]

Interestingly, Flint's response to the pandemic was relatively sensitive to the needs of its residents. In addition to a previously enacted ban on shutoffs, when the pandemic began the city government ordered restoration of disconnected water service and financial aid for low-income residents to pay water bills, policies that were later adopted statewide. Pauli writes that Flint's actions "may offer inspiration to struggling public water operators in the United States and beyond."[92]

What has the pandemic meant for the fortunes of bottled water? The industry responded by promoting its product as being essential in an emergency. A March 2020 IBWA press release stated that "bottled water is always there when you need it—in good times and bad, such as public health emergencies."[93] In the first months of the pandemic, sales of bottled water soared as panicked consumers stockpiled supplies.[94] The industry hoped to turn this temporary spike into a market opportunity, as a 2021 report indicated: "The panic buying solidified bottled water's status as a necessary product to keep stocked within the home. . . . The lingering psychological effects of the pandemic along with the loss of safe tap water in Texas during the 2021 winter storms will install [*sic*] some consumers with a desire to keep even more packaged water on-hand 'just in case.' Brands and retailers have an opportunity to facilitate these stocking up behaviors."[95]

A UNIQUE THREAT

Recall the bottled water industry's argument from the previous chapter that it is not in direct competition with tap water but rather in competition with other packaged beverages. The use of bottled water as a short- and medium-term substitute for toxic tap water in Flint and other communities whose water has been compromised—and the fact that packaged water firms tout their donations of bottled water in response to those crises in their advertising—is further evidence that packaged water is indeed being utilized (and viewed) as a direct replacement for the tap, despite claims to the contrary. Its ready availability creates a permission structure for governments to postpone making direly needed but costly investments in public water infrastructure.

This contradiction highlights perhaps the most unique attribute of the commodity of packaged water, and a key way that it effectively encloses drinking water. Bottled water is fundamentally equivalent to tap water in one utterly obvious yet crucial way: it is nothing but water.[96] In contrast with all other beverages, it directly substitutes for a good that is essential to life, vital for public health, and a key function of local government. The unspoken but fundamental claim of bottled water is to be a perfectly acceptable—indeed, a superior—replacement for tap water. Unlike soda, milk, juice, or beer, it purports to replicate a vital service provided by government for over a century—delivering a life-sustaining substance essential for maintaining public health—which public water utilities do for a minuscule fraction of the cost and environmental impact.

What will happen if "competition" from packaged water is allowed to undermine the maintenance of water infrastructure for the great majority of the population in the global North who still receive high-quality drinking water? Maude Barlow tells me this is one of her big fears: "The longer we leave undertaking the necessary upgrades, the more the public will lose faith in public water. As they turn to private water, many will not support their tax dollars going to upgrade a system they don't use except for toilets and showers. . . . So it becomes a vicious cycle: 'I don't trust the water; I will pay for an alternative source; but now I don't want to pay again to make sure the public system is safe because I don't use it.'"[97]

In settings where tap water remains almost universally available and safe to drink, by increasing public distrust of tap water and thereby diminishing the pressure on government to reinvest in drinking water infrastructure, the bottled water industry creates its own market. Put another way, if enough of the public is persuaded to *never* drink from the tap—to shun it as a source of drinking water altogether—then an increasing share of our tap water may indeed become less reliable or safe to drink. This dynamic is the essence of accumulation by dispossession. It is also a savvy marketing achievement, but one with troubling implications for society and democracy.

FROM FEAR TO RESISTANCE

Widespread dependence on packaged water serves as a key marker of water injustice. Its substitution for tap water in contexts of austerity such as Flint, Benton Harbor, Jackson, or schools in Baltimore and other cities—not just during an immediate emergency but for the medium or long term—is a glaring symptom of the abandonment of the social compact by governing elites in a context of increasingly harsh neoliberalism. In some predominantly non-White cities, low-income rural areas, and Indigenous communities in the United States and Canada, as well as in many parts of the global South, residents may find they have no options other than either to buy costly packaged water day after day, drink contaminated water, or go thirsty.

It is also evident that a large and growing swath of the populace in the global North is concerned, in some cases justifiably, about the safety of their tap water. Falling public trust in tap water is one facet of a more generalized loss of faith in government to solve societal problems—but this is itself a symptom of four decades of neoliberalism and austerity, whose advocates have promoted a narrative of state failure and starved the public sector. And the antithesis of trust is fear, which leads even more of society to turn to the inverse quarantine solution of bottled water.

However, this distrust in the tap, like bottled water consumption and spending, is not evenly distributed. In the United States, both are higher and growing faster among low-income people and communities of color, especially Blacks and Latinos/as, who on average can least afford the exponentially higher cost of relying on this product for all their drinking (and sometimes cooking) needs.[98] Bottled water is embroiled in a complex dynamic of social inequality and uneven threats to tap water quality caused by deteriorating public water infrastructure and dwindling public investment.

Bottled water and packaged water are also playing a key role in the much larger struggle over the future of public goods—in this case, our vast, elaborate public water treatment and delivery infrastructures, built over more than a century with the taxes paid by many generations. Pulido writes that "infrastructure is the manifestation of past wealth and capacity, and its eroding status . . . signifies the politics of abandonment. Crumbling infrastructure is where past economic regimes meet the present. . . . Infrastructure maintenance is a form of social investment. The decision to neglect infrastructure so that it becomes toxic must be seen as a form of violence against those who are considered disposable."[99] The disproportionate distrust in tap water among communities of color and low-income people in the United States is at least in part an artifact of this abandonment—a response both to specific present instances of environmental injustice such as Flint and Jackson and to a far longer history of systemic racism, redlining and residential segregation, disinvestment, and environmental injustice on a national level that predates the current era of neoliberal austerity.[100]

The near-universal availability of packaged water, and its displacement of tap water for drinking, weaken the political pressure needed to force government officials to properly fund, repair, and maintain our deteriorating public water systems, and to ensure that tap water continues to be treated to the highest drinking water standard for everyone. It defuses the urgency for change. Starving public water systems of vital funding corrodes not only pipes but our belief in collective solutions to water needs. And a war on tap water is also a war on public water utilities—indeed, on the very notion of the public sphere.

In this chapter I have argued that unlike any other beverage, the private good of bottled water directly substitutes for a vital function performed by the public sector and that in the process it works to undermine that function—the essence of accumulation by dispossession. That is what makes the threat posed by bottled water unique, and it is why this commodity must be treated as more than merely another consumer good.

However, these urgent threats can be resisted. In many places, that is already happening. The multiracial urban-rural Michigan alliance described above is one provocative example. But other forms of resistance to the rapid growth of plastic bottled water, and to the "war on the tap" and its threat to public water systems, have also been building over the past two decades—especially in places where safe tap water is widely available. The next chapter examines one key facet of this opposition: a constellation of efforts by local governments, citizen and consumer groups, environmental activists, and others to reverse the shift to bottled water from the consumption side. They too are working to problematize this commodity and advocating for alternative models that include restoring deteriorating water infrastructure, making drinking water far more accessible in public places, promoting and rebuilding trust in tap water, and ensuring safe and affordable public water for all.

CHAPTER 4

Reclaiming the Tap

The water bottle has in some ways become the mink coat or the pack of cigarettes. It's socially not very acceptable to the young folks, and that scares me.

—John Caturano, Senior Sustainability Manager, Nestlé Waters, 2019

It is an unbearably hot and muggy early August day in Ottawa. The temperature at 11 a.m. is already well into the nineties (over 34 °C). By the time I climb two flights of stairs in an older office building in the central city, I'm sweating heavily. I am welcomed into a rambling suite with boxes, books, and papers everywhere. My host tells me that the organization is packing up to move to another office. Richard Girard is the executive director of the Polaris Institute, a Canadian think tank founded in 1997. I've come here in search of the origins of the movements challenging the bottled water industry, a trajectory in which both Polaris and Canada figure prominently. We sit down at his large desk to chat. "I remember . . . back in the days when the bottled water movement was just starting," Girard tells me, "and I would always tell people . . . how it became blown open in the media, and it was because some restaurant in California, or somewhere, was the first restaurant to ban bottled water. It was such a big deal—like, 'Why are you selling Perrier?' And then all these other restaurants started to do it, and then somebody else paid attention, and then maybe San Francisco did something, and then the municipal thing started, and then it kind of exploded."[1]

The roots of Polaris's involvement with bottled water, Girard explains, date back to the turn of the twenty-first century and the global justice movements spawned by the expansion of so-called free trade agreements and institutions, including the World Trade Organization (WTO) and the scuttled Multilateral Agreement on Investment (MAI):

> Tony Clarke—the president and founder of the Polaris Institute—was a protagonist in the civil society actions in Seattle [against the WTO]. . . . Then they started campaigns against the MAI and other nefarious trade agreements, all from the position of corporate power. And that led to looking into the issue of the trade in services . . . and then water privatization. . . .
>
> In the 1990s the big water companies like Suez and Vivendi, Veolia, were really focused on trying to get contracts in the global South and expand as much as possible. That didn't really pan out for them, but during that time Polaris started doing a lot of strategic corporate research, writing corporate profiles of the big water services companies. And then . . . Tony and some others started thinking about, well, bottled water is exploding right now. Is this another form of privatization? We really need to focus in on what is going on in the bottled water industry, who are the main players? So that led to him wanting to write a book about it, led to *Inside the Bottle.*

The publication of *Inside the Bottle*[2] in 2005, a book one observer described as "a mixture of detailed corporate and market analysis, industry exposé, and political advocacy,"[3] coincided with Polaris's launch of a nationwide campaign to take on both the industry and the commodity of bottled water—the first effort of its kind. The book made two intriguing and non-obvious connections: it linked the emerging bottled water industry with the global corporations involved in water system privatization around the world, and it framed the rapid growth of bottled water as a peril to the public provision of drinking water. The campaign worked to involve Canadians as civic actors—rather than consumers—in reclaiming their public drinking water systems, as part of a broader process of rescuing democratic institutions from corporate influence.[4]

Polaris aimed to activate that resistance in two arenas where public voices still retained at least some influence on decision-making: municipal governments and university and college campuses. Girard describes the campaign's evolution:

> I started then doing research into the big companies, so Coca-Cola, PepsiCo, Nestlé. . . . And then Polaris started doing an actual campaign. So we had done all this research, and then we started the Inside the Bottle campaign in Canada, but we did have partners in the U.S.: Corporate Accountability International, Food and Water Watch, and others.
>
> In Canada we really focused on . . . getting municipalities to ban the purchase and sale of bottled water on city property. . . . And the vast majority of those victories that we had are still in place. . . . Some [city councils] would ban it completely on city property; others were just like, it's not going to be in the vending machines, or we're going to reinvest in tap water. The key for us was access to publicly delivered tap water and trying to hit the

corporations as much as possible. . . . And then we had another arm, which was focused on campuses.

Polaris developed a highly effective approach to winning policy change at the municipal level, including Canada's largest city. In 2008, Toronto banned the purchase and sale of bottled water on city property, despite vociferous opposition from the Canadian Bottled Water Association, Nestlé Waters, Refreshments Canada, and lobbying firms hired by the industry.[5] The Toronto campaign, says Girard, illustrated the group's potent organizing model:

> There was a critical mass of people working on the issue in this country, and we basically had it down to a science of how to approach municipalities. We worked really hard with the FCM [Federation of Canadian Municipalities] and others to get them to actually [send] communiqués to their members about the issue. . . . We were very strategic about who would be a champion of the issue, and getting Toronto to do that was really important. Having that mayor say the stuff that he was saying, and making comments like that, that was pretty big. . . . So, of those municipalities our biggest victory was in Toronto.

• • •

This chapter explores the first of two major facets of the movements challenging the industry: activism targeting bottled water at the consumption end, as opposed to the sites where it is extracted and produced. These demand-side challenges focus on the commodity of bottled water generically, largely without regard to brand names or whether the water originates from already-treated municipal water supplies or from springs and aquifers.

These campaigns consist of a diverse mélange of efforts that defy easy categorization. They range from the community to the national (and even transnational) levels and span the public, private, and nonprofit sectors. While the motivations of the community residents, activists, students, environmentalists, local officials, and others involved are highly varied, most of these initiatives share a few key aims. First, they seek to reduce or eliminate the sale and consumption of packaged water, because of either its negative environmental and economic effects or the threat it poses to public water. Second, they aim to substantially improve access to tap water in public (and sometimes also private) spaces. Third, they work to promote increased consumption of tap water and to stigmatize packaged water—to make it a contested and even a reviled commodity.

In concrete terms, "reclaiming the tap" has taken two main forms over the past two decades: on the one hand, hundreds of cities and localities banning the purchase of bottled water for governmental use, and often prohibiting its sale on public property, such as parks, beaches, zoos, or museums; and on the other, thousands of universities and other public and private institutions prohibiting or restricting sales of bottled water in vending machines, coffee shops, cafeterias, dormitories, or classroom buildings. In almost all these cases, eliminating bottled water is accompanied by simultaneous efforts to promote tap water use and make it far more accessible and appealing, by replacing aging water fountains with shiny new filtered refilling stations, distributing refillable water bottles, or educating people about tap water quality. Most of these initiatives have emerged organically, whether from engaged students, local green groups, or city councilors, but many are also inspired or supported by national and international campaigns coordinated by a handful of major organizations.

On top of formal government and institutional policies, more recently there is a burgeoning range of efforts by nonprofit groups, small businesses, and others that strive to make tap water available in more private spaces, especially for refilling reusable bottles, and to enable people to find those refill points easily—thereby undercutting the bottled water industry's key argument that only its single-use product can meet the need for convenient hydration on the go.

While this element of the reaction against packaged water emerged and has spread mainly in places with widespread access to high-quality tap water—primarily in privileged nations and communities not affected by water safety crises—that is now beginning to change as the plastic pollution crisis has energized the demand to reclaim the tap on a global scale.

FIVE MOVEMENT PROTAGONISTS

The earliest major research about bottled water's impact emerged in the late 1990s—well before *Inside the Bottle* appeared and just a few years after the industry widely adopted PET plastic, Pepsi and Coke introduced Dasani and Aquafina, the Big Four consolidated their market control, and sales mushroomed. The initial concern was the environmental impact of billions of single-use plastic water bottles. Girard recalls that "the environmental stuff . . . originally, that was the big thing that led restaurants to start banning bottled water." The Natural

Resources Defense Council's blockbuster 1999 report, "Bottled Water: Pure Drink or Pure Hype?,"[6] put hard numbers on the industry's growth and impact for the first time, contributing a few years later to the first efforts to ban bottled water by universities and local governments.

Douglas Holt writes that "while climate change was far and away the focal environmental issue of this period, bottled water was arguably the most intensive and best-organized environmental campaign focused on a particular issue. The core argument of all of these efforts, regardless of creative spin, was to switch from bottled water to tap. The lead argument came straight from the ethical values paradigm: bottled water is a big environmental problem and it is such an easy one to fix, so you should do something about it."[7] Several major environmental organizations, including the Sierra Club, Greenpeace, and World Wildlife Fund, have been involved in anti–bottled water campaigns at various times.[8]

Tracing the ideological and historical DNA of the anti–bottled water movements on the consumption side is like peeling back the layers of an onion. One origin story seems to lead to another. In broad terms, these efforts emerge from a twin set of ideological and institutional roots. On one hand, they are an outgrowth of environmental groups opposed to the huge energy, water, and waste footprints of what they view as an unsustainable and unnecessary product. On the other, they stem from organizations that have been involved in anticorporate and global justice activism since the late 1990s, as Polaris illustrates, but also to earlier consumer and anticorporate advocacy dating as far back as the 1970s. The most significant and long-standing actors in opposing bottled water at the consumption level represent the latter set of roots. They are a quartet of advocacy organizations—two based in Canada and two in the United States—that have also been active in fighting tap water privatization: Polaris Institute, the Council of Canadians, Corporate Accountability, and Food and Water Watch. A fifth major player, the Story of Stuff Project, is a newer arrival that straddles these two camps (see table 5).

MORE CANADIAN ROOTS

Returning to the second, anticorporate set of roots, Polaris's Inside the Bottle campaign proved highly successful—winning bans on bottled water in ninety-two municipalities, on twenty-six university and college campuses, and in eleven school districts in just a few years.[9] However,

TABLE 5 MAJOR NORTH AMERICAN ORGANIZATIONS WITH BOTTLED WATER CAMPAIGNS

Organization	Headquarters	Year founded	Previous organization names	Earliest work on bottled water
Polaris Institute	Ottawa	1997	—	2005
Council of Canadians	Ottawa	1985	—	2009
Corporate Accountability	Boston	1977	INFACT; Corporate Accountability International (CAI)	2005
Food and Water Watch	Washington, DC	2005	—	2007
Story of Stuff Project	Berkeley, CA	2007	—	2010

by 2011 its work on the issue had begun to taper off. "We got to a point in Canada," Girard explains, "where we had a huge amount of success in getting municipalities to ban the purchase of bottled water, and campuses, and it was a big issue, and then a couple things happened and our funding went away. And then that led to us stopping the campaign, and I think it coincided with a bit of, you know—people knew the issue."

However, another major organization was present to pick up the baton. The Council of Canadians was founded in 1985 by several prominent cultural and political figures, including authors Farley Mowat and Margaret Atwood, environmentalist David Suzuki, and author and activist Maude Barlow, who became the group's chairperson. The Council's founding concerns included defending Canadian economic and environmental sovereignty in the impending Canada-U.S. Free Trade Agreement[10] and advocating fair trade policies. Its work expanded to include the fate of the Great Lakes and protecting Canada's freshwater supplies in global trade agreements, along with opposing privatization of public services, particularly water. It now has more than fifty local chapters nationwide.[11]

In 2009, the Council launched its "Blue Communities" initiative, in collaboration with the Canadian Union of Public Employees. Barlow recalls the moment: "[Conservative Party leader] Stephen Harper was prime minister. He was promoting public-private partnerships on municipalities that wanted to increase their infrastructures; maybe they were growing, or the pipes were one hundred years old; they needed money. And he said, if you want federal funding you have to move to a private model. And we were fighting bottled water. . . . So instead of being against everything, or against bottled water, against privatization, we decided to be *for* something. To have a vision. We wanted to think about becoming aware, municipalities taking a stand."[12] In order to become a Blue Community, municipalities must pass resolutions committing themselves to three specific actions: first, to "recognize and protect water and sanitation as human rights"; second, to "protect water as a public trust by promoting publicly financed, funded and operated water and wastewater services"; and third, to "ban or phase out the sale of bottled water in municipal facilities and at municipal events."[13]

The Blue Communities effort explicitly linked the bottled water industry with the privatization of public water systems, a connection Polaris's campaign had also made, although not explicitly in its demands of municipal governments. The third requirement—banning governmental purchases of bottled water and its sale in public facilities—extended the ambition of the project to match the actions that Toronto and other Canadian cities had already taken. Starting with the Vancouver suburb of Burnaby in 2011, the list of municipalities signing onto the initiative grew slowly at first but has surged in recent years to include Montréal, Vancouver, and dozens of smaller communities, as well as universities and school districts. The campaign now also encompasses cities, campuses, and churches in the United States, Europe, and Latin America (see table 6 later in chapter).

Barlow explained to me the role that bottled water plays in the Blue Communities initiative:

> We launched this project in Canada a decade ago to fight the privatization of water in its two most prominent forms—private operations of water services and bottled water—although there are many other forms of water commodification. We now have over fifty towns and cities[14] representing twenty-five million people that are official Blue Communities, and they take it very seriously. Most supply glass carafes to offices and for council meetings, and many, including Berlin, Munich, and Brussels, have all put funding into installing many new drinking water stations in their cities, in parks and community centers, et cetera.[15]

Barlow's framing of water services and bottled water as the two major facets of water privatization is noteworthy. It underscores the connection between packaged water's growth and global struggles over privatization and the human right to water. It also implies that these two facets of commodification are less distinct from one another than many observers believe.

FRUIT OF A BOYCOTT

Across the border in the United States, the bottled water issue was similarly kicking off during the first decade of the 2000s. The first national group to plunge into these waters in a major way was Corporate Accountability International (CAI), an advocacy organization founded in 1977 as INFACT (Infant Formula Action Coalition) by student activists at the University of Minnesota to target the Nestlé corporation's highly contentious infant formula marketing practices in the global South. It was one of many groups in the United States, Canada, and Europe involved in the campaign, which raised widespread awareness of the infant formula controversy through an international boycott against Nestlé products that lasted from 1977 to 1984. One observer writes that the Nestlé boycott "generated the largest support of any grass roots consumer movement in North America, and its impact is still being felt in industry, governments, and citizen's action groups around the world."[16] The boycott led to the creation of a code of industry marketing rules by WHO and UNICEF, which Nestlé agreed to adopt, partially changing its practices.

In the 1980s and '90s, INFACT took on other industries and corporations, including General Electric for its involvement in nuclear weapons manufacturing, and the tobacco industry over its health impacts. The group changed its name to CAI and in 2005 turned its focus to the bottled water industry, soon launching the "Think Outside the Bottle" campaign. Ironically, the group had come full circle back to targeting its former nemesis: Nestlé. Penny Van Esterik, a professor of anthropology at York University who was active in the Nestlé boycott, told a 2019 gathering of bottled water activists that she saw important parallels between the two campaigns: "We're addressing the same company, Nestlé—making an obscene profit, selling back to the public what it doesn't need. Water, like human milk, is a unique product and there's no substitute for pure water or human milk. Both are incommensurate, critical for human societies' survival, and they're necessities, not lifestyle choices. No one has the right to privatize and profit from a human need. And now Nestlé has been doing for water what it did for human milk over these last few decades. . . . So I think we should work together."[17]

Through Think Outside the Bottle, CAI supported a growing network of student activists aiming to persuade their universities and colleges to ban bottled water sales. It also jumped into campaigns to convince numerous cities to ditch bottled water, lobbying city councils and butting heads with the IBWA and other proindustry groups. In 2008 it supported a successful effort led by the mayors of San Francisco, Salt Lake City, Minneapolis, and New York City to get the U.S. Council of Mayors to pass a resolution calling on cities to "phase out, where feasible, government use of bottled water and promote the importance of municipal water."[18] In 2007, pressure from CAI led PepsiCo to admit publicly that the water source for its Aquafina brand was municipal tap water supplies and to print that fact on its labels.[19] Reflecting on this period, the group's water campaign director Lauren DeRusha tells me, "We found that talking to folks and organizing with folks about bottled water was an entry point into talking to people about the privatization of water, which at that time was not really an accessible topic to folks in the United States, which has completely changed now in 2019. But at that time it was kind of an entryway into having this conversation . . . as we were seeing the bottled water industry shift the public climate, spend a lot of marketing dollars intentionally to shift the public climate towards mistrust in tap water."[20]

CAI changed its name again in 2017 to Corporate Accountability. More recently the focus of its water campaigns has shifted, emphasizing threats to public tap water in a post-Flint era, the U.S. water affordability crisis, and advocacy for funding to restore public water infrastructure. Consistent with its historical work on multinational corporate practices, the group is involved in coalitions opposing water privatization plans in cities including Lagos, Nigeria, and World Bank support for water privatization across the global South in general. However, it still maintains an emphasis on the bottled water industry, particularly its erstwhile target, Nestlé, as indicated on its water campaign website:

> The bottled water industry plays an insidious role in driving the corporate control of water. It has spent millions of dollars on misleading marketing to change the way people think about water. Corporations like Nestlé want to turn water into a commodity to be bought and sold for profit, rather than an essential resource to be shared and managed democratically. . . . But tap water isn't the only enemy for corporations like Nestlé. Public control of water resources also limits their quest for massive profits. . . . And when public water crises like those in Flint and Detroit hit, the bottled water industry jumps in and positions itself as a solution.[21]

Overall, then, for Corporate Accountability (International), challenging the bottled water industry represents only one element of its efforts to oppose and reverse corporate control of water in the United States and globally, and to advocate for water justice.

LINKING FOOD AND WATER

The second major U.S. player in the anti–packaged water movement is Food and Water Watch (FWW), founded in 2005 by former staff members of Public Citizen, a consumer and environmental advocacy organization launched in 1971 by Ralph Nader. Before FWW's creation, Public Citizen was involved in opposing municipal water privatization in U.S. cities including Stockton, California. In addition to food-related activism on issues including genetic engineering and factory farming, FWW's water campaigning has run the gamut from opposing municipal water privatization, to advocating more federal investment in water infrastructure, to taking on packaged water. It has produced a series of heavily researched reports on these issues, including several on the bottled water industry.[22] While largely U.S.-focused, FWW's work now extends to Europe as well.

Through its Take Back the Tap campaign, FWW was long involved in institutional efforts to ban packaged water. "We've mostly worked with college campuses," the group's water campaign director, Mary Grant, tells me. "We've [also] done some work . . . around encouraging cities to not use bottled water in municipal buildings . . . and eliminating waste. It's more of a consciousness-raising effort . . . making people more aware of these single-use plastics, as well as [reducing] bottled water consumption in places where tap water is safe."[23]

FWW's emphasis has shifted lately, says Grant, away from a primary focus on bottle bans. "There is value in institutional bans," she continues. "I think we need policies to encourage people to drink tap water, and to shift off of bottled water. [But] most of our work has shifted towards focusing on policies, instead of institutional bans." FWW has also supported local community groups opposing the extraction of local spring water by commercial bottlers, as the next chapter describes.

TELLING THE STORY OF BOTTLED WATER

The fifth major organization active in campaigning against bottled water is The Story of Stuff Project, which emerged after the viral success of founder Annie Leonard's 2007 animated video about overconsumption titled *The Story of Stuff*.[24] According to executive director Michael O'Heaney, in an effort to harness the overwhelming popular interest generated by the film, the new group began targeting several other commodities and industries:[25]

> We made this series of first videos—*Story of Bottled Water, Story of Electronics,* and *Story of Cosmetics,* which were basically efforts to sort of take everyday consumer products, but use those as totems for bigger truths about our economy. . . . *The Story of Bottled Water* is really about manufactured demand. How do you convince people to buy something that they can get virtually free from the tap? . . . [The video] was produced with Food and Water Watch, Corporate Accountability, Polaris Institute, Pacific Institute, [and] Council of Canadians. . . . [We] got all these incredibly wonky but smart people around the table and hashed out this script. . . . But basically, we made this video, and it blew up . . . in a way that I think we weren't really anticipating.[26]

Up to this point, the organization had been solely a producer of media to facilitate social and environmental activism. However, after the Flint disaster, when the other national groups began shifting away from supporting local campaigns against bottled water extraction to focus more on water injustice and threats to public water systems, Story of Stuff began to fill that gap, contributing to what O'Heaney calls "site fights":

> We got more engaged, at a very sort of hyperlocal level, first in San Bernardino [California], then in Cascade Locks [Oregon]. And through that [we] started to meet this network of other grassroots organizations around the country, and in Canada as well, working on these issues. So as Food and Water Watch and Corporate Accountability were kind of waning in terms of their support for these grassroots communities . . . we took on some of that mantle, I guess, at some level. And in particular, [we provide] two things . . . financial support, [and] tactical and strategic advice. We made this series of three videos about San Bernardino, Cascade Locks, and then *A Tale of Two Cities,* about Evart, Michigan, and the contrast with the crisis in Flint.

While producing these site-specific videos to support and amplify grassroots campaigns, the group also tackled issues of water justice, deteriorating infrastructure, privatization, and shutoffs. "Bottled water as a product, we think, is a stupid product," O'Heaney continues. "[But] we also think it's a great way to enter into conversation about bigger issues of water justice, water access. . . . *A Tale of Two Cities* . . . was our most explicit, 'Hey, see, what's happening here.' Where you have this privatization of a water source and you have an underinvestment in public water infrastructure . . . you see Nestlé then taking water from Evart and giving it away free to the folks in Flint."

The organization was also well positioned to respond to the explosion in concern over single-use plastics in recent years. According to

O'Heaney, "We've been pretty instrumental in the birth and growth of the Break Free From Plastic movement, which is the more edgy, justice-oriented, plastic pollution movement in the world. . . . Bottled water is a plastic issue. And so there's been a good overlap there, but fundamentally it is about water justice, privatization, and basically who controls the commons." In other words, Story of Stuff framed bottled water as both an environmental and a social justice problem, while creating a bridge between activism targeting the industry's water extraction on one hand and opposing the consumption (and disposal) of this commodity on the other.

These five North American organizations, then, have been the key protagonists in what amounts to a loose social movement coalition contesting the bottled water industry and its product. For each of the groups, that work is an outgrowth of deeper concerns with resisting the growth of corporate power. Aside from the Polaris Institute, they all continue to work on the bottled water issue today to varying degrees, but as only one element of larger efforts to oppose water privatization, defend public water systems against disinvestment and deterioration, and address injustices in access to clean and affordable tap water domestically and globally.

FIGHTING (WITH) CITY HALL

Of the various elements of this consumption-focused activism, the efforts to win municipal-level policy changes are the most visible and have generated the most significant and lasting impact. Because the vast majority of water utilities in the United States and Canada are public entities, bottled water is almost always competing directly against cities' own public tap water, which they deliver using ratepayer and taxpayer funds. Municipal governments are also usually responsible for garbage and recycling services, and they bear the cost of dealing with the massive increase in unrecycled and discarded plastic bottles, especially since China's National Sword policy banning plastic waste imports.[27] If city hall is buying bottled or packaged water to serve in its offices instead of tap water—or if city council meetings feature plastic water bottles at every seat—public funds are effectively endorsing a product that works against the municipality's efforts on both of these fronts. As DeRusha explains, "A lot of the narrative at that time was around how as stewards of the public water system, it doesn't make any sense for the municipal government to be basically endorsing the bottled water industry, and subtly undermining their own service that they are charged with providing. Why wouldn't you provide tap water to your employees at a meeting? What message are you sending by lining up the bottled water to give them? And definitely that message was really resonant with cities." For these reasons, anti–bottled water campaigners have often found local governments to be fertile terrain for policy change, and they have worked with local residents and sympathetic public officials to great effect. The appeal—and the impact—of these policies can be simultaneously fiscal (reducing cities' expenditures on unnecessary bottled water and on waste management), political (using public procurement power to make a statement along with other cities), and cultural (changing norms by revalorizing tap water and reducing bottled water consumption).

In these efforts to challenge the bottle at the local government level, it is useful to think of at least three tiers of policy change. The lowest-hanging fruit is the most common: ending public purchases of packaged water for local government use and by public employees—in other words, no more taxpayer funds spent on Dasani, Aquafina, or Poland Spring bottles in the office fridge or at city council meetings or public hearings, with pitchers of tap water taking their place. Beyond this lie a range of midlevel opportunities that involve making bottled water hard to access on public property: restricting or banning its sale at public parks, sports fields, beaches, museums, and the like. This can include all food carts, cafés, concession stands, and vending machines located on municipal property, potentially a large market. Big public events are another half-step—think concerts in city parks drawing thousands of people, where pricey single-serve bottles are typically the only available water source, and leave a mountain of plastic waste behind. Banning bottles and requiring event organizers to provide free tap water is an alternative that reclaims public spaces for public water. On the third tier, a more radical model of banning the bottle is to do just that: outlaw *all* sales of single-use bottled or packaged water in a community, whether on public or private property. This controversial approach has been implemented in only a few communities so far, but it appears to be growing. Finally, local taxes on bottled water, such as the nickel-per-bottle tax enacted by Chicago in 2008, are another way to reduce consumption and fund public water improvements.

Let's examine where and how the first two tiers have been realized. Los Angeles was likely the first city to ban municipal purchases of bottled water, back in 1987, but Mayor Tom Bradley's executive order was later ignored.[28] When L.A. refreshed its policy in the mid-2000s, it was part of a surge of local governments across the global North taking similar steps. "The movement is spreading," wrote Peter Gleick in 2010. "Cities like San Francisco, Vancouver, St. Louis, Ann Arbor, Urbana, Santa Barbara, Manly, Toronto, Ottawa, Rome, Florence, Liverpool, and others . . . are moving to ban government purchases of bottled water and to endorse campaigns to promote local tap water."[29] Indeed, the first decade of the century saw a huge wave of cities end municipal buying of plastic water and expand drinking water access—sometimes on the initiative of progressive mayors or city councilors, other times pushed by local activists and/or national advocacy groups. The motivation for these cities is often fiscal as well as environmental. Cook County, Illinois, for example, saved over $400,000 annually on bottled water purchases when it enacted its ban in 2007.

Some cities come back for a second or third pass, strengthening these policies along the way. New York City ceased buying bottled water for city offices in 2008, but in 2020 Mayor Bill DeBlasio signed an executive order expanding the ban to all city-owned and -leased properties. According to journalist Justine DeCalma, "The move would eliminate at least 1 million single-use plastic beverage bottles that the city buys each year. . . . It could also have wider-ranging effects since the city owns or leases over 17,000 properties spread over an area about twice the size of Manhattan . . . [including] The Trump Organization's two skating rinks in Central Park."[30]

REVIVING FOUNTAINS

Banning bottled water sales, however, can be problematic if there is nowhere to find safe, convenient tap water. Even in wealthy nations, public drinking water infrastructure in some places has been defunded and neglected to the point where it functionally no longer exists. An investigation by *The Guardian* of public drinking water access in Britain, for example, found that in the entire Manchester metropolitan area—with a population of 2.8 million—there was *not one* working publicly maintained water fountain. "The same situation was found across Merseyside," writes correspondent Nicola Davis, "with all five councils replying that they have no working fountains, and in South Yorkshire, where a spokesperson for Sheffield council revealed that all fountains 'were taken out a few years ago due to health risks and damage.'"[31]

Thus there is an obligatory two-step dance here: policies banning or restricting bottled water need to be accompanied—and often preceded—by corresponding efforts to restore or improve access to tap water in public places, to restore not only trust in tap water but often the lost cultural practice of drinking it.

Nonetheless, the renaissance of public fountains is now well under way. Kendra Pierre-Louis writes that "some cities are slowly bringing back—or at least increasing maintenance of—water fountains. In 2013, Los Angeles put together a comprehensive plan to upgrade and restore public water fountains. Minneapolis spent $500,000 on 10 new fountains designed by local artists."[32] New York City has committed to installing or repairing five hundred fountains with bottle fillers in underserved areas, city parks, and other public spaces by 2025 (complementing nearly two thousand existing fountains), a goal explicitly linked with waste reduction.[33] And Paris, whose remunicipalized utility Eau de Paris is expanding drinking water infrastructure, has even outfitted several of its public refilling stations with spigots for *sparkling* water.

Even some cities that have not enacted bottle bans are joining in. London's mayor Sadiq Khan made headlines with announcements of new public water fountains to be installed across the city, in 2018 declaring one hundred new refilling sites at a cost of 5 million pounds.[34] Khan's office solicited input on the locations for the fountains, stoking public interest.

These efforts are typically accompanied by campaigns to promote access to public tap water, with local government and public water utilities frequently working in tandem. In Australia, Sydney Water's website features a searchable map to its hundreds of public bottle refilling stations (see figure 10), plus over five hundred restaurants and other businesses that make free tap water available. Sydney Water also provides portable refill stations free of charge to organizers of public events over one thousand people, provided they do not sell bottled water.[35]

In Europe, the EU has taken major steps to improve access to tap water through its recently revised Drinking Water Directive, which was pushed by water justice groups. By 2025, all EU countries will be required to make free drinking water available in public buildings and to provide access to water fountains or refill stations in other public spaces. According to the European Greens, this decree "will lead to new rules in many Member States, where the construction of public water dispensers is not yet required by law."[36]

FIGURE 10. Bottle refilling station, Manly, Australia. Photo: Author.

RECLAIMING THE TAP IN SAN FRANCISCO

San Francisco, which until recently was the largest U.S. city to ban both public purchases and sales of bottled water, exemplifies this movement to reclaim the tap. Its policy now prohibits the sale of single-use packaged water in any container under twenty-one ounces on city property—including parks, beaches, food carts, and city-owned venues. Organizers of large events in public spaces must hook up to the city water system and provide free tap water. A staff member with the San Francisco Department of Environment outlined the policy's evolution:

"We had the mayor's [Gavin Newsom's] executive order on bottled water, which passed in 2007, which prohibits any department from using any city funds to purchase bottled water. We have the Bottled and Packaged-Free Water Ordinance, passed in 2014, and that restricts the sale [of bottled water]. . . . And we also have the Drink Tap ordinance, which requires that any new buildings that are going to have drinking water fountains have refilling stations."[37] These efforts to eliminate bottled water are a key part of the city's larger Zero Waste policy, which the staffer says has already reduced the volume of waste it sends to the landfill by over 75 percent: "Our overarching goal is how do we get to zero waste. And this is just the first line of attack here, about banning the most unsustainable single-use plastics."

The landmark 2014 ordinance expanding the bottled water ban was spearheaded by then-city council member David Chiu. "I want to remind people," he told the council, "that not long ago, our world was not addicted to plastic water bottles. Before [the 1990s], for centuries, everybody managed to stay hydrated."[38] Chiu's campaign to pass the ordinance was supported by CAI, the Sierra Club, and the Surfrider Foundation, among other groups.

The Department of Environment staff member tells me the motivation behind the policy involves not only environmental and cost considerations but also concern over packaged water competing with San Francisco's high-quality tap water, which originates in Yosemite National Park: "We have such pride and joy in where our tap water comes from. . . . We have this pristine water from Hetch Hetchy [Reservoir] that exceeds state and federal regulations for ways that it needs to be filtered. And yet we have people in the city here who feel compelled to buy plastic water."

In 2017, the city expanded the prohibition to include all packaged water containers under one liter in volume—including bags, cans, and boxes.[39] And in 2019, the ban was extended even further to the city-owned SFO airport. With nearly fifty-eight million passengers moving through SFO annually, this move greatly increases the visibility of the city's anti–packaged water efforts. Travelers caught unawares can purchase reusable bottles or mugs and fill them at over one hundred hydration stations.[40]

But it is on the supply side—expanding access to tap water—that San Francisco has perhaps been most innovative. Its "Drink Tap" initiative cuts across several city departments, reflecting a broad view of tap water access as an economic, environmental, and public health issue. As of

2021, the city's Public Utility Commission (PUC) had installed 173 Drink Tap stations—drinking fountains with bottle fillers—in public locations (not counting SFO) and inside nearly all of the city's public schools, with more in the works.[41] How does the city address the problem of public distrust in tap water quality? "Whenever we install a new Drink Tap station," explains PUC manager John Scarpulla, ". . . we do a robust water quality test of the station. . . . [By] showing everybody, 'Yup, this passed the test, you should feel totally free to drink safe, delicious tap water from here,' that helps."[42] The PUC also promotes the city's water quality with advertising on public transit and other venues.

All these efforts—from installing refill stations to the educational efforts around tap water—are funded by San Francisco's penny-per-ounce tax on soda and sugar-sweetened beverages, approved by nearly two-thirds of voters in 2016.

REALLY BANNING THE BOTTLE

While all of the above local efforts represent meaningful change, they leave unquestioned one basic premise: that these policies should apply only to public property—that telling private businesses what not to sell on their own premises is a bridge too far, or at least politically impossible. Yet governments regulate private sector actions and prohibit products that harm the environment or impose high costs on society all the time.

A few communities have taken a far more ambitious step: banning all private (as well as public) sales of single-use bottled water. This micromovement began in 2009 when residents in the town of Bundanoon, Australia, were fighting a proposal for a water-bottling plant on a nearby aquifer. As local business owner Huw Kingston told the *New York Times,* "The idea was floated that if we don't want an extraction plant in our town, maybe we shouldn't be selling the end product at all."[43]

Interest in this approach crossed the globe, surfacing in historic Concord, Massachusetts, where after multiple attempts, octogenarian activist Jean Hill persuaded residents in 2010 to approve a ban on all sales of single-use plastic bottles. The measure was eventually validated by the state's attorney general and took effect in 2013. Since then, Massachusetts has become the epicenter of this more ambitious model, with over a dozen communities around the state following suit. And since 2019 a wave of bans on single-use water bottles has swept across Cape Cod, spearheaded by Sustainable Practices, a local environmental group led by Madhavi Venkatesan. The organization is shepherding its "Commercial Single-Use Plastic Bottle Ban" resolution through the town meetings of all fifteen townships on the Cape. As of this writing, it has succeeded in seven.[44]

Taken together, the twenty-one communities that have implemented such full private and public bans on single-serve plastic water as of this writing are home to nearly 210,000 people. That's a modest number, and so far all are modest-sized towns. Could this more radical approach spread to bigger cities and even become the norm? If that does happen, it will be because more governments manage to include water and beverage bottles as part of their broader efforts to tackle the problem of single-use plastics.

ADDING IT UP

Ultimately, how significant is the bottle ban phenomenon in all its various forms? On one level, this strategy can be conceptualized as a form of boycott—with governments, rather than individuals, as the actors refusing to purchase the product. This stands in contrast to other forms of consumer activism. Rather than a "buycott" that advocates consumption of greener products or more ethical brands, an approach many critics deride as compatible with neoliberalism, this is an *anti*-consumption strategy. It involves rejecting an unnecessary market commodity and replacing it with a free, uncommodified, public good.

This approach also embodies a form of policy contagion—the conscious spread of parallel policies among cities and municipalities, which in the neoliberal era have often proved to be the most receptive venue for implementing progressive social change. Social movements frequently turn to local governments to enact innovative policies that would be stymied at the state, provincial, or national levels by entrenched interests.[45] In the case of bottle bans, the larger the population such policies apply to, the greater the impact should be. This happens both directly, through reduced bottled water purchases, and also indirectly, shifting cultural norms by offering better access to water at new fountains and refill stations, so that people feel less compelled to consume plastic water. These are localized, tangible instances of decommodification, but if implemented across hundreds of communities large and small, their effects can multiply.

Reliable data on the total number of water bottles eliminated by such bans are unfortunately not available. Even San Francisco does not track how many bottles its Drink Tap program and packaged water ban have saved. One for-profit company, Tap, maintains a website that it says aggregates data from over 255,000 refilling stations around the world, claiming a total of nearly fifty million plastic water bottles avoided to date.[46] Yet this is clearly just the tip of the iceberg.

In the absence of broad quantitative data, let's look at the total population living in communities covered by bottle ban policies. In Canada—the epicenter of this movement—over 7.1 million people live in the fifty self-declared Blue Communities,[47] and another 6.7 million live in other municipalities that have declared bottle bans, largely during Polaris's earlier campaign. Thus as of 2022, almost fourteen million Canadians reside in places where local governments have declared some form of water bottle ban—nearly 37 percent of the national population.[48]

Adding in the twenty-seven cities in Europe, the United States, and Brazil that are Blue Communities brings the total population covered by that initiative to nearly twenty-four million people globally. Stepping back further, all types of municipal government bans on bottled water now cover at least thirteen million people in the United States and a total of over forty-five million people worldwide. This includes the 2019 ban enacted by the South Delhi municipal council in India, with nearly three million inhabitants.[49] Table 6 shows the largest cities that have enacted some form of bottled water ban.

To increase these numbers further, one can factor in all the state, provincial, county, and regional governments that have adopted their own bottle bans. Because these include the huge governmental apparatus of Maharashtra State in India—population 116 million—which banned single-use plastic water bottles in 2017,[50] these supralocal bans have now been implemented in jurisdictions totaling 171 million people. Worldwide this means that over 216 million people live in jurisdictions with some form of bottle ban. Finally, if we add the people in communities with bans on the private sale of bottled water, and a few million students worldwide attending colleges and universities that prohibit its sale, the total rises further still.

These numbers require a few provisos: there are likely additional bans that were missed in this count, and some older bottle ban policies may no longer be actively enforced. Nonetheless, they make it possible to draw a few conclusions. First, there was a big initial wave of municipal bottle bans enacted between 2007 and 2011. In subsequent years the level of activity ebbed somewhat, but new communities have continued to declare these policies, and there has since been another major

TABLE 6 LARGEST CITIES ADOPTING POLICIES BANNING BOTTLED WATER, AS OF AUGUST 2022

Municipality	Nation	Population**	Year(s) policies adopted	Bans or phases out gov't purchases of bottled water?	Bans or phases out sales of bottled water on municipal property?	Signatory to Blue Communities Initiative?
New York City	U.S.	8,230,290	2008; 2020	Yes	---	---
Los Angeles	U.S.	3,983,540	2007; 2019	Yes	Yes	Yes
Berlin	Germany	3,426,354	2018	Yes	Yes	Yes
South Delhi	India	2,733,752	2019	Yes	---	---
Toronto	Canada	2,731,571	2008	Yes	Yes	---
Paris	France	2,148,271	2016	Yes	Yes	Yes
Brussels-Capital Region	Belgium	2,095,688	2019	Yes	Yes	Yes
Hamburg	Germany	1,852,478	2022	Yes	Yes	Yes
Montréal	Canada	1,704,694	2019	Yes	Yes	Yes
Munich	Germany	1,260,391	2017	Yes	Yes	Yes
San Jose, CA	U.S.	1,009,340	2008	Yes	---	---
Edmonton	Canada	932,546	2016	Yes	---	---
San Francisco, CA	U.S.	883,255	2007; 2014; 2017	Yes	Yes	---
Seattle, WA	U.S.	776,555	2008	Yes	---	---
Vancouver, B.C.	Canada	631,486	2009; 2020	Yes	Yes	Yes
Sacramento, CA	U.S.	525,398	2008	Yes	---	---
Liverpool	U.K.	*498,042*	2007	Yes	---	---
Miami, FL	U.S.	478,251	2008	Yes	---	---
Halifax, N.S.	Canada	448,231	2009	Yes	---	---
London, Ontario	Canada	383,822	2008; 2021	Yes	Yes	Yes
Vaughan, Ontario	Canada	331,572	2009	Yes	---	---
Thessaloniki	Greece	325,182	2018	Yes	Yes	Yes
St. Louis, MO	U.S.	294,890	2008	Yes	---	---
Gatineau, Québec	Canada	290,239	2007–11*	Yes	---	---
Augsburg	Germany	259,900	2019	Yes	Yes	Yes
Longueuil, Québec	Canada	252,828	2007–11*	Yes	---	---

Burnaby, B.C.	Canada	*247,336*	2011	Yes	Yes	Yes
Freiburg im Breisgau	Germany	*230,940*	2022	Yes	Yes	Yes
Móstoles	Spain	210,309	2019	Yes	Yes	Yes
Richmond Hill, Ont.	Canada	208,052	2007–11*	Yes	---	---
Salt Lake City, UT	U.S.	200,831	2007	Yes	---	---
Burlington, Ontario	Canada	193,668	2010	Yes	---	---
Bath	U.K.	193,282	2008	Yes	---	---
Oshawa, Ontario	Canada	178,893	2008	Yes	---	---
Sherbrooke, Québec	Canada	171,158	2007–11*	Yes	---	---
St. Catherines, Qué.	Canada	141,490	2012	Yes	Yes	Yes
Pasadena, CA	U.S.	141,029	2008	Yes	---	---
Trois-Rivières, Qué.	Canada	140,420	2019	Yes	Yes	Yes
Kingston, Ontario	Canada	135,707	2011	Yes	Yes	Yes
Bern	Switzerland	134,591	2013	Yes	Yes	Yes

* Estimated date range (Polaris Institute 2014).

**Current population estimate, only within city or jurisdictional limits.

SOURCES: Council of Canadians n.d.-a; Polaris Institute 2014.

surge in bottle ban policies, largely but not only in new Blue Communities. In fact, well over half of the total population covered by these bans lives in places that have enacted or strengthened them since 2017. The pace is picking up, not slowing. Second, it is evident that packaged water has now become incorporated into the larger international movement against single-use plastic, which has also spread across the global South. We can expect to see more local governments in both South and North join this reaction against plastic water in coming years.

LEARNING (TO LIVE) WITHOUT BOTTLES

Colleges and universities have been especially fertile ground for challenging packaged water: they are quasi-self-contained communities, largely in charge of their own physical facilities and with governing bodies at least ostensibly responsive to the needs of students. The first

U.S. campus to formally exile bottled water was private Washington University in St. Louis in 2009. Also that year, the University of Winnipeg and Brandon University in Canada did the same. Since then, hundreds of colleges and universities have followed suit, from Australia's Canberra University and the University of Ottawa to the Ivy Leagues (Harvard, Dartmouth), big U.S. public universities, and community colleges. Leeds University in the U.K. eliminated over 180,000 water bottles per year when it banned their sale in 2008.[51] Banning the bottle became one of the major student environmental causes of the decade. In North America, many of these initiatives were supported by one or more of the advocacy campaigns described above. Food and Water Watch in 2016 claimed seventy-three U.S. campuses with either full or partial bans in its Take Back the Tap network.[52]

DeRusha recalls how she initially became involved with CAI as a student at Stonehill College in Massachusetts:

> I was a freshman in college having a social justice awakening—a very rude one, about just how . . . much of [U.S.] foreign policy is really done at the behest of transnational corporations. . . . And it was right in that pretty dark period that I actually found Corporate Accountability and I interned on the Think Outside the Bottle campaign. . . . And the bottled water campaign felt like something really tangible that I could do as a student. And so I ended up organizing my campus too. And we didn't end the full sale [of bottled water], but we ended the school's budget spending on bottled water for events and things like that. And that was a really powerful experience for me. . . . That was like a vehicle for me to both educate people, but also to move them along and to enter this conversation about water as a human right and how incompatible that is with treating water as a commodity.

Almost all such campus bans are coupled with expansion or replacement of fountains and refilling stations, to make it easier for students to fill reusable bottles or mugs with clean water. Distributing university-logo metal bottles to all new students at first-year orientation is a common practice, helping shape new drinking water norms. Even where bottle ban efforts fail, they often cause administrators to improve campus drinking water infrastructure.

What is the effect of all these campus bans? Does removing commodified water actually cause students to drink more tap water, and perhaps develop a lifelong aversion to plastic water? Or, as the packaged water industry insists, do students merely substitute less healthy soft drinks or carry in bottled water bought off-campus? The evidence is scant and mixed. A 2015 study on Washington University's first-of-its-kind ban documented a 39.4 percent decline in purchases of all bottled beverages and a 46.3 percent drop in sales of fountain drinks.[53] While the study didn't track students' bottled water consumption specifically, it is clear that the alleged postban shift to soft drinks did not occur. On the other hand, a 2015 study of the University of Vermont's 2013 ban, enacted by a student vote, found that consumption of less healthy beverages like soda and sports drinks had actually increased, and the number of plastic bottles discarded on campus had not fallen.[54] While opponents seized on this as evidence that bans are counterproductive, university officials vowed to increase their efforts to expand tap water access.

Where is this campus bottled water activism heading? The initial surge of campus bottled water activism that grabbed headlines in the late 2000s—with students holding blind taste tests between bottled and tap water outside student unions and launching loud campaigns to win campus bottle bans—has slackened somewhat. Nonetheless, the list of campuses adopting these policies continued to grow throughout the 2010s. And more recently, the movement to eliminate bottled water from campus is back with a vengeance—but this time as a by-product of the global movement against single-use plastics, and on a scale that's harder to ignore. By far the biggest moves come from California, where in 2020 both UCLA and the entire twenty-three-campus California State University system adopted policies that will see plastic beverage bottles vanish entirely by 2023. The same will happen across all ten campuses of the University of California system by 2030.[55] The combined enrollment of those two higher-ed systems is enormous—nearly eight hundred thousand students.

UNFAIR TREATMENT?

All of these governmental and institutional policies represent a lot of activity directed against a single commodity. How do its producers respond? Publicly, the bottled water industry brushes off these developments, but it also claims it is being unfairly singled out. A bottled water firm representative I interviewed argued that "we essentially want to be treated equally with all other packaged beverages. So [if] these organizations and universities and etcetera are banning bottled water, they should also be banning soda and fruit beverages and everything else in the vending machine on their properties." In a 2013 press conference, Nestlé Waters CEO Kim Jeffery predicted that the growing number of bottled water bans would not stop the product's steady growth but would merely drive people to drink more unhealthy beverages. He also attacked Food and Water Watch and CAI, calling them "not solution-based organizations. They would prefer that we not have a license to operate our business. We have nothing to talk about with these two organizations."[56]

The industry frequently lobbies and testifies against proposed bottle bans, along with some individual bottling firms. However, the IBWA would much prefer to have local residents be the face of opposition to such policies, according to the group's director of government relations:

> At the local level it is preferable for a local citizen (not a national nonprofit) to discuss with their neighbors and friends the many benefits of bottled water to a community because their voice will have a stronger impact. IBWA's involvement (and sometimes the mere mention of the organization) can tilt arguments in favor of a ban or restriction simply because local citizens do not want what they perceive as a large, multinational organization or business trying to influence a local issue. In those instances, it is better to have a local voice introduce any pro-bottled water messages that IBWA would have delivered on this topic.[57]

While in some cases such pressure has worked, the industry's efforts have largely failed to stop municipal-level bans, in part because the cost savings from ending bottled water purchases are so appealing to local governments.

PARKS' PLASTIC PURGE PREVENTED

However, the industry had more success in fighting a ban by one high-profile institution: the U.S. National Parks. "In 2010," writes journalist Adele Peters, "days before the Grand Canyon planned to stop selling bottled water, Coke pressured the National Park Service director to block the ban. Their leverage? Coke is a huge funder of the National Park Foundation. The ban was eventually allowed, after a public outcry. At the Grand Canyon, like many other parks, bottled water made up almost a third of the trash that the park had to haul away."[58] In 2011, the National Park Service (NPS) approved a plan allowing individual parks to stop selling single-use plastic water bottles in gift shops and concessions and to increase the use of water fountains and refillable bottles.

The IBWA, however, hated the new approach. "These misguided bans on the sale of bottled water are not likely to reduce the presence of plastic bottles within the recycling streams of our national parks," read one press release.[59] The result was one of the most public and bare-knuckled confrontations ever seen between the industry and its opponents, much of it taking place in the halls of Congress. In 2015, DeRusha remembers, "we began working with parks and people around the country who were calling on their park to go bottled water free. Because the Park Service is such an important cultural marker—they're such an emblem of sustainability. . . . And to have park rangers hanging out by the bottled water machine really sends the wrong message. . . . And so people around the country were really, really passionate about this . . . and many parks went bottled water free over the course of our campaigning." Only 23 out of 417 national parks adopted bottle prohibitions, but these included some of the most iconic and heavily visited sites. Derusha continues, "The International Bottled Water Association . . . began interfering at that time through intensive lobbying which resulted in a rider amendment getting introduced onto a spending bill that would . . . rescind the policy that allows parks to go bottled water free. And we organized, basically went toe to toe with IBWA, organizing our membership to call on their members of Congress to stand up to this corporate interference, and stand behind the Park Service, who clearly wanted to do these policies. . . . And ultimately we won every round."

A study found that the parks that adopted bans had collectively avoided as many as two million plastic bottles, leading the NPS to reaffirm its policy in 2016. However, this was a pyrrhic victory, says DeRusha: "It was within a few weeks of David Bernhardt becoming deputy secretary of the Interior under Trump's administration—and Bernhardt came directly from a firm that had legally represented Nestlé Waters. . . . Within a few weeks of that [in August 2017], the bottled water policy was rescinded."

Yet why would the industry expend such great effort to quash the bottle ban option, which affected a tiny slice of revenue? A 2020 IBWA report admitted that "banning the sale of bottled water in national parks would not significantly impact the industry's sales figures. However, it would set a very bad precedent and would be used by bottled water critics in their efforts to ban the sale of our products in cities, colleges, zoos, and other public places."[60]

In June 2022, after prodding from hundreds of environmental groups, the Biden administration took action on this issue. U.S. Interior secretary Deb Haaland announced that the federal government would completely phase out procurement and sale of all single-use plastics, including water bottles, on 480 million acres of federal lands—including national parks—within ten years.[61]

REFILL MADNESS

The final element of the demand-side pushback against packaged water lies beyond the realm of government and educational institutions. This is a panoply of mostly local efforts by nonprofits and businesses to increase tap water consumption by making it easier to find in both private and public spaces. In the process, they are partly blurring the line between those two spheres.

This latest burst of action is all about refilling. In the past several years there has been an explosion of refilling initiatives that enlist private business to commit to serving free tap water to customers and noncustomers alike. They are enabled by two key developments: the rapid spread of reusable water bottles and refill stations, and the advent of mobile phone apps allowing people to easily find those refill points. Listed businesses and other entities (libraries, museums, etc.) typically identify themselves with branded window stickers, contributing to a sense of movement building (figure 11). In the U.K., the group Refill—part of the nonprofit City to Sea—was founded in 2015 with a pilot program in Bristol. Its app now lists over thirty thousand sites all the way from Guernsey to northern Scotland, and it claims to have kept over one hundred million plastic bottles out of the trash in the U.K. alone since 2019. The group encourages activists to form local refilling schemes in their communities and currently counts over four hundred such efforts. Its efforts in London are financially supported by the mayor's office and the (private) water utility Thames Water, and it aims to create a global network of similar initiatives.[62]

Analogous efforts are popping up elsewhere: Free Tap Water NZ in New Zealand, Choose Tap in Australia, WeTap in the United States, and Blue W in Canada. Evan Pilkington, the founder of Blue W, tells me his goal is "that wherever you are in the world, you won't have to pay for a drink of water when you're thirsty." He continues,

> There are people who wake up in the morning, they grab their keys, they grab their lunch, they grab their work and their laptop and head out the door, and they grab their [refillable] water bottle, right? So the businesses that post to Blue W are trying to attract those people who would have walked by otherwise looking for a place to refill, so they'd head to a library or maybe the bus station . . . and find a fountain if they're lucky. But now,

FIGURE 11. Free refill logos on café door, Dunedin, New Zealand. Photo: Author.

> because there's a blue W in this window, they walk in when they knew they didn't want tea or coffee, but they know they can come in, refill the bottle. The owners recognize that . . . it brings in foot traffic and it aligns them with something positive.[63]

These efforts have apparently caught the attention of not only small businesses but large retailers. In the U.K., the grocery store chain Asda and the restaurant chain Pret a Manger have installed free filtered water stations in hundreds of stores in collaboration with Refill.

The packaged water business has taken note of the environmentally motivated refill movement and is becoming worried, as the epigraph to this chapter illustrates. Bottled water sales in the U.K. fell in 2019 for the first time since the Great Recession, and refilling was one major cause. According to industry publication *The Grocer,* "The bad news doesn't stop there. Bottled water has also continued to bear the brunt of the environmental concerns, as a growing number of consumers opt for refillable water bottled over single-use plastic. . . . So momentous is the movement that Richard Caines, senior food & drink analyst at Mintel, names reusable bottles 'one of the biggest risks to sales of bottled water.' That risk is only set to grow as filling up from the tap becomes increasingly easy on the go. . . . All in all, the picture is looking pretty bleak."[64] The industry's concerns are validated by polling. In the U.K., 63 percent of bottled water purchasers and 72 percent of people who carry reusable bottles say that they would use refilling stations even more if their availability increased further.[65] The movement to reclaim the tap appears to be flying with a strong tailwind.

TAKING STOCK

What is the significance of these developments? Has this loose amalgamation of efforts in the public, private, and nonprofit arenas actually managed to reduce bottled water consumption or hurt the industry's prospects? Over a decade ago, during the Great Recession, bottled water sales actually fell for two years running in most of the global North. At the time, some observers commented that opposition movements were likely a big factor, along with sharply reduced consumer income. Yet the downturn was ultimately short-lived. "Drinking bottled water did become stigmatized on some college campuses and in some niche cultural elite circles," wrote Douglas Holt in 2012, "but the campaign's impact on the mass market was negligible. Sales predictably contracted during the acute recession in 2008–2009, but then grew again."[66] Peter Gleick tells me that at that time, "I sort of felt . . . [that] this [fall in sales] possibly reflects the social movement against bottled water, . . . [but] you know, the demand has ticked up again, and the social movement has diminished to some degree."

However, while activism challenging bottled water consumption did ebb somewhat for several years, in hindsight that appears to have been a lull between two large waves. We are once again witnessing a major upsurge in governmental and institutional policies, including new bottled water bans in several major cities and substantial expansion of access to tap water, particularly since 2018 and 2019. Those were the years that China began rejecting rich nations' recycled plastic exports and the climate justice movement galvanized millions of youth in worldwide protests. In the interim, this facet of the movement has been swept up into two larger currents: the global backlash against single-use plastics, and the post-Flint fight to defend and reinvigorate public drinking water.

While it is difficult to prove causality—and the pandemic also affected the industry—those movements do appear to be playing a key role in reducing sales. Bottled water sales in the U.K. in 2020 were still below their 2018 levels—a remarkable change in fortunes for a market that was growing unabated just three years earlier.[67] The industry views these developments as the result of a pro-refill, antiplastics movement, especially among young people, and it is concerned. This falloff in sales seems to reflect a cultural shift—evident in opinion polls—toward renormalizing drinking tap water, with people demanding its increased availability in more convenient forms in more public and private places.

What does the history of these consumption-focused efforts reveal about the activist DNA of the broader movements challenging bottled and packaged water? The genealogy of the "bottle ban" strand of activism helps to address one of the questions I posed in the Introduction, regarding the relationship between the movements contesting bottled water and those fighting tap water privatization and advocating global water justice. The trajectory of what might be called the "big five" North American organizations described in this chapter shows that their work on bottled water—which for most of them dates back nearly two decades—grew out of earlier commitments to anticorporate and global justice activism, whether against unfair trade and investment agreements or transnational corporate practices. Their work on bottled water is also an extension of prior or ongoing work to oppose municipal water privatization and to strengthen public tap water as part of a reinvigorated public sector. At the same time, for most of these groups bottled water has never been a sole, or even the primary, focus. Rather, it has usually played a supporting role in their overall challenges to water commodification and corporate control.

What role do concrete efforts to ban bottled water, to reduce its ubiquitous presence in society and stigmatize its consumption, play in decommodifying water? In direct, material terms, these efforts (strive to) reduce the sale of bottled water not only by government bodies but often by private businesses on public property. The corollary to bottle bans—investing in expanded tap water access infrastructure—further helps reduce packaged water consumption by undermining the industry's one often-valid, if sometimes implicit, claim for its product: that it is just too hard to find reliable public drinking water outside of the home. New shiny refilling stations and water fountains in visible public places weaken that argument, at least theoretically leveling the playing field between public and private drinking water, reducing packaged water's necessity and therefore its consumption. In more indirect terms, these efforts to restore a collective practice of public tap water consumption—enabled through a new infrastructure of omnipresent refilling stations, reusable bottles, window stickers, and smartphone apps—operate on the terrain of culture to shift societal norms around drinking water.

Taken together, these consumption-side efforts involve a wide range of both old and new modes of accessing free—uncommodified—drinking water. They include not only city-funded water fountains and refill stations but also private businesses serving up free tap water to customers and noncustomers alike, and a logistical and technological apparatus allowing people to find all these sites while on the go.

What's being constructed here, it seems, is far more than the sum of its parts. It amounts to a new kind of hybrid physical-and-virtual water access infrastructure, whose presence makes a provocative assertion: that access to drinking water should be cost-free, universal, ubiquitous, unstigmatized, and devoid of single-use packaging. In short, this is the reassertion of a right to public drinking water in society.

These developments dovetail well with the youth-led global movements against plastic pollution and climate change. The resulting convergence works to denormalize and stigmatize bottled water among important sectors of the public, making it less appealing, more fraught, and uncool. If the growth of bottled and packaged water is a wedge that accustoms or inures the population to accept private control over water more broadly, then conversely, slowing and reversing this commodity's growth constitutes an important form of decommodification.

Finally, this suite of efforts to ban the bottle and reclaim the tap is a specific way to oppose austerity and defend public goods. It leads to reinvestment in our shared but often dilapidated water infrastructures. It asserts that the state has an obligation not only to keep water safe for drinking in the home but also to make it once again widely accessible in public places. This in turn helps to reinforce a broader awareness of the vital importance of a shared public sphere.

HOW CAN WE RECLAIM THE TAP IF IT ISN'T SAFE TO DRINK?

However, there is one important objection to these arguments: that this consumption-side activism—the bottle bans, the buildout of refill stations—applies only in places with widespread access to safe water supplies. How can the tap be reclaimed if that tap water is unreliable or unsafe? Is this aspect of packaged water activism essentially a protective action available to relatively well-off communities and nations that enjoy high-quality drinking water—an artifact of water privilege? Or can reclaiming public tap water and shifting drinking water culture generate political pressure that can be harnessed in the service of water justice for all?

Some observers argue that this consumption-focused anti–bottled water activism is mainly a product of places that already have nearly universal potable tap water, primarily in the global North. "Most of the opposition to bottled water," write Gay Hawkins, Emily Potter, and Kane Race, "is confined to places where state provision of water is both expected and emblematic of citizenship."[68] While valid in broad strokes, this argument needs nuancing. For one thing, it is possible to improve or restore water quality and (re-)establish trust in public tap water in places where it was previously untrustworthy. The former UN Special Rapporteur on the Human Right to Water, Catarina de Albuquerque, told an audience at the World Water Forum in Brazil: "In my country, in Portugal, when I was younger, people were afraid of drinking water from the tap, because [it was] not very safe. And there was a huge campaign, huge investments made, to make sure that people realized the best water we can drink is the water coming out from the tap. It's cheaper, the quality is better, it's much more controlled, and it's environmentally friendly. But then for us to achieve that, information has to be available, transparent, so that people gain confidence."[69]

This massive public investment in Portugal's drinking water systems is a good example of the kind of infrastructure restoration that water justice groups in North America and worldwide are demanding today. Yet when applied to the real-life context of communities with unsafe water today, it raises difficult questions. This is true not only in much of the global South but also in settings within Northern countries that mirror those conditions. For example, dozens of Indigenous communities in Canada are under long-term unsafe water advisories and are forced to rely on bottled water. Some have expressed concern at the prospect that proposed governmental bans on single-use plastics might include bottled water. As the CBC reports, "June Baptiste is a councillor for Lhoosk'uz Dene Nation in B.C. which currently relies on bottled water brought into the community for clean drinking water. Any ban on single-use plastics that would affect access to bottled water would not go over well in the community, she said. Lhoosk'uz Dene Nation has running water connected to its homes, but Baptiste said it is contaminated with heavy metals that leave the water yellow and smelling like sulfur."[70] The Blue Communities project acknowledges this dilemma and has modified its requirement regarding bottled water bans for communities whose tap water is unsafe.[71] Clearly such situations necessitate a nuanced approach to challenging packaged water consumption.

However, the argument that it is only in places where the water is largely safe (or in wealthy nations with the capacity to make huge expenditures on infrastructure) that this mode of bottled water activism appears—while largely true until recently—is now no longer accurate, as anti–packaged water sentiment merges with the global antiplastics movement. Local institutional bans on single-use bottled water are now being adopted even in places that do not have universal access to safe drinking water. The governments of South Delhi and Maharashtra State in India are two recent examples, the latter having a population larger than most nations on earth. Maharashtra officials said they planned to replace plastic bottles in state offices with water treated by reverse osmosis and served in glass bottles or dispensers.[72] Universities in the global South are also getting in on the act: the Federal University of Lavras in Brazil has become a Blue Community, and Silliman University in the Philippines—part of the Break Free From Plastic movement—has halted the sale of single-use water bottles. These examples are likely just the first of many such efforts to come.

FROM CONSUMPTION TO EXTRACTION

In closing, it is worth making a final observation about this facet of the movements confronting bottled water. The consumption-focused, reclaim the tap form of activism challenges the commodity of plastic packaged water (and the bottled water industry) in general and even abstract terms. The groups demanding that local governments ban its purchase and sale typically focus on the negative impacts of bottled water as a generic commodity, rather than on specific corporate actors, and without regard to the original source of the water they contain. This makes strategic sense. When a city bans municipal purchases of bottled water, it is nixing bottles containing both refiltered tap water and groundwater extracted from springs or boreholes. In doing so, these efforts target 100 percent of the bottled water market, without distinction by manufacturer or source. They do not usually focus on brand names, and the opposition to their work in North America has mostly come from industrywide lobby groups, as opposed to individual bottled water firms.

That differs distinctly from the battles over the industry's extraction of spring water and groundwater (which in the United States constitutes just over one-third of the bottled water market, but the majority in Canada and many other nations). Such conflicts at the production end are highly focused on particular corporate actors, with opponents naming names and working to protect groundwater in specific locations. It is to those deeply local extraction struggles that I turn next.

CHAPTER 5

Cascade Locks: A Decade-Long Struggle

I follow Anna Mae Leonard up the gravel road behind the state fish hatchery, set at the base of a steep slope on the edge of national forest land. We enter a forest of Douglas firs, then fork right onto a short trail that dead-ends at a large, deep pool, lined with ferns and filled with crystal clear water. Spring water gushes into the pool from a brick-lined opening in the hill. A metal catwalk with a wooden railing leads along the outer edge of the pool, perched over the steep slope. The only sound, aside from the faint noise of the freeway, is the steady stream of water that flows out from the pool on a concrete sluice, falling twenty feet onto moss-covered rocks in the thick forest below. If a proposed water exchange between the state of Oregon and the nearby town of Cascade Locks is approved, Nestlé will build a large bottling plant, and 118 million gallons per year of this spring water will be packaged into Arrowhead brand water, distributed, and sold throughout the Northwest.

There is nowhere to sit, so we grab some decaying cut sections of log for makeshift stools. A few beams of sunlight reach us through the forest canopy. Neither of us can know that in two weeks, this area will be partly burned in an enormous forest fire, started by a teenager throwing fireworks. Leonard is a grandmother in her late fifties, with hair worn in a single braid and a gentle smile. She is a member of the Confederated Tribes of Warm Springs, which along with three other Columbia River tribes holds substantial rights here on their ceded lands under the Treaties of 1855, including to their traditional salmon fisheries. "The spring here where we're sitting is, in the English language, Oxbow Spring," she tells me.

> In our native tongue, it's *K'uup Waniitcht*. . . . From ancient times, this is an area where people came to pray, to practice our spiritual ways, our traditional ways, and that might involve songs, but it's also a place where people would stop to eat lunch and drink the healing water, this water here. . . . Water is a human right. It's a human right. In traditional Native societies, it is inconceivable to sell water, like it's inconceivable to sell the air. The ultimate goal of the Wanapum Fishing People Against Nestlé is to stop the use of bottled water in Indian country.[1]

. . .

In 2008, Nestlé Waters announced a proposal to build a large bottling plant in Cascade Locks, Oregon, a town of 1,100 people located in the scenic Columbia River Gorge. The plan would give Nestlé a fifty-year guarantee of access to Oxbow Spring's water, which it would purchase as a customer of the local water utility, doubling the cash-strapped town's revenues and promising to create nearly fifty jobs. Cascade Locks city officials, Oregon's governor, and the state fish and wildlife agency all lined up behind the proposal, continuing to back it during the long approval process. For several years, very few voices within the town publicly opposed the plan. It looked like the deal's approval was a foregone conclusion.

Yet as I sat with Leonard under the trees more than nine years later, Oxbow Springs still flowed untapped (figure 12). The bottling plant had still not been approved, and it appeared further than ever from being built. Four months later, the proposal was officially dead. How was a group of community members and activists able to hold off the world's largest food and beverage corporation for so long, and eventually scuttle its water bottling plans entirely? This winding story involves an array of regional and national organizations, a prolonged campaign directed at state government, and ultimately the emergence of a homespun local movement that assembled a diverse coalition including Native activists and tribal governments, culminating in a precedent-setting countywide public vote on water bottling that contributed to the proposal's eventual demise. However, the broader tale both predates and follows Cascade Locks, incorporating half a dozen other Northwest communities where Nestlé and other global bottled water firms have attempted to gain a toehold and access spring water. The details of this decade-long set of struggles offer valuable lessons, not only for

FIGURE 12. Oxbow Springs, Columbia Gorge, Oregon. Photo: Author.

residents of other rural communities facing similar proposals, but for water protection movements more broadly.

REPLACING MOTHER MCCLOUD

To understand what drew Nestlé to the Pacific Northwest, and eventually to the rainy heart of the Columbia Gorge, we need to step back several years to the company's previous major attempt to bottle spring water on the West Coast. That six-year conflict took place in the far northeastern California town of McCloud, nestled in dense forests at the base of towering, glaciated Mount Shasta. The unincorporated hamlet of 1,200 residents was long a company town owned by the McCloud River Lumber Company, colloquially known as "Mother McCloud." The timber industry faded in the 1980s and the mill closed entirely in 2002, pushing the unemployment rate over 20 percent.

In September 2003, officials of the sole local government body, the McCloud Community Services District (MCSD), approved a contract with Nestlé that was negotiated behind closed doors. The contract allowed the company to build what would have been the nation's largest water bottling plant on the former mill site, giving it access to 520 million gallons of spring water annually for ninety-nine years as a customer of MCSD. Although Nestlé would have paid the district only $0.00008 per gallon (one cent for each 123 gallons of water), the contract would have generated $350,000 annually for MCSD, whose total revenues were less than $1 million, plus a promised 250 jobs.[2]

How did Nestlé choose this community for its proposed bottling plant? According to journalist Michelle Conlin, "Nestlé employs 11 water hunters around the U.S. Besides monitoring water supplies, they search for new sources, typically in remote, pristine places like McCloud."[3]

When the service district board voted to approve the contract—for which Nestlé covered all legal costs—on the same day it was made public, many residents were outraged.[4] Local opposition to the plan coalesced into the grassroots Concerned McCloud Citizens, which joined with McCloud Watershed Council (MWC) and later two angler groups, California Trout and Trout Unlimited, who were worried about the plant's impact on the area's blue-ribbon fisheries. MWC later also collaborated with Food and Water Watch (FWW).

Nestlé's proposal split the town down the middle. Supporters of the plan, who had earlier tried to court other bottlers, argued that commercializing McCloud's abundant spring water was a no-brainer, a way to fix its deep economic woes. "This is the only thing the service district has to sell," one resident told me. "We have water—pure, simply, end of the story, water. That's all we have to sell, other than what you get from me every month [in taxes]. . . . So we thought, 'Gee, who could complain about a clean industry, a water bottling plant?'"[5] As tensions rose, an opponent's tires were slashed and a proponent's fences torn down.[6]

Local critics, however, saw the plan as a serious threat to the town's future access to drinking water. A member of the McCloud Watershed Council said, "the main reason . . . that the people who switched from pro-Nestlé to anti-Nestlé [did so] was . . . that in the contract, it stated that if we [were] to go into drought . . . we would have to dig the wells, we would have to drink the well water, and Nestlé would get our pure spring water."[7]

Legal challenges by opponents forced Nestlé to prepare an environmental impact report under the California Environmental Quality Act (CEQA), slowing approval significantly. Then in 2008, facing a nationwide recession, Nestlé announced it would greatly reduce the facility's size.[8] *BusinessWeek* described this development as a "cautionary tale for any company. [Formerly], multinationals could arrive in economically depressed communities and pretty much have their way. But . . . residents in McCloud were able to turn their issue into an international sensation. Now Nestlé has capitulated."[9] Finally, in September 2009, Nestlé revoked its McCloud proposal entirely. Only a few months earlier, the firm had reached agreement with local officials in Sacramento, California, to build a large plant there to bottle the city's municipal water.

In analyzing Nestlé's retreat from McCloud, activists said they had prevailed because of a mix of effective messaging, legal challenges, and public relations harm to the company from unfavorable media coverage. "We framed it with two things," a California Trout staff member told me: "that really we are sort of interwoven, that water is more valuable than this [Nestlé plant]. And that . . . Nestlé is going to make billions of dollars . . . and you got treated like a Third World nation by them. And you deserve better than that. If selling water is a good idea, then think about what you just did—you just sold the farm for almost nothing."[10] Nestlé Waters' CEO, however, described the decision as purely a business move: "The Sacramento plant will allow us to serve our Northern California customers with lower distribution costs and a reduced environmental footprint. . . . We no longer have a business need to build a new facility in McCloud and we are withdrawing our proposal."[11]

BOTTLING CITY WATER

A plant that bottles municipal water is difficult to find. Often located in industrial parks, they look like any other large warehouse. Bottling firms simply hook up metered connections to the city water supply, refilter the water, add proprietary minerals, bottle it in plastic, and haul it out in trucks to stores. It's a bargain—bottlers usually pay commercial water rates, sometimes at a discount, and permitting and siting hassles are almost nonexistent. They fly under the radar, avoiding the controversy of many spring water extraction plans. Given all this, the protests that erupted at Nestlé's Sacramento plant starting in 2014 were remarkable.

According to Dan Bacher, a local independent journalist, Nestlé's 2009 proposed bottling facility in Sacramento was heavily promoted by then-mayor Kevin Johnson, a former NBA basketball star and establishment-aligned Democratic politician. Activists had "kicked Nestlé out of McCloud, and then they came down here because we had a mayor that wanted them there. . . . This is 2009. Economic collapse, the housing bubble burst the year before. So that was essentially Kevin Johnson's thing: 'Hey, these are jobs . . . so what's wrong?'"[12]

Nestlé's plant was capable of bottling at least eighty million gallons per year, although city officials say it usually extracts less. For this water, the company paid less than two-tenths of a cent per gallon, the same rate paid by residents and other businesses.[13] As of 2015 Nestlé also bottled tap water from three Southern California cities, plus spring water from eleven sites statewide.[14]

In 2014, drought hit nearly all of California, and by 2015 it was far more severe. Governor Jerry Brown mandated emergency cuts in residential water use of at least 25 percent.[15] "[California] Department of Water Resources [is] telling people to cut back on showers, to let their lawns get brown," Bacher told me. "They're touting as heroes people that do that and use less water in their showers. . . . Meanwhile, Nestlé is pumping out all this water and selling it back to the public, making an enormous profit. . . . And that's what I think caused the real outrage."

After revelations in 2015 by the *Desert Sun* newspaper that Nestlé was extracting water from a spring in the San Bernardino National Forest on a permit that had expired twenty-seven years earlier,[16] Nestlé Waters' new CEO, Tim Brown, defiantly told a radio interviewer that "if I could increase [water bottling in the state], I would."[17] This juxtaposition—commercial bottlers unapologetically pumping scarce groundwater without restriction amid a historic drought, while residents had to slash their own water use—made the issue an international news story.

In this heated context, a group of Sacramento activists, humorously calling themselves the Crunch Nestlé Alliance, organized some of the very few public protests of facilities bottling only municipal tap water. The first, in October 2014, blocked the sidewalks outside the plant gates. Bacher recalled, "Nestlé had already called its people off. . . . They were smart and they avoided the conflict. . . . Crunch Nestlé Alliance was the main group. But at the protest there were people from the peace community, Indigenous community, Chicano community, like the Brown Berets were involved in it." In March 2015, the Alliance again blocked the entrance to Nestlé's facility, this time outfitted with mediagenic costumes and props. "People made these pitchforks and there were barbarians at the gate," said Bacher. ". . . And that Nestlé's

Crunch sign—all the TV stations loved that." The protest received international media coverage. "We never thought this was going to go global," one of the Crunch Nestlé members told me.[18]

The local activists' efforts gained support from statewide and national groups, including the Courage Campaign and Corporate Accountability International, which launched online petitions that gathered over 160,000 signatures in favor of halting water bottling in the state.[19] While they did not win their demand, the fact that Nestlé's extraction of tap water became a major focus of contention at all is significant. It suggests that piggybacking on public water infrastructure may no longer be the reliably discreet and uncontroversial market strategy it once represented.

Sacramento gave Nestlé a new, stable source of municipal water for its low-cost Pure Life brand. Crucially, however, it was not *spring* water, which sells under different brand labels for a higher price and is more profitable.[20] Thus the company continued searching for new spring water sources, casting its gaze north to the wetter Pacific Northwest.

HUNTING FOR WATER IN THE NORTHWEST

As you drive eastward out of Portland, Oregon, along the wide Columbia River, the city drops away suddenly. Deep green forested hills along the river rear up higher and higher, becoming mountains so high you're craning your neck to see the tops. Forty-five miles from Portland, in the heart of the Columbia Gorge, basalt cliffs rise up from the river, only one hundred feet above sea level, to the Benson Plateau at over four thousand feet.

This is a pinch point—the narrowest part of the passage between the arid interior and the rainy Willamette Valley, and the only nearly sea-level mountain crossing on the entire West Coast. A massive landslide about five hundred years ago created the original Bridge of the Gods, a natural dam that blocked the Columbia, eroding to leave a series of rapids—the Cascades for which the range is named—that posed a major obstacle to Euro-American settlers and led a town to grow here. A set of shipping locks built in the 1890s eased travel until they were flooded by Bonneville Dam in 1938.

Above and below Cascade Locks, massive dams generate part of the Northwest's abundant hydroelectric power, which today is counted as renewable energy. Yet the dams have wrought great ecological and cultural damage—especially the Dalles Dam that in 1957 drowned Celilo

Falls, a vital ancestral fishing site and trading center for the Columbia's Indigenous people, and the longest continuously inhabited community in North America.[21] The dams have also effectively severed the river, heating its water in summer and causing a dramatic decline in spawning steelhead and chinook, coho, and sockeye salmon, now at only 5 to 10 percent of their historic levels.[22]

The economy of this area ran on fishing and timber for more than a century after European settlement dispossessed the Indigenous inhabitants. Large timber mills employed dozens of people in Cascade Locks until the last one shuttered in 1988. The Columbia Gorge National Scenic Area Act, signed by President Reagan in 1986, enacted protections against development on almost 300,000 acres (121,000 hectares) of public and private land in an eighty-five-mile-long swath on both sides of the Gorge, with exceptions carved out for the few towns and cities. A recreation, tourism, and service economy began to replace the industrial one, faster in some places than others.

Here in Cascade Locks, hikers along the famed Pacific Crest Trail drop down sharply to cross the Columbia over the "new" Bridge of the Gods—a narrow, century-old metal toll bridge lacking a pedestrian walkway but offering stunning views up and downriver. In this monumental landscape, Cascade Locks is squeezed between the mile-wide Columbia and the verdant mountain wall. Just above the river runs the Burlington Northern Railroad, whose long trains constantly roar through town. Next comes the town's one main street, and a block further up is the concrete barrier of Interstate 84 and the city limits. Above lies Mount Hood National Forest land, most of it in the Hatfield Wilderness.

This is a wet place, eight months of the year. With average annual precipitation of 77 inches (195 cm), the forests are a lush, moss-draped, dripping wonderland with waterfalls huge and small. On the mountain slope above town, dozens of springs seep out of the ground. One of them is Oxbow Spring, which after leaving its containment pool, running down the mossy hillside, and passing through a small state fish hatchery joins Herman Creek, a cold stream that quickly dumps into the Columbia.

Just upriver from Cascade Locks, precipitation tapers off dramatically in the abrupt rain shadow of the Cascades. Twenty miles east in sunny, piney Hood River, a major agricultural valley and now a recreational mecca, only thirty-one inches per year fall, and twenty miles further the mountains have dropped away, depositing you into grass- and sage-covered desert fed by thirteen inches or less.

"A COMMUNITY ON HARD TIMES"

In 2010, at the depths of the Great Recession, Cascade Locks had an official unemployment rate of over 18 percent.[23] Like many rural Western communities, it exhibits tensions between a history of reliance on natural resource extraction—in this case logging—and a regional economy increasingly oriented around recreation and tourism.

On a chilly April day in 2010, with snow still lingering on the cliffs towering over the town, we[24] sat down in Cascade Locks to talk with Jerry Rogers,[25] a local official. I asked him to describe the town. "We're a community on hard times," he responded, "and we're a community in transition, where our economy died with the timber industry, and we really haven't come out of that." He continued,

> In general, I think you could say that this is a conservative town. It's a guarantee that they vote Republican, it's definitely red. . . . It's relatively low-income . . . small town, rural, so in general there's a distrust of government. . . . I bet you that we're pro-gun rights here, and I would offer that it's a more Christian or [a] type of a religious mindset, but it does have its Oregon flavor. . . . There's an environmental ethic floating around here as well—I mean you've got people very distrustful of government, but say they want this to be a green community. . . . Most people would tell you, I support the Scenic Act and its protection of these beautiful mountains. Nobody in town thinks that we should be clear-cutting. Where they get upset is like, [within city limits] we're not in the Scenic Area . . . but there's a mentality in the Gorge that says anything that happens in the Gorge is subject to review by environmentalists and advocacy groups like Friends of the Gorge, and it's just flat-out unfair.[26]

Other residents echoed this sense of lost economic vitality. "If I could have something here," Emily Caples[26] told me, "it'd probably be a bank first, or a pharmacy, or a bigger grocery store, or just the basic living things would be nice. But we don't, so we adapt."[27]

Deanna Busdieker, a city council member, said that many town residents were hoping to return to an older, extractive economic model: "I think the mill closed in about 1980, and it has felt for me, personally, that time stopped for a lot of people when that mill closed, and they just want to go back to being a company town and selling your natural resources is just what you do, . . . and that's not the way things work anymore."[28]

One element that arose in most of the conversations in Cascade Locks was the town's relationship with its neighboring city—the much larger Hood River. By a stroke of political fate, Cascade Locks ended up within Hood River County, but the two communities differ drastically in demographics, economic base, political outlook, and cultural orientation. Many residents insist Cascade Locks has been shortchanged as a result. "Our county money focuses into Hood River," said Robert Hanford.[29] "They're really an agricultural community, but we're in their county. We pretty much have to live off of them too and we're dictated a lot of things by them, and we don't feel that we get . . . everything we should from them because we're two different entities." Another local official, Caleb Townsend,[30] told me that "Hood River has a tendency to be a lot of kind of liberal, Birkenstock-wearing environmentalists . . . Subaru-driving, you know, and a lot of that 'Not in Our Gorge' type attitude, and some of them have views that yes, stopping bottled water will save the planet." Les Perkins, manager of Farmers Irrigation District and also a Hood River County commissioner, confirmed that "there's very much a feeling that the residents of the city of Hood River dictate what happens in Cascade Locks. . . . There's just this deep-seated feeling that they are neglected and left behind."[31]

This sense of neglect increased for many in Cascade Locks when the town's high school and middle school closed in 2009. According to Nancy Marquardt,[32] a former teacher, "We don't have enough kids to keep a high school here or a middle school, so that's why it's up at Hood River now. That's a major issue with a lot of the old timers . . . It's strange, you used to see kids just running around." Brent Foster, a lawyer living in the Gorge, told me, "You have to start from the perspective of Cascade Locks, [which] like tons of rural Oregon towns has been totally left behind in the kind of neoliberal vision of success. They've not benefited from the economic growth that the neighbors to the west see in Portland and many other places. And they have every right to be totally pissed off."[33]

This is the political and cultural milieu into which Nestlé's bottling plant proposal landed at the depths of a national recession: genuine concerns about a withering local economy and institutions, a deep desire for viable employment options, and widely felt grievances toward the town's thriving neighbor city, grounded in divergent political and cultural orientations and a perception of being blocked from pursuing its own economic development visions.

The decade-long tale of this fight over water bottling in the Columbia Gorge features several distinct episodes, unexpected plot twists, and a continual series of new characters. It can best be recounted as a drama in six acts.

ACT ONE (2008–10): SWAPPING WATER

In a cramped office in a fading turn-of-the-century building in downtown Portland, Julia DeGraw—an energetic twenty-something Portland native and the newly hired Northwest organizer for Food and Water Watch—explained why her organization decided to establish a presence in the region:

> [Nestlé's] water sources in California are drying up, and they just got kicked out of the Mount Shasta area. . . . They don't have a single water bottling operation in the Northwest, and they know that this is a wet region and they want to get a foothold. . . . Nestlé specifically has this track record of going into rural communities that have been hit with hard economic times, and . . . taking advantage of that, and extracting this resource, often with adverse impact, in order to make a large amount of profit. And what has happened in Michigan and Maine and other locations has made it very clear that once they set up shop, it's very, very hard to undo what has been done. So the way we'd have the largest impact on the industry in general . . . [is] to stop these water bottling facilities from going in in the first place.[34]

This argument—that spring water bottlers like Nestlé prey on hard-hit towns with a history of dependence on resource extraction—is commonplace among the industry's critics. However, a bottled water company staff member I interviewed denied that the industry intentionally seeks out economically distressed rural communities for its water sources: "We don't . . . look at a map and say, 'Okay, where are there communities where the industry is leaving and they need somebody to come in and backfill?' That is not part of our decision process. The primary, first and foremost important things are, is there a water source that meets the FDA definition of spring water and . . . does the water quality meet our requirements and [is] the quantity at a level that could sustain a project like this?"[35]

After shopping around the region in 2007 and courting officials in several Washington state towns—Enumclaw, Orting, Black Diamond—without success, Nestlé set its sights on Cascade Locks. Rogers described how the company first approached local officials: "So this [Nestlé representative] invited. . . . us to meet . . . and just said, 'Hey, we've been looking at siting a plant in the Northwest and we discovered that you guys have two things we're looking for. You've got excellent springs, and you've got industrial land here, and those actually are hard to find together.' . . . We actually did a tour and . . . we went from basically west to east, and when we were done, their favorite spring source . . . was the Oxbow Springs and Oxbow Hatchery."

However, local officials explained to Nestlé that the city did not own the Oxbow water.

> And then they said, "Well, who does?" Well, the state does. And then they said, "Well, there's an awful lot of water coming out there; do you think they would allow us to take a small amount?" . . . So we put together a meeting with all the key [state agency] people . . . and we just posed this question: What's the possibility of the city taking a small portion of your water right and giving it to Nestlé for them to bottle up? So they're going, "Man, we love this idea," and then we turned to WRD [Oregon Water Resources Department] and [said], "Okay, do you see a problem with an exchange of spring water rights with well water rights?" And they said, "No . . . you can do that," so that meeting was like, "Holy crap, this is a doable project!" And we've been running ever since.

What was the state's interest in this arrangement? Democratic governor Ted Kulongoski, who left office in January 2011, instructed the relevant state agencies to pursue the deal on economic development grounds. The governor's support for this bid to bottle Cascade Locks spring water may or may not have been related to Nestlé's expenditures on lobbying Oregon politicians, which totaled $35,000 in the 2009 legislative cycle alone.[36]

Doug Bochsler, program manager with the Oregon Department of Fish and Wildlife (ODFW), explained: "We were introduced to this process through the governor's economic development group and asked to participate through the governor's office. That's how it all started. And we told them that as long as we can benefit, the fisheries of the state of Oregon and the mission of ODFW, we'll talk, we'll move along. . . . We're looking to protect the interest of our hatchery and improve the interest of our hatchery—that's what we're going to do."[37]

The result was a complex three-way proposed water swap—an exchange not of water *rights* (despite Rogers's comment above) but of water. Townsend explained the logistics of the deal:

> What we are doing is we're swapping with Oregon Department of Fish and Wildlife, essentially exchanging our well water for their spring water. . . . Nestlé's going to put in all the infrastructure, all the plumbing. . . . [ODFW] likes it because it gives them then the opportunity to supplement their water resource; in the summer sometimes they have lower flows and they are unable to expand. . . . Essentially it is a gallon-for-gallon trade—we essentially get the spring water and then replace it with well water. And we in turn sell that spring water to Nestlé just like any other industrial water customer.

This deal was unusual because it involved a local government, a state agency, and a private corporation. Proponents framed the plan as a win-win-win scenario. It was also atypical in another way: the spring water source was owned not by a private landowner or the town but by the state of Oregon. It would have been far easier for Cascade Locks simply to sell its well water to Nestlé, but the firm was seeking valuable spring water instead. If the spring it chose had been located on privately owned land, the approval process would have been much quicker as well. But because Nestlé settled on a state-owned water source, the path to sealing the deal was longer and more complex. This key fact—that the water right was effectively owned by all Oregonians—would turn out to be pivotal in the outcome. Not only did it mean that a major additional layer of regulatory approvals was required, but critically it enabled opponents to frame the plan as a corporate grab of public water.

The remaining details were more familiar. Nestlé planned to build a large $50 million bottling facility on land it would purchase in the city's industrial park. This factory would bottle up to 118 million gallons of Oxbow spring water per year as Arrowhead water, plus an undetermined but substantial volume of well water for its Pure Life brand. For both water sources it would be a customer of the city's water utility, paying less than one-fifth of a cent per gallon for water it would eventually sell for thousands of times that amount. Up to two hundred trucks per day would haul the water away—over a billion bottles a year—the vast majority of it to stores in western Oregon and Washington. The company claimed the plant would employ almost fifty people full-time, but it did not promise those jobs would go to local residents.

Most attractive to local leaders was the revenue the plant would generate. According to Townsend, between projected tax revenues and the sale of electricity to Nestlé, "We're looking at . . . very close to a million dollars a year in [total]." Nonetheless, when I first spoke to local officials in 2010, Nestlé had neither put its promises into a contract nor purchased the land for the bottling plant. Seven years later, it still had not done so.

LONELY LOCAL OPPONENTS

Organized resistance to Nestlé emerged almost as soon as its plans became public, but over the course of nearly ten years it surged, ebbed, and was reshaped substantially. Within Cascade Locks itself, very few residents were willing to publicly oppose the bottling plan in the early years, while boosters of the proposal were loud and numerous. According to Perkins, "There were a lot of vocal folks within the community that were saying, 'Hey, we want the economic development [from the Nestlé proposal]. . . . It's something that this community needs.'" Rogers argued that local support was near-unanimous: "I literally only know one person who's opposed to it. . . . She is a newcomer and she's one of those classic people who comes in the community, they want to shut the door and say nothing should change from this point out."

Indeed, in 2010 only one local resident opposed to the plan was willing to be interviewed. Karla Newton[38] recalled that she first learned about the proposal when the city manager announced at a meeting "how [the company] really wanted our water, and how they were going to do this right this time, that we had this great and wonderful opportunity to sell water to Nestlé." She continued,

> Where I used to live is [the] Calistoga [California] area. . . . A lot of people there believe that Saint Helena Creek was drained dry from them. . . . That's partially why I started by going on the StopNestlé blog, and said, "[Can] anyone give me some advice about how this has happened?" I also knew about McCloud from being from California. And evidently [a statewide water organizer] was googling or got something . . . and that's how I connected with them. I just threw it out there, like, "How do I deal with these people? It's Nestlé."

Newton described how local officials worked quietly to build early community support for the plan. "What they decided to do then was hold some meetings, that were just with the movers and shakers of the community. . . . And they had two meetings . . . so that they didn't have a quorum for the council. . . . What they were trying to do was get a few people on board with it, agreeing with it." She attended one of these gatherings: "I sat right next to [a Nestlé representative] and kind of hit him with a nice quiet little question about what happened in McCloud. . . . He wasn't expecting [many people], he didn't advertise it anywhere, but I got ahold of everyone I could and let them know. That's why [Columbia] Riverkeeper was there, that's why Friends of the [Columbia] Gorge . . . sent out a whole batch of 'Please come to the meeting, ask the hard questions.' And Nestlé was surprised." Thus regional environmental and water advocacy groups were alerted to the proposal early on, becoming involved from the very start.

However, the claim that there was only one local opponent did not square with the concerns I heard expressed in other formats, even in the proposal's heady early days. At a Nestlé-organized "town hall meeting" in 2010 attended by about seventy-five people, a company spokesperson opened the proceedings and framed the agenda, followed by supportive comments from city and Port of Cascade Locks officials. Despite a tightly constrained format that offered no opportunity for open discussion, many town residents found ways to interject their concerns about the plant's possible harmful effects, especially the high volume of truck traffic and its effect on air quality, traffic safety, and the condition of local roads. One man asked the Nestlé representative who would bear the cost of maintaining local roads with hundreds of truck trips daily. Another resident, an Indigenous fisherman, voiced fears about the impact on salmon in Herman Creek and the Columbia: "Will Nestlé clean up the mess? You'll be soaking my nets with oil." A woman from nearby Stevenson, Washington, urged attendees to drink tap water and shun bottled water. Finally, a community member stood up to express his fears about groundwater depletion: "After you tap all our water out, are you going to tear everything out, or are you just going to leave it, like back east?"

"WE CONTROL THE VALVE"

Nonetheless, many residents of Cascade Locks voiced a strong belief that local water—both groundwater and surface water—was exceptionally abundant, and they were highly skeptical of arguments that the bottling plant would ever threaten water availability. According to Dierdre Jefferson,[39] opponents of the bottling plant "don't want to see the water leave this area [pointing out the window at the Columbia River]—there it goes! It flows into the Columbia and there it goes. All the water's leaving this area. What's the difference if it goes out in a bottle or out in that huge river? . . . We get eighty inches of rain a year. We've got one thousand people here. And we're surrounded by Scenic Area national forest. Who's gonna use our water? Nobody." A longtime resident, Lorraine Harmon,[40] similarly argued, "[Nestlé opponents are claiming that] 'We're going to run out of water!' Give me a break, you know. It's just not very realistic to think that a town in the Columbia River Gorge where it rains is going to run out of water. . . . If they're talking about two [bottling] lines, then they're talking about 10 percent of our water capacity being used. Well, I just don't think it's reasonable to think we'd run out of water at that point."

Another major point of contention was the question of who would exercise ultimate control over water extraction. Here, as in McCloud,

Nestlé would not hold ownership of the spring or the water rights. The complex water swap between local officials and the state of Oregon technically excluded the corporation from the legal arrangements to move the spring water from the public sector to the private sphere, yet Nestlé was the stated beneficiary of the water. If the deal were approved, the company would access the spring water (and additional well water) as a customer of the local water utility, but a strongly preferred customer with fifty-year contractual rights.

The ODFW's Bochsler insisted that the state would not be ceding any rights or protections: "We can protect ourselves in this agreement and turn our water back on from the spring [if needed]. By holding our water right, we maintain the ownership—the property right ownership of that resource. And then we're exchanging the water, kind of like in a business contract with protections in it for us." Likewise, many local supporters of Nestlé's plan emphasized a belief that the community would hold the reins in its dealings with this multinational firm. "I think a lot of people think that they've got a wild card running around here," said Kevin Cardozo,[41] another local official, "that's going to do whatever they want to do and they're big and they're strong and powerful. But . . . the city has the ability to write resolutions and procedures and things of that nature to control these people. . . . So this whole thing will be controlled basically by the city. . . . We're going to write right in the contract that we control the valve."

The bottled water company staff member I interviewed echoed the notion that local officials ultimately call the shots in such arrangements: "As far as water withdrawals, it would be like any other industry coming into town. . . A community would have a commercial. . . rate in their industrial plans as far as how much volume could be set aside for industrial customers, and [bottling companies] would have to work within their plan—within their limits of supply."

However, aside from the questionable premise that a small community could effectively regulate a large multinational firm, distinctions between loss of control and loss of rights may be largely semantic. "Even though Nestlé isn't getting water rights in this particular case," argued DeGraw, "they're gaining a lot of control. Once Nestlé invests millions of dollars building a huge water bottling facility and has a contract in which the city is obligated to give them 118 million gallons of water a year. . . . You don't need explicit water rights in order to gain control of a resource." In other words, the lack of a legal exchange of property rights does not mean the market has not exerted control over water. Commodification can occur in the absence of privatization.

ACT TWO (2010–14): BUILDING A COALITION

By 2010, a statewide alliance had been assembled to fight the bottling proposal, titled the Keep Nestlé Out of the Gorge Coalition. This alliance eventually grew to comprise sixteen organizations, including environmental, consumer, and faith groups such as Environment Oregon, Alliance for Democracy, Columbia Riverkeeper, the Sisters of the Holy Names, and Oregon Physicians for Social Responsibility. However, says DeGraw, four groups were central to the active coalition, participating in regular strategy sessions: Food and Water Watch, the forest advocacy group Bark, the regional Sierra Club chapter, and the public employee labor union AFSCME—all based in Portland.

Yet while these groups could fairly easily mobilize large numbers of people to send emails to state officials or attend a Portland rally, cultivating local opponents willing to speak out in a small community with a vocal pro-Nestlé majority was another matter entirely. A Sierra Club volunteer described these challenges: "We've taken twenty-six-some [Sierra Club members] in the area and we've mailed to them, we've emailed to them if that was possible, we've called them if possible, and of that we've got one member who is actually taking action," she admitted. "And our goal is . . . [to] make it possible so that when you take action, you're bringing your neighbors with you or your friends or your family . . . so that [people] have a little bit of cover as well, because we understand it's a small town, and as soon as you're out on an issue it's everybody's business. . . . We definitely need stronger communication with folks in Cascade Locks."[42]

A LONG SLOG

For a state with a vaunted green reputation, Oregon's water protections are surprisingly weak. A 2016 investigation by *The Oregonian* revealed that the state's groundwater regulations are feeble and outdated, the water agency allows unsustainable rates of extraction, some aquifers are being rapidly depleted, and it is "virtually impossible for the state to enforce its pumping limits."[43]

Nonetheless, the fact that Oxbow Spring was already state-owned water turned out to be good news for opponents: it ensured a long, numbingly complex approval process, offering many opportunities for activist groups to intervene. The coalition mobilized a record-breaking number of citizen comments to the state Water Resources Department opposing the water swap and made a formal submission arguing the deal was not "in the public interest"—the key legal metric for approving the arrangement.[44] Looking back after the multiyear effort, DeGraw observes,

> It was kind of a gift that ODFW was being approached to give water—publicly owned water—away so that Nestlé could bottle it . . . [so] we were able to run a statewide campaign. . . . [We generated] seventy-seven thousand comments to Governor Kitzhaber, right? And so that got national press, and the first year I got interviewed by the *Wall Street Journal.* So it was big. It was big news, and then it became this big long fight, and it's really significant that it became a big, long fight because it showed that we had staying power, and that we weren't just going to walk away from this, and that we were winning.[45]

However, for almost five years, the choice to focus the campaign on the state approval process meant this struggle would play out in slow motion, moving forward almost imperceptibly, punctuated by an occasional rally or press conference. In 2012, despite the huge volume of opposition comments, the Water Resources Department approved the water exchange. Crag Law Center filed a lawsuit protesting the decision, stopping the clock for two more years. Finally, in 2014, their appeal was denied. Officially this was just another modest setback for opponents—there were still more approvals required and likely years to go before the deal could actually proceed. Yet it turned out to be a pivotal moment, because it was at this point that state and local officials and Nestlé apparently lost patience with the sluggish bureaucratic process and decided to change tactics.

ACT THREE (2015): WATER CATCHES FIRE

Suddenly, after four years of almost imperceptible progress, in 2015 the Nestlé bottling proposal exploded into public view, rapidly becoming a huge media story and generating a slew of new opponents. Three major developments explain this change.

First, the water exchange plan was dramatically altered by the state, Cascade Locks, and Nestlé, shifting it onto a fast track for approval. "Now," wrote the *Oregonian*'s Kelly House, "Nestlé wants to scrap the existing permitting process for an approach with the potential to cut the remaining wait time in half. Instead of obtaining the water through a gallon-for-gallon trade between the state and Cascade Locks city government, which would then sell the water to Nestlé, the company wants the state to trade its legal right to some of the Oxbow water. The new tactic would also eliminate a key sticking point in the current permitting process: [consideration of] the question of whether the deal would negatively impact the public."[46] Gone was ODFW's insistence that it would never surrender its water right at Oxbow Spring in order to protect the fish hatchery. Instead of a reversible exchange of water, this would be a permanent transfer of water *rights* to benefit one multinational corporation. "This starts us down the road. . . . It narrows the scope of opposition," said Cascade Locks city manager Gordon Zimmerman.[47]

According to DeGraw, this decision was decisive in fanning the flames of local and regional opposition. "All of a sudden, because of lawsuits and how long it was taking, and Nestlé being like, 'We want to get going,' . . . ODFW cried uncle and they said, 'We're going to do this fast-track process.' And that fired the tribes up, even fired the locals up, fired everyone. . . . That was absolutely the fast track proposal [that] exploded it."[48]

The switch to a water rights transfer galvanized two Columbia River tribes into taking action, framing the move as a threat to salmon and thus to their treaty rights. E. Austin Greene Jr., chair of the Confederated Tribes of Warm Springs, wrote to the new governor, Kate Brown, with a concise history lesson:

> The Cascade Locks area is a part of the Tribe's aboriginal title lands. The Tribe transferred title to the lands to the United States government as a component of the Tribe's 1855 Treaty with the United States. However, the Tribe has never left its aboriginal home on the Columbia River and takes an active role in exercising its treaty rights in the Columbia River area. . . . The Tribe urges the State to reconsider its process, withdraw the current [water rights] transfer application, and . . . requir[e] a public interest review as part of the water exchange.[49]

The Confederated Tribes of the Umatilla Indian Reservation followed suit. "Our concern is that there would be enough water still being sent down through to ensure the salmon are protected," said Umatilla spokesperson Chuck Sams.[50] "We don't have that info yet."[51]

The second game-changer was drought. In spring 2015, it became clear that much of the western U.S. was in the grip of a major drought. By July, twenty-three of the state's thirty-nine counties had been declared in a drought emergency, including Hood River County. This declaration arrived at the worst possible time for backers of the water rights transfer.

Pamela Larsen is an educator and a longtime Hood River resident. "I was born and raised watching Mount Hood," she tells me. "The summer I had come back from [living in] Asia—I was like, [gasps] . . . I had never seen it so dry. And I had seen it for forty-some years. I was like, 'What's going on?' . . . So it was on my radar, this thing about water."[52]

Larsen cofounded what eventually became the Local Water Alliance (LWA)—a new grassroots group that first emerged in Cascade Locks, putting an end to the claim that almost no genuine opposition to Nestlé existed in the town. The drought "was a huge part of our success," she says, describing a county commission meeting where an amendment was introduced to exempt Cascade Locks from the countywide drought declaration. Scientists testified about the devastating impact of climate change on the Mount Hood glaciers that local farms and orchards depend on for irrigation. "I had doom and gloom for twenty-four hours after hearing those speakers," she recalls. The amendment, supported by Cascade Locks officials and Nestlé, failed. Meanwhile, on the town's five-member city council, things were beginning to shift.

AN ALLIANCE EMERGES FOR WATER

Busdieker, a Gorge resident for over twenty years, served on the town's planning commission and was elected to city council in 2014. In March 2015, she was the first local elected official to publicly voice doubts about Nestlé's proposal, a risky move that exposed her to reprisals but also catalyzed the creation of what became the LWA. "Once I said something," she recalls, "all of a sudden, things started happening. And Pamela put this meeting together to decide how they would be able to support me. Actually the first meeting we had, we even had it over in Stevenson because [laughs] we didn't want anyone in Cascade Locks to know what we were doing. But there were ten or twelve people at that meeting, and that was really encouraging. . . . [Then] there were like twenty, twenty-two people at the second meeting, and we were all like, 'Wow, hey, maybe something's happening here.'" Larsen, the only non–Cascade Locks resident at the initial gathering, describes it from another vantage point. "I didn't go in there like I knew anything, 'cause I didn't know what I wanted to do," she says.

> . . . I just knew we had to do something. So, I went in with the approach of "What do you love about living here?" And we brainstormed. . . . What do they want in the future? And through that exercise, they all came up with that they didn't want something like Nestlé, this huge international

> corporation. They wanted local, sustainable, all these other things. And so we didn't come at it as "We hate them," you know? It was a very different approach, and I think it just gave the whole movement a different feel. . . . It's not just about battling something you don't like. It's really saving what we love and cherishing what we love, and how do we do that?[53]

Soon a second chapter of the group formed in Hood River. Crucially, though, the LWA offered strength in numbers for Cascade Locks residents who did not want the Nestlé deal to go through but were only just now finding one another, and finding their voice.

DEFENDING WATER AND SALMON

The third element that led this issue to erupt in 2015 was the involvement of several vocal independent Native activists, who played a fundamental role in the coalition fighting Nestlé's plans—a role distinct from, but complementary to, that of the official tribal councils.

Anna Mae Leonard is one of a small number of Warm Springs tribal members who live permanently in Cascade Locks. She explains how she first found out about Nestlé's plans:

> Well, my son was at me for two years: "Mom, they're going to take the spring up here, this beautiful spring, Oxbow Spring." "Who's gonna take it?" "Nestlé." I didn't know they were one of the world's largest food and beverage corporations. Didn't matter, I would have still taken it on. Doesn't matter who you are—if you're doing wrong, then something's got to be done about it. So one day, one morning, he'd just gotten on me again about it: "What are we gonna do? We've got to do something." He went up and down the river to the in-lieu [tribal fishing] sites . . . the Oregon sites and back to the Washington sites, asking the fishermen if they knew about this proposal, and no one knew about it.

The bottling plant proposal had been announced back in 2008, but none of the four Columbia River tribes—Yakama, Umatilla, Warm Springs, and Nez Perce (Nimiipuu)—had taken a public stance on the issue until the Warm Springs Council formally opposed the water rights transfer in May 2015. Leonard continues her story:

> Two days before July Fourth—July second [2015], I went to get a newspaper and a cup of coffee, and I saw these three ladies standing in the entrance, and they had their "No Nestlé" signs, and I was like, "Oh, no. There's only three women out here? That means we're losing. We're not winning." I had hoped . . . we were like, "Okay, the environmentalists, they got this." I should know better . . . but I stopped and talked to the ladies, and they were with

> the Local Water Alliance, and they ran up to me and they're like, "What are the Indians doing, what are the tribes doing, are you guys going to do anything?" And so, we're not winning yet. "You're not winning?" I told them. Something to that effect. They're like, "No!"
>
> . . . So that's where it started. We did our first rally on July Fourth, and I was surprised with how many—there wasn't a whole lot of Natives. There was probably thirty, but that was pretty good for pulling it together in two days. . . . And then we collected about, oh, between Warm Springs and Yakama . . . a couple thousand [petition] signatures.

Anna Mae Leonard is not actually her real name—it is Klairice Westley. She took on the pseudonym when she staged a hunger strike to protest the water bottling proposal, paying homage to two prominent American Indian Movement activists: Anna Mae Aquash and Leonard Peltier. "After the first rally on July Fourth, we realized that we needed to reach tribal members from all four tribes, and we needed to do it right now, so we needed something that's going to get attention." She continues, "I [had done] a protest fast before in my thirties about youth detention issues . . . so I decided I'm going to do a fast. I'm going to go without water and food, and I'm going to challenge the [Cascade Locks] City Council to try the same, to stop and think about what the earth would be like, all of life, without water. If we allow our water to be controlled, owned by a foreign corporation, they can do whatever they want with it." Word of her action spread rapidly among tribal members:

> At the same time, there was a big fire in Warm Springs, and one elder lady was driving by, taking things to the fire, you know, food, donations and stuff. And she stopped, which, I counted on something like this happening. Somebody's going to come by and they're going to throw it on the Facebook, and then that's what she did. She said, "We'll get this out there. Come on, take a picture with me." She's a real respected elder, you know. So she took a picture with me and it just flew. It went viral . . . and I was surprised how many tribal members showed up that week . . . even though the fire was going on at home, a lot of things happening, and that started it from then out. Each rally, we had more and more tribal members showing up. . . .
>
> But it was hard dealing with a lot of people, and there was a situation. A man here in Cascade Locks tried to run our feet over. . . . It would've broken our feet, our ankles. It would've broken our feet all the way off. It was less than an inch from the bottom of my foot hanging over the side of the curb.

Leonard's five-day fast in August 2015, as she sat dressed in traditional longhouse regalia in front of the Cascade Locks City Hall, drew a huge amount of news media attention to the battle over Nestlé's plans.[54] It also brought her into contact with new Indigenous activists.

I meet Whitney Kalama in the community of Warm Springs, one hundred miles southeast of Cascade Locks, up and over Mount Hood on the dry side, set in a canyon among sagebrush-covered bluffs and scattered juniper trees. The 650,000-acre Warm Springs reservation is one of the fragments of land remaining in the hands of the Columbia River tribes, after the forced surrender of the Warm Springs' nearly ten-million-acre ancestral homeland under the 1855 Treaties.[55] We sit down in a café, with loud music playing, and she talks about her connection to the site of the proposed bottling plant (see figure 13):

> I actually grew up in Cascade Locks, on a platform, in the summertime. That's where I caught my first fifty-five-pound salmon as a child, on a platform with a big hoop, with a long stick. I pulled it up myself. My dad was standing by, but I did all the work. My dad was a fisherman. That was the preserving of salmon, and he would do some resale to the tourists that would come, and friends he knew, he would bring it back. We'd go fish, we'd fish for weeks at a time, in the summertime when the reds came. We'd live at one of the in-lieu [fishing] sites for months during the summer and fall. And for a long time I grew up going to the East Wind drive-through, the Char Burger, and going to get ice cream at the Inn; we'd always buy dinner from there. . . . I spent my summers there. I grew up there. I used to walk up the hill across the railroad tracks from the marina park, which is a little beach that's on the far end. . . . We all, each family has their platform, their scaffold. . . . Warm Springs has fished along that river since I don't know what year, when the water was high, before the dams came—there's a lot of history down there.[56]

Kalama spent much of her youth living away from Oregon but returned to Warm Springs in 2013. "What drew me back is . . . you can't get your own culture, your own language [elsewhere]. . . . That's a big thing that I hold sacred is our language, which we're losing, and if we don't try to continue, then it will be lost." She became deeply involved in the struggle over Nestlé's proposal when she heard about Leonard's hunger strike:

> I saw a Facebook post of this lady, older lady, and in our language, a grandma's called a *kathlah* or *alla*—depends if it's your maternal or paternal. So I saw somebody posting . . . whoever was getting it out there on Facebook. And I noticed it and I said to myself, "If this *kathlah* can do it, why can't us younger ones go down there and support her?" She's sitting there, I saw the post about it—it was the day before the fast started for her. And she was prepared to fast. And I said, "I want to go down there." And my husband just kind of looked at me like, what are you talking about? . . . I reshared it, and I said, anybody want to go with me?

Another woman from Warm Springs accompanied her to Cascade Locks:

FIGURE 13. Tribal salmon fishing scaffold, Cascade Locks, Oregon. Photo: Author.

> We got dressed, we braided our hair and took our shawl and took our traditional regalia, which is called a wing dress in English, and I wore my moccasins and put on some beaded earrings and took my Pendleton shawl with me, and we got down there. And I saw her, and she was just sitting there quietly and praying to herself and she had her eyes closed. And [when] she opened her eyes . . . her face lit up, and she was like, "I'm so glad you girls are here." She didn't know us. . . . We didn't even know her real name until I think it was like a month down the road. And she was so excited. . . . She was so happy that we were there to sit with her. And I told her, "We're here to sit and visit and pray." And we sat all day with her.

Kalama continues, "Before we left we actually went down to the [Oxbow] spring and washed up and washed our faces, and we prayed and I put my feet in. . . . The water was so cold, and in our *Ichishkíin* [language], it's *Chúush*. And I had a Coke bottle, because it was floating around in my car, and I went to fill it up and . . . I drank it right away. And . . . it kind of got to me. I was like, I'm filling up a bottle with this water that we're trying to protect."

Kalama organized a series of protests by tribal members in Cascade Locks and raised awareness about the Nestlé proposal among the four tribes, launching a Native-specific petition against the plan. She pushed

FIGURE 14. Rally against Nestlé proposal, Oregon State Capitol, Salem, September 2015. Photo: Will Doolittle, whereismyriver.com.

the Warm Springs Tribal Council to take a more vocal stance, leading it to send a letter to the Cascade Locks City Council opposing the water exchange. She also helped plan a major public rally against the Nestlé proposal at the state capitol steps in Salem that September (figures 14–16).

Neither of these two activists formally joined the LWA, but they collaborated with its efforts. Simultaneously, they also created independent Native-led organizations: Westley cofounded Wanapum Fishing People Against Nestlé with her son Aushwol and two Yakama tribal members, and Kalama helped create a social-media group called Columbia River Treaty Natives Against Nestlé in Cascade Locks.

Ironically, at the time I speak with Kalama in 2019, the Warm Springs reservation is in the midst of a major water crisis. Lacking funds to replace its forty-year old drinking water system, and with insufficient help from the federal Bureau of Indian Affairs, which is responsible for upkeep, the 3,200-person community has experienced a series of boil-water orders due to bacterial contamination since at least 2017.[57] "They declared us a state of emergency," Kalama says. "And it was really hard to not go and turn the faucet on and boil potatoes or make hot tea, or just drink water. It wasn't even safe to do the dishes unless you boiled it first. . . . I go out and fill up five-gallon sports jugs out of a spring that

FIGURE 15. Rally against Nestlé proposal, Salem, September 2015. Photo: Will Doolittle, whereismyriver.com.

FIGURE 16. Anna Mae Leonard speaking to rally at Oregon State Capitol, September 2015. Klickitat River Chief Wilbur Slockish, Jr. on right. Photo: Will Doolittle, whereismyriver.com.

we have and it's called Rattlesnake Springs. And that's the only water I trust."[58] The emergency obliged many tribal members to turn to bottled water. "People would donate bottled water, and Nestlé [water] was coming in. And I was so angry. . . . I was like, 'Arrowhead—that's Nestlé.' . . . I saw a lot of people saying [in] a couple articles, . . . 'Why the hell are they giving away Nestlé water when we just had a big, huge battle about that?'"

ACT FOUR (2016): BOTTLING ON THE BALLOT

After the LWA was born, events moved quickly. As Busdieker recalls, "We put together a group and had a bunch of protests and got our nonprofit status and put together a ballot measure, all between April and early September."

The decision by the LWA to pursue a ballot measure—a citizen-initiated referendum put before voters—to ban water bottling was a major departure from the previous tactics that either the statewide or local opponents had pursued. Brent Foster, a lawyer who had worked on other local ballot measure campaigns in Oregon, entered the picture at this point. "I was certainly aware of the [Cascade Locks] fight," he says,

> [but] I hadn't been very involved in it other than to kind of watch. It just seemed so crazy—at some point I thought, "Well, someone's going to stop this. The state will come to its senses. Kate Brown will ultimately put a kibosh on it." But as the thing went further and further, I was like, "Oh my God, no one's stopping it!" And I'd met one of the chief petitioners, Pamela. And she . . . talked with Julia and some of the other folks early on about whether they'd considered local ballot measures as an option. You know, local ballot measures are one of the few tools where you can inject true or honest democracy into a completely hijacked system. . . . So I think that was kind of my entrance, was just watching for years as no one—none of the electeds—did what they should do.[59]

Foster was also persuaded by the profile of the local opposition in Cascade Locks and the organizing they had already done:

> They were business owners. These were not radical leftists by any means. They were just worried about what life was going to be like in their town if they had Nestlé come in. That was really powerful to me, and kind of listening to them and meeting with the Native American folks. Klairice Westley had just had a hunger strike. And seeing how passionate these folks were about it, it was just about, from a legal perspective, how do you take what their interests were and merge it—make it in a form that's legally defensible and logical for people to vote for? And that's what this was.

The choice to push for a ballot measure was also backed by data. "We were able to pay for a poll," recalls DeGraw. "The reason why [the ballot measure] was crucial was [that] it was so clear from the way that the fast-tracking proposal went—and how uncertain winning our legal challenge [to the water rights exchange] would be—that a ballot measure would . . . send such a strong message. It'd be this really powerful hard 'No' to Nestlé. And it felt winnable."[60] The poll showed that approximately 70 percent of county residents would support the measure.

The second decision—which jurisdiction to pursue the ballot measure in—was easier: it would be at the Hood River County level. This choice rested on several fortunate circumstances for bottling opponents. First, Oregon is among the states that allow citizen-sponsored initiatives on local election ballots. Second, Hood River is one of only nine "home rule" counties in the state, a status that gives residents the power to change the county's governing charter—its basic laws—by a vote. Third, organizers knew the county's electorate would be far more receptive to banning water bottling than the voters of Cascade Locks alone.

Yet this was not merely a choice of convenience. Foster says he became involved because the organizers "realized this *was* a Hood River issue." He continues,

> This isn't just Cascade Locks, right? Like, what happens to Cascade Locks happens to Hood River. If they put this giant facility in Cascade Locks, it's going to have huge ramifications for Hood River, because what Nestlé does is, it's not like they just put up a plant and take one water source. . . . They send their trucks out, like the giant tentacles of an octopus, to suck up water wherever they can get it. And [in] the community where I live, I know multiple people whose wells have gone dry in the last few years. When you look at Hood River and you look at the future of the snowpack and everything else, it's like we are going to be living in something that is basically a desert in the not-too-distant future.

Kristen Robison, a campaign consultant, proposed the ballot measure to the statewide coalition, collaborated on a strategic plan and messaging, and managed and trained LWA members. The LWA, working with Foster and Crag Law Center, drafted a text for the ballot measure they felt would likely withstand an eventual legal challenge. In September 2015, a brigade of volunteers fanned out to collect signatures to qualify the measure for the ballot. In this county of just over twenty-two thousand people, the organizers needed to gather only 664 valid voter signatures for what they called the "Hood River Water Protection Measure." Three months later, however, Larsen and her fellow organizers marched

FIGURE 17. Local Water Alliance members submit ballot measure signatures, Hood River, December 2015. From left: Pamela Larsen, Aurora Del Val, Molly Kissinger, Heidi Jimenez, and Melody Shapiro. Photo: Blue Ackerman.

into the county courthouse with boxes containing over three times that number (figure 17). Eventually the measure was placed on the May 2016 election ballot and given a number—Measure 14-55—and an official text: "Shall the County Charter be amended to prohibit the commercial production of bottled water and the transport of such waters?" The measure also defined "commercial" bottling as more than one thousand gallons per day and allowed a temporary exception for emergencies.[61]

In the meantime, the heat had continued building—literally and figuratively—around the state's decision to pursue the fast-track water rights exchange. Not only had the drought of 2015 dramatized the threat of climate change to water availability, but that summer's extreme heat killed nearly half of the Columbia River salmon run—as many as 250,000 fish.[62]

By late fall, pressure from the tribes and the public yielded results. On November 6, 2015, just before the agency was about to approve the water rights exchange, Governor Brown directed ODFW to abandon the fast-track process and return to the previous, slower, water exchange that included a public interest review.[63] What had happened? Westley says that tribal pressure was decisive: "When Umatilla [tribal leaders] met . . . with the governor's representatives—that's when, that very next day, the transfer of water rights was withdrawn."

LAUNCHING A CAMPAIGN

The governor's concession did not deter the ballot measure campaign from moving ahead. The LWA activists and their allies now had five months to win a high-stakes struggle against a deep-pocketed multinational firm and its regional backers. They established an office in Hood River, with an unpaid staff member to direct the campaign: Aurora Del Val, who had lived in Cascade Locks since 2003. She was later supported by a paid organizer, Molly Kissinger. The Alliance rapidly gathered endorsements from over one hundred businesses, including more than forty farms and orchards. They found that the campaign had broad appeal, even among constituencies not typically associated with environmental issues. "We had this total cross section," recalls Foster. "I mean, we had libertarians . . . you had Republican business owners. I mean, anyone who's really thinking about . . . 'How do we continue to farm for the next twenty, fifty, hundred years?' . . . they've got to be concerned about their water supply. And the true kind of corporate ideologues? Yeah, we're not going to get them. But you know what? They don't even make up 15 percent of the population, so who cares? That was how we approached it." In this small, ideologically diverse county, transcending political divides was essential for the campaign. "We wanted this to be nonpartisan," says Del Val, "so we had one mailer that said, 'What do Republicans, Dems, and Independents all have in common? . . . They want to protect their water.'"[64]

The highly grassroots campaign relied on dozens of volunteers. Heidi Jimenez is a nurse who had moved to Hood River from Kansas a few years earlier. Although she had no political experience, she ended up volunteering nearly full-time as the LWA's office manager. "I remember telling Pamela, I don't know if I can go collect signatures," she recalls.

> First of all, it didn't sound very fun and appealing to go stand out and get signatures. . . . I had a little child, really small, like four, and my partner's away for long periods of time. . . . So it's just the two of us. We don't have family in Hood River. And so I would have to bring him, but we did it. And I think it's just a matter of starting the process, just taking baby steps in doing it and then just realizing, Oh, it's not that bad. I can go downtown and go collect signatures and hound people.[65]

Larsen describes the supportive ethos that the LWA tried to cultivate among volunteers: "We did different things that were nurturing and fun so that people felt that we really built community within our movement. . . . Grassroots, but doing it in a very nurturing way."[66]

On the other side, Nestlé established a pro–bottling plant political action committee (PAC), the Coalition for a Strong Gorge Economy, which hired a top political consultant, Rebecca Tweed, who had previously led successful efforts to defeat statewide ballot measures to raise business taxes and label genetically modified foods.[67]

CRAFTING THE MESSAGE

Central to a campaign like this one are choices about how to effectively characterize the issue for voters and convey what is at stake, involving what social movement scholars and many activists call message framing or messaging.

One challenge for ballot measure advocates was making the issue locally relevant. Nestlé's bottling plant was proposed for an isolated part of the county, with a different economic and cultural profile and no obvious hydrologic connection to the surface water used for irrigation by most Hood River Valley farmers and orchardists. So the LWA faced a dilemma: How to dramatize the potential impacts of water bottling for the larger county? The campaign chose to describe bottled water extraction as water *exportation*—permanent removal from the region. Although the bottles were unlikely to literally cross international borders, this framing proved compelling to county voters. As DeGraw recalls, "Water exporting immediately sounded like a terrible idea for farmers, or even local people who just understood that the county depends on agriculture. . . . The way we got past the laissez-faire bent of libertarians is that water is different. Your neighbor selling out to Nestlé lowers your water table. Any neighbor making a deal with Nestlé lowers the water table."[68]

That focus on the threat of potential future water scarcity, specifically to the county's prime industry—irrigated agriculture—was effective. Del Val explains, "Knocking on doors . . . I'd hear from some of the farmers, 'How is this connected with us?' . . . I'd say, 'How do you tell a huge corporation to turn off the water when we're in a drought? What happens if you have a neighbor, an aging farmer who doesn't want to do it anymore and ends up selling their water rights to Nestlé? What's going to happen to your water?'"

This central message was augmented with two corollary themes. The first was the foot in the door: the idea that once Nestlé opened the tap, other extractive users would follow, and it would be impossible to shut the tap off. As one LWA campaign mailer stated, "We just can't afford to open our county to the long line of water bottlers looking for water. There's no way to let in 'just one' water bottler" (figure 18). The second was to frame the decision as a zero-sum choice between corporate water extraction and local water availability. "Last year's emergency drought was a reminder that our water is limited," the mailer continued, calling Measure 14-55 "a common sense way to protect . . . our limited water supply for uses that actually matter—our orchards, homes, and small businesses."[69] These themes reinforced official messages that residents were already hearing. Del Val recalls that "the irrigation district was telling people, 'Don't water your lawns. Let's save the water for the farmers.'"

But how significant a threat really was the Nestlé bottling proposal to the water that farms and orchards depended on? I travel up into the Hood River Valley, with pear and cherry orchards on all sides, framed by dramatic views of two snow-capped peaks—Mount Hood to the south and Mount Adams to the north—to meet Les Perkins, the manager of Farmers' Irrigation District. If anyone is in touch with the dynamics of farming and water in this region, it is Perkins. Sitting at a big round table in his busy office, he explains that almost all agriculture in the valley is irrigated by river and stream flows, but drought and climate change are now altering those flows, with far less water in summer and autumn.

Would additional spring water extraction sites in the valley pose a threat to farmers' water access? "The supposition is that in the Hood River basin we're highly connected to surface water," he tells me, "so that if you start pulling off of groundwater, it's likely to have an impact on surface water . . . but we just don't know enough yet."[70] Perkins was not so concerned about the impact of Nestlé's pumping on the fish in Herman Creek over in Cascade Locks—"It's a very small amount of water"—but he says arguments about future groundwater extraction sites for bottling in the Hood River valley did make a big impact on local farmers: "There were certainly some local folks who were against [the ballot measure] who just looked at it and said, it's just a bunch of liberal wackos trying to control private enterprise. But I would say more common was people just saying, yeah, it just doesn't seem like a good idea. Like, that's a resource that we use heavily here for a long-standing economic activity. And so just putting in the bottle and shipping away doesn't seem like the best use."

Perkins says he supported Measure 14-55, but not because of the water availability issue. "It comes down to that privatization of a public resource that I philosophically have an issue with," he tells me. ". . . I think you're way better off having local ownership and local control of those resources. . . . If it's owned by a corporation that is not in the

Now is Our Chance to Protect Our Limited Water Supply From Bottled Water Exports

Last year's emergency drought was a reminder that our water is limited, and a secure water supply is critical for thousands of existing local jobs. It doesn't make sense to export bottled water when experts say droughts are increasing, and we don't have enough for our farms, homes, and small businesses.

The Water Protection Measure is a common-sense way to protect our local water for use by our families, farms and fishery, and to protect our water from getting trucked away in over 1.6 billion plastic bottles every year.

But the Measure isn't just about protecting our County from Nestlé's plant, it's about making clear that we want to protect our limited water supply for uses that actually matter—our orchards, homes, and small businesses.

Local Farms, Orchards, and Hood River's Rockford Grange Support Measure 14-55

Over 40 local farms and orchards have endorsed the Water Protection Measure. As Hood River's Rockford Grange explained:

"We just can't afford to put every good water source in the County in the sights of water bottlers."

Visit **LocalWaterAlliance.org** to see our farm endorsers.

Voter Alert

Watch out for Nestlé's Political Hype!

Nestlé, one of the largest corporations in the world, recently formed a political front group called "Coalition for a Strong Gorge Economy" which is headed by a high-priced political lobbyist from Salem.

Don't be tricked by a local face fronting Nestlé's slick advertising efforts. Setting the precedent that our County will give away our water for industrial-scale water bottling puts our water security at risk and is not good for our economy.

"We used to support the Nestlé project, but wher researched their claims, realized it was just smol and mirrors"

— Kathy and Bob Tittle, longt Cascade Locks residents an business owners

Get Ready. We'll soon be bombarded by Nestlé's political hype, intended to scare and mislead voters. Here are just a few of the claims you're likely to see:

CLAIM # 1: "Bottled water is good for Hood River County's economy and jobs"

REALITY CHECK: Water bottling plants are highly automated, and provide a small number of generally low-paying jobs. **Nestlé won't commit to offering even one job to a local.**

Don't buy Nestlé's inflated jobs numbers. For years they claimed their plant would bring 50 jobs, then this month, Nestlé magically doubled this already exaggerated number. But they still won't disclose real job numbers at their existing plants.

Water is the new gold. We can't afford to open our county to the long line of water bottlers looking for water. There's no way to let in "just one" water bottler.

Setting the precedent that Nestlé can truck away 200 million gallons of our water a year puts the water supply we need fo agriculture, our homes, and small businesses at risk.

CLAIM # 2: "Cascade Locks needs Nestlé"

REALITY CHECK: Despite Nestlé's exaggerated promises of jobs, they won't commit to giving even one of their low-paying jobs to a local resident.

Many Cascade Locks residents oppose the Nestlé plant an the 200 extra truck trips a day that will rumble through their small downtown, hurting efforts to attract businesses with the type of family-wage jobs Cascade Locks needs.

FIGURE 18. Yes on Measure 14-55 campaign mailer, 2016. Design: Katsandesign.com.

Nestlé's Water Trucking Operations Put the Whole County at Risk

Iestlé's big truck traffic and water bottling operations hreaten our entire County and put every good water source into Nestlé's sights.

Iestlé plants around the country commonly increase heir water supplies using water tanker trucks. For example, Nestlé's Denver plant trucks in water from sources 100 miles away, and there's nothing stopping hem from doing the same here.

The risk that Nestlé will target water sources across he Gorge to truck into a Cascade Locks plant hreatens our already limited water supplies needed or our orchards and farms.

Why Do Over 100 Local Businesses Support Measure 14-55?

Water is the foundation of our local economy.

Bottled water plants are highly automated and produce only a small number of low-paying jobs.

Measure 14-55 protects the water security local businesses and agriculture need to thrive now and into the future, when water supplies will be even more limited.

It's just common sense—Hood River County can't afford to become the biggest bottled water exporter in Oregon.

Nestlé trucking in water at Denver plant

CLAIM #3: "It will help the tax base"

REALITY CHECK: The City of Cascade Locks and Nestlé have long assumed Nestlé would not pay any property taxes for at least 5 years under the standard Enterprise Zone tax abatement program.

Nestlé's 200 truck trips a day would heavily impact our roads, and they would take advantage of our fire and police protection, but taxpayers would be footing their bill.

CLAIM #4: "It's illegal and will cost the County"

REALITY CHECK: This is a classic political scare tactic, used when an opponent is trying to distract voters from the merits of a measure, and it's just not true.

"As a Hood River Valley business owner for 34 years, I'm voting yes on 14-55 because our water supply is just too important to waste on bottled water."
— Butch Gehrig, *owner Odell Chevron*

Counties like ours have the right to protect our water for uses that really matter, such as our orchards, families and small businesses. The corporate water bottlers may not like that, but it doesn't make it illegal. And the County has no obligation to spend a dime to legally defend the measure.

CLAIM #5: "It's just a little bit of water"

REALITY CHECK: Nestlé wants to truck 230 million gallons of water a year out of Hood River county: That's over 1.6 billion plastic bottles of water and almost 10,000 gallons a year for every person in Hood River County.

Want more information? Like to volunteer or donate to help pass the Water Protection Measure? Visit **www.LocalWaterAlliance.org**

community, the drive is for profits, and the need of the community is really pretty far down the list." This message—the loss of control over local water to a huge, distant, global corporation—was another framing by the Yes campaign that resonated with voters. "I think part of it [was], it's a corporation, a large international corporation coming in," says Perkins, "that's really the sole beneficiary in most ways . . . so I think that David versus Goliath side kind of comes into play."

When approaching voters in the town of Cascade Locks itself, the campaign stressed a different set of themes. "When we were talking with Cascade Locks people, we'd keep it very local," Del Val tells me.

> [We said,] "If you think truck traffic is bad now, imagine what it's going to be like." And we'd also do some education: "Did you know that Thunder Island Brew[ery] hires like thirty-five people?" . . . We'd talk about how Nestlé never, ever promised a job to a local. [And] having conversations with the locals about fishing, because there are a lot of fishermen. Like, "Well, what's the water level been like?" It was so low, it was like mudflats over in lower Herman Creek, and that's where the municipal water was going to be drawn for the whole water exchange thing.

Del Val also stresses the importance of running a genuinely local, grassroots campaign. "What really made a huge difference and what made our campaign successful," she reflects, "is that it had the look and the stamp of locally run. . . . We were just very careful, on the campaign front, making sure that we had people who lived in-district, in the county, because people can smell bullshit, you know?" DeGraw elaborates: "We made Cascade Locks-specific handouts that focused on the trucks, and safety. . . . And the people in Cascade Locks were willing to reach out to their friends and family in Cascade Locks. It's really crucial that the contact with the community happens with other folks within that community."[71]

Yet not all ballot measure supporters in Cascade Locks were happy with the strategic choices that were made. Busdieker says she felt the campaign was "focused on Hood River, because that's where most of the people live and that's where most of the votes are. And that did cause some divisions, because this is Cascade Locks, and this is our town and our community, and a lot of people felt like it was pulled out of our hands." She also disagreed with the focus on the threat to farms and orchards, because "they're not getting their water from Cascade Locks."

The campaign also decided not to emphasize some messages that might have seemed like obvious choices. "[The result of the polling] was pretty essential, in terms of how the messaging was arranged," Del Val recalls. ". . . Instead of using kind of the classic, more environmental language, it was 'Go with the economics.' And . . . the polling showed that using the [issue of] bottled water and plastic water bottles is not an effective one. You're not going to win with this. That's what inspired *me* . . . [but] I really had to put aside my personal motivation and really pay attention to what's going to be the winning strategy."

The opposing camps in the ballot measure campaign also threw into sharp relief two divergent models for economic development for Cascade Locks: on one hand, the traditional approach of landing a single large industrial employer; and on the other, a vision of a highly diversified local economy with many small service employers catering to the region's recreational users. By this point, the town's demographics and economic base were shifting. "Housing is insane price-wise [in the Portland area]," Del Val tells me. "That's why people are looking to Cascade Locks for more affordable housing. . . . it's changing the demographics. . . . They're a lot of younger twenty-something, thirty-something people." Busdieker adds, "We've got the brewery. They've been here for three years, they're already winning awards. . . . We've got the fish market which is a local, Native family. . . . The bike trail was opened all the way through from Troutdale to Cascade Locks . . . so that increased bike tourism, and then the Pacific Crest Trail movie [*Wild*] came out. . . . Things are happening."

In April 2016, the pro-Nestlé PAC filed a campaign finance report, declaring that its only contribution had been $35,000 from the International Bottled Water Association. However, LWA organizers were angered when just a week before the election in May, the PAC amended its report to show the contribution had actually come from Nestlé Waters, which had also made two other donations, for a total of $105,000—four times more than the Yes on 14-55 campaign ended up spending.[72]

What did Nestlé get for its expenditures? "I don't remember their campaign," Perkins tells me in 2019. "That probably says something right there. . . . I don't think they did a good job of positioning and framing things. . . . They took how things had worked in other communities and directly translated that to Hood River without really understanding [the] differences. It's a different place with a different perspective and very strong feelings about water." Foster recalls that "the biggest thing [was] the slick advertising. . . . Every day, people open[ed] up their mailbox and there's a new three-part ad for Nestlé. I just think, one, people were smart enough, they didn't buy it. . . . The other big thing is the campaign did a good job with telling people, 'You know what? Nestlé's about to lie to you.' Right? And that's an important thing."

THE HOME STRETCH

In the final month of the campaign, dozens of lawn signs, almost balanced between pro- and anti-Measure 14-55, lined the streets of Cascade Locks. Local opponents of Nestlé's proposal, formerly silent, had chosen to show their colors. In Hood River, the LWA held frequent rallies along the city's main street, waving signs and getting supportive honks (figure 19).

Five weeks before the election, a city council meeting took place in Cascade Locks. On the agenda was a resolution declaring the town's opposition to Measure 14-55, which it termed "a direct threat to the city's home rule."[73] A large contingent of ballot measure supporters, including Warm Springs and Yakama tribal members, turned out to protest and testify (figures 20–22). The environment was tense. "That night that we gave the testimony," recalls Kalama,

> they had called the meeting. . . . It was totally dark out, and we had been outside all day. There were drummers, there were big drums set up across the street. A lot of people with signs. And one of the [slogans] that we came up with was "Vote for 14-55—to support the Treaty of 1855." . . . We were shouting, "Water is sacred, not for sale!" . . . And a lot of the townsfolks were dead set on if Nestlé came to town, that everybody would have a job. . . . And they were saying, "Oh no, we're going to have all these jobs." . . . And we said, "At what cost? At what cost to the river? To the salmon?"

In the packed room, Warm Springs tribal councilor Orvie Danzuka testified that "the Treaty of 1855 guaranteed access to our usual and accustomed places, this area being one of them. Something like this could have a severe detrimental effect on our first food. We want to be in line to make sure this doesn't happen."[74]

The Cascade Locks City Council voted 6–1 to oppose the ballot measure, with only Busdieker dissenting. The tense debate had shown the opposing sides in stark relief, but ultimately the council's vote was merely symbolic.

On election night—May 17, 2016—LWA volunteers and supporters gathered in Hood River to watch the election returns come in. Before long they were celebrating jubilantly. Measure 14-55 had passed by an overwhelming margin: 69 percent in favor. Jimenez recalls, "It was like, holy shit—all of this work that we just did actually worked! Like we made a huge change. Nestlé's not coming. Like we don't have to worry—we can rejoice! And it was an incredible feeling of just having everything come to fruition." Just as remarkable as the resounding victory was the fact that the level of support had not fallen since the poll conducted months earlier,

FIGURE 19. Yes on Measure 14-55 rally, Hood River, Spring 2016. Photo: Blue Ackerman.

FIGURE 20. Warm Springs and Yakama tribal members rally to support ballot measure, Cascade Locks, April 2016. Photo: Blue Ackerman.

FIGURE 21. Cascade Locks City Council meeting, April 2016. Photo: Aurora Del Val.

FIGURE 22. Warm Springs and Yakama tribal members rally, Cascade Locks, April 2016. Whitney Kalama is second from left. Photo: Blue Ackerman.

despite exposure to the opponents' advertising—a rare outcome. Nestlé Waters' natural resource manager released a statement: "While we firmly believe this decision on a county primary ballot is not in the best interest of Cascade Locks, we respect the democratic process."[75]

The election result also set another precedent. This was the first time that water bottling had ever been banned outright by a vote of the public. "It's the first ballot measure of its kind," DeGraw tells me, ". . . [and] definitely the only charter amendment ballot measure that put into the constitution of [a] county that commercial water bottling is illegal."[76]

Notably, however, the ballot measure failed in one precinct: the town of Cascade Locks itself, by a margin of 58 to 42 percent. While legally irrelevant, this statistic became central to the fact that the election result did not actually put a stop to the bottling proposal. "We could've gotten more," reflects Del Val, "but the fact that we got 42 [percent in Cascade Locks]—I think that's pretty darn good, when we got 69 percent of the entire county."

ACT FIVE (2016–17): A HOLLOW VICTORY?

However, supporters who expected this to mark the end of the saga quickly had their hopes dashed. In the election's aftermath—despite the decisive result—both local and state officials vowed to push ahead with the water exchange and the bottling plant proposal anyway. Nestlé kept its office in the town open. "Because it did not pass in the Cascade Locks precinct," Busdieker tells me in 2017, "the rest of [city] council believes that they are doing the will of the people . . . and [that] it's their duty to keep moving forward." Hood River County officials also refused to implement the ballot measure, despite the fact that it had changed the county's charter to explicitly ban water bottling. Busdieker recalls, "I called the executive director of the county commission, and he basically told me that they have no intentions of enforcing it, and they expect every water district in the state to back them on it . . . because they don't want people telling them what they can and can't do with the water." Cascade Locks city manager Zimmerman told a Portland TV news channel that "we're still following what we believe our citizens want, and what the city council wants."[77] Opponents countered that since Oxbow Spring did not even lie within city limits, the county charter amendment should clearly apply.

A document drafted by Cascade Locks officials after the election outlines a series of potential legal strategies for combating or invalidating the ballot measure, the last of which is titled "Do Nothing." It continues, "This is always an option. We could work with NWNA [Nestlé Waters North America] to reach a contract then defend the contract for all of the above reasons in court. . . . Per the ballot measure the enforcement is left to a citizen or group of citizens, not the County. Let 'them' sue us."[78] The bottling proposal appeared to be stuck in a game of legal chicken, with all sides anticipating a lawsuit.

DeGraw recalls that "I thought that Governor Brown would say no to the whole project if we won the ballot measure, because she was willing to say no to [the water rights exchange]."[79] But in the face of intransigence by local and state officials, and silence from both Nestlé and the governor, bottling opponents girded for another long slog. In September 2016—four months after the ballot measure passed resoundingly—Klairice Westley was again sitting on the steps of the state capitol in Salem, staging another five-day hunger strike, this time without water as well. "I didn't expect that they would care, or the governor would care," she remembers:

> I sat, for the second fast, in front of the state capitol, and [the governor] never did come to speak with me, although some of the legislators did, and . . . the first thing they would say is, "No, this is all done. You can go home now." You know, teasing. "Yeah, you won, what are you doing? Nestlé's done, the people voted, in Hood River County." They would all say the same thing to me, every last one of them, and I would shake my head, "No." And they would go, "What do you mean, No?" And I'd tell them, "No—they're moving forward. Go ask the governor about it, then. Go ask Fish and Wildlife and Oregon Water Resources. The people voted to protect the water, [but] they're moving forward and it's official."[80]

Westley continues, "So, the [tribal] councils have come to the rallies, and the fasts, and the city council meetings. JoDe Goudy [council chair of the Yakama Nation] made it very clear to the governor that if the water exchange application goes through, that the tribe will take legal action. . . . So that's why it's critical to stop the application before it gets signed, sealed, delivered. Before it gets to a Standing Rock place."[81]

On September 21, dozens of tribal members and leaders, some in traditional dress, along with many supporters, joined her for a large rally on the capitol steps (figures 23 and 24). JoDe Goudy delivered a forceful speech tracing the history of the dispossession of the Columbia River tribes, the significance of treaty rights, and the meaning of water:

> I've come here to stand today in solidarity with this effort . . . on behalf of the *Chúush,* the water. The Yakama Nation is one of the four treaty tribes that's located on the Nch'i-Wàna, what others know as the Columbia River. On my father's side, my blood comes from Wenatchapum. . . . On my mother's side, my blood comes from Dog River, what most of you would know as Hood River. . . .
>
> Article 3 of our Treaty, and all the reserved rights . . . stand up in the hierarchy of law much higher than the state of Oregon. Article 6 of the U.S. Constitution, Clause 3, states that treaties are the highest law of the land and all states [are] thereby bound to [them].
>
> And I will tell you this: that the original peoples of the Nch'i-Wàna are not going to stand for Nestlé. It is not happening. It's not happening. I've come here today as the chairman of the Yakama Nation to tell Governor Brown, to tell Cascade Locks City Council, and to tell Nestlé, and whoever else is advocating for this corporation to come in and do this to this water—it is not happening. . . . [If] a water right is going to be issued by the Oregon

FIGURE 23. Columbia River tribal members and supporters at state capitol rally, Salem, September 2016. Photo: Blue Ackerman.

FIGURE 24. JoDe Goudy, chairman of Yakama Tribal Council, at state capitol rally, Salem, September 2016. Photo: Blue Ackerman.

> Department of Fish and Wildlife, you can better believe Yakama Nation is immediately filing suit.[82]

Several attendees described this event as being especially powerful, saying they sensed that a major development was imminent. However, for more than a year after the rally, everything around the water exchange proposal remained in a state of legal and administrative limbo.

A RESOLUTION AT LAST

Finally, on October 27, 2017—seventeen months after voters approved the ballot measure—Governor Brown asked the Fish and Wildlife Department to cancel the water exchange process completely. "ODWR's processing of the proposed exchange would take significant staff resources and legal costs," she wrote. ". . . This is of particular concern given the uncertainty around the City's plans for a Nestlé plant. In 2016, 69% of Hood River County voters passed a ballot measure. . . . This law makes the ultimate goal of the proposed water exchange uncertain."[83] The department complied. Nestlé closed its Cascade Locks office. Nearly a decade after it was first announced, this was the final nail in the coffin of Nestlé's proposal. For the LWA, the statewide coalition, the Indigenous activists, the tribes, and their supporters, it was a complete victory.

What finally caused the governor to stop the plan? None of the officials or activists I spoke with had had any advance warning of the decision. "They held their cards very close to the vest all the way to the end," Perkins says. "I think probably, realistically, it was [the governor] looking at, 70 percent of the voters in Hood River County voted against this idea of privatization of water. You take that to the west side of the state, which is really the power center of the state, and that's a big way to go up. Hood River still has a fairly conservative bent to it—we're not as liberal as the city of Portland or Eugene or Salem. So I think it was probably just a read of politics in the state." The political implications of the decision were an obvious explanation, since Brown was approaching a reelection campaign the following year. According to DeGraw, "It did not come out of the blue. A lot of folks think that Governor Brown felt threatened by her Republican challenger . . . and that she [was] going to have a competitive race. . . . I think she just knew that that would be a hit she couldn't take while trying to win statewide office."[84]

Foster attributes Brown's move to two factors: "One was the ballot measure, and so I do think that it was a sign, belated as it was, that the ballot measure was a valid expression that Hood River, including a large number of Cascade Locks [residents], was against this. The tribes [also] had a big role. . . . Part of it is just like, what side of history do you want to be on? Do you want to be on the side that sends away a sacred tribal spring from the Columbia River Gorge to a big corporation?" Kate Brown went on to win reelection in 2018.

ACT SIX (2017–20): THE HUNT BEGINS AGAIN

The bottled water company staff member whom I interviewed told me that spring water bottlers typically "like to have multiple projects on the drawing board at different stages, so that in case . . . something about one of them causes us to determine that it's not a viable option, we still have others." Indeed, even before the ballot measure passed resoundingly in May 2016, Nestlé had begun looking at other communities in the Columbia Gorge. It did so under the radar, beginning with the town of Waitsburg, Washington. According to the *Walla Walla Union Bulletin,* "[Waitsburg City] Council members . . . expressed their shock and disappointment that [Mayor Walt] Gobel and City Administrator Randy Hinchliffe had already met with a Nestlé representative several times since February [2016] without their knowledge . . . [and] that Nestlé's contractors had already been in the city's watershed scoping out the city's springs to determine whether such a project in Waitsburg had merit."[85] Ultimately, DeGraw tells me, "Not only did Nestlé get disinvited from bottling water in Waitsburg, but . . . the mayor of Waitsburg was forced to resign just for talking to Nestlé in private."[86]

In November 2016, Nestlé tried again to site a bottling plant in the larger community of Goldendale, Washington. By this point, the company's efforts were being closely followed by the activists and tribes who had won the ballot measure and were determined to help other communities. Del Val recalls,

> [We organized in] a coordinated effort with the residents of Goldendale, neighboring towns, and tribal members a very successful showing of over 150 people at a packed Goldendale City Council meeting once the news broke that Nestlé was talking to the mayor. . . . How powerful it was to hear thirty-five speakers eloquently and respectfully asking the city officials of Goldendale to learn from Cascade Locks and the successful water protection county ballot measure results. . . . A city council member who came to the meeting predisposed to support talks with Nestlé was so moved and shaken by the "No Nestlé" speakers that his level of support needle moved dra-

matically. I even remember one speaker who . . . signed up to speak in favor of a plant but over the course of the evening switched sides.[87]

Yakama Nation attorney Ethan Jones testified that "threats to water are threats of genocide to Yakama culture and way of life. . . . We will do whatever it takes to stop a Nestlé facility from being built in Goldendale."[88] Several residents formed the Goldendale Water Coalition, which filed a public records request to unearth details of city leaders' negotiations with Nestlé. After they revealed their findings in a newspaper op-ed, alleging that officials had worked with business leaders to stack the deck in favor of the deal, the mayor shelved the plan.[89]

In both towns, says DeGraw, "The city leadership and mayors realized this is way more trouble than it's worth. . . . And all of those communities referenced Cascade Locks. . . . And so, we really did keep Nestlé out of the Gorge."[90]

A LEAKY GEYSER

It was not only Nestlé that viewed the Pacific Northwest as fertile terrain for water bottling. Crystal Geyser, owned by a Japanese pharmaceutical corporation, owns several springs in California but was seeking another site. The company bought groundwater-rich land near Randle, Washington, on the Cowlitz River, and in 2019 approached Lewis County officials for permits to build a major bottling plant on the site. Area residents rapidly organized to oppose the plan, and the local Cowlitz Tribal Council also voted unanimously against it.

Several organizations and activists that were involved in the Nestlé struggle in Oregon reappeared here. Kalama worked to raise awareness among tribes in Washington: "Warm Springs and also the Yakama Nation have fishing rights on the Cowlitz River. . . . I did a lot of emailing to the Chehalis tribe, Yakama Nation . . . six or seven [tribes], Puyallup, Nisqually . . . [asking], can you take this to your general council or your tribal council? . . . Suquamish, Quinault, I tried to reach out as far as I could to other tribal nations."

Then, because of an astonishing mistake, the company's strategy for this community was leaked to the media. Crystal Geyser's chief operating officer, Page Beykpour, wrote an email to his boss Ronan Papillaud, president of its affiliate company, CG Roxane, but inadvertently sent the message to *The Chronicle,* a local newspaper that had been critical of its bottling proposal:[91]

Dear Ronan:

Obviously the Chronicle will not print a story about our community involvement in all of our other locations. They are in bed with the opposition, which is fine. . . . As this project stands it is dead because the media has successfully convinced officials and the media against us. . . .

The only possible chances I see are:

1. Hire a PR firm solely for the purpose of gathering grassroots supporters. . . . Once we have gained enough people, we mobilize them in the same fashion as the opposition (reach out to County officials, media, governmental agencies). Along these lines, I've already lined up local contractors and their subs to gather their employees. . . .
2. We sue the subdivision for their damage to the (aquifer) due to irrigation and septic system failures. Hopefully, this gets them to the table and they are prepared to have an open minded communication about the site being off property but piping distance away. When I'm [*sic*] Washington we looked at a good location piping distance away, but even if you hide the building, we cannot escape the truck traffic complaint for others who have joined this group.
3. A combination of one and two.

All of the above are super long shots, but from my perspective worth it. We will face the same all over Washington and Oregon, unless we find the unicorn site.

All we lose if we pursue this strategy is time and internal resources, and some minimal costs associated with the PR firm and filing a lawsuit. The biggest risk is negative PR from a lawsuit, but frankly, if substantiated we have something to rest on.

Otherwise, I say we dump this site. Please let me know your thoughts.[92]

The leak outraged opponents and gave their efforts a major boost. In February 2020, the Lewis County Board of Commissioners approved an ordinance to ban all water bottling in the county—the same result that Hood River County had achieved through the ballot measure. But bottling opponents now had their sights set higher than a countywide ban. According to DeGraw, who left FWW in 2018 but helped advise the Lewis County effort, "Those organizers in Randle have met with the fricking governor [of Washington]—they're not messing around. I think they might actually succeed in passing statewide legislation that's going to make it harder for water bottlers to get permits."[93]

In fact, an unprecedented bill in the Washington legislature to do just that—ban all permits for bottling spring water (but not municipal water) in the state, retroactive to 2019—passed the state senate in early

2020.[94] Washington would have become the largest jurisdiction in the world to ban this industry, but the bill died in the state house after fervent opposition from the IBWA, the Washington Beverage Association, and other industry groups. Its sponsor, Democratic state senator Reuven Carlyle, wrote that "the lethal, mercenary, icy hand of lobbyist bill assassination slaughtered SB 6278 to ban new state permits for commercial bottled water extraction."[95] Nonetheless, energized by this near win, bottling opponents say they are considering a statewide ballot measure in Washington to accomplish the same goal.

LESSONS FROM CASCADE LOCKS

As of 2021, eighteen years after it first approached McCloud, Nestlé had still not sited a bottling plant anywhere in the Pacific Northwest. That year, facing falling sales for spring water, the company sold its entire North American bottled water business to BlueTriton.

After the ballot measure victory, DeGraw wrote that the outcome "set an important precedent for other communities to follow if they, too, are determined to protect their water."[96] But what exactly enabled residents and activists in the Columbia Gorge to hold off the combined efforts of Nestlé, local officials, and state government for a decade, and how transferable are these lessons to other settings?

A handful of factors make the case of Cascade Locks unique, or at least atypical. First was the crucial detail that the spring water Nestlé sought was state-owned—it was literally public water. This fact combined with Oregon's water laws to ensure a complex, years-long path to approval, as well as a greater degree of sunshine on the process. It also gave bottling opponents the ability to frame Nestlé's bottling plans as a private grab of public water, an opportunity they seized to cast the issue as relevant to all Oregonians.

Related to this was Nestlé's choice not to purchase water rights but instead to become a customer of the local water utility, albeit with a long-term contractual supply guarantee. This strategy mitigated some risks for the company but introduced others. Nestlé used this approach in both McCloud and Cascade Locks but not in some of its other bottling sites, including Maine, Michigan, and Ontario. Had the company chosen to acquire a privately owned spring in Oregon or Washington instead, its bottling plant might well be operating today.

The second factor was the particular opportunities created by local political structures. The LWA and its allies had access to the citizen-initiated local ballot measure, a tool unavailable in many places. This recourse to the ballot, plus the county's small population and a low signature threshold for qualifying the measure, enabled the LWA to run a local, volunteer-powered campaign at modest cost. The home rule provision in Hood River County allowed a vote to change county law to ban a specific industry, which superseded Cascade Locks' city government, although the ballot measure was never tested in court. Finally, the boundaries of the county were favorable to the 14-55 campaign, allowing them to choose an electorate receptive to their message.

However, many of the lessons from the battle over bottling in the Columbia Gorge are more widely applicable to analogous struggles elsewhere. The LWA and its allies assembled an extremely broad coalition centered on the defense of local water, harnessing a shared interest in protecting that water in order to transcend political and cultural polarization. They drew on a deep understanding of the community and the particular meanings of water to its multiple constituencies—farmers, orchardists, small businesspeople, recreationalists, tribal members, and fishers, among others. Rather than focusing on the commodity of bottled water itself, they evoked the potential threat to water availability to galvanize community support. Reflecting on the outcome, Foster tells me:

> Look, water is the issue that ties everyone together, and so when you've got an outside corporate interest like Nestlé threatening the region's water supply, and just threatening to do something entirely stupid with it, then you couldn't ask for better glue to pull the community together. I mean, this thing passed by 69 percent of the vote and, despite one of the biggest corporations in the world pouring everything into it, they couldn't stop it.
>
> And I remember some people, even within the environmental community, said, "Oh, gosh, how are you ever going to beat Nestlé?" And I said, "How would we *not* beat Nestlé?" . . . Who the hell wants to give away the most precious resource that almost any community has, which is their water supply? For nothing. . . . I mean, you have to be completely high to be able to think that this is going to provide some meaningful number of jobs or economic benefits.

The multiple campaigns against the bottling plan—the longer-running statewide coalition, the LWA, and the grassroots work of Native activists—also highlight the importance of understanding and using available political structures, tools, opportunities, and vulnerabilities. Among those opportunities, the ballot campaign seized on the reality of a Democratic governor, disinclined to take action to stop the water exchange but sensitive about her environmental image and tribal relationships, facing an election. But the governor's hand needed to be forced, and neither statewide nor local legislation was in the cards. The campaign picked the most favorable strategy and constituency—a county-level ballot measure—for achieving democratic change.

Even more significantly, the LWA and the ballot measure campaign managed to avoid being pulled into the culture war. "The most important thing is being able to talk about water in a way that is accessible and relates to people from all sides of the political spectrum," says Foster: "There wasn't uniform agreement among everyone in the world [on] the fact that, for example, water as a commodity alone was an issue, or the fact that it came in these terrible plastic bottles. . . . They focused on the core agreement, which was that water—just the presence of it—was so critical locally that we couldn't watch hundreds of millions of gallons of it get shipped off."

This highlights the key issues of framing, messaging, and strategy. The LWA read the constituency and the moment well in crafting its messages. They were aided by the severe multiyear drought. Regardless of what individual voters might have believed about climate change, everyone could see the drought—and the specter of present and future water scarcity that it dramatized—which amplified support for the water bottling ban. In a county whose economic lifeblood is agriculture, with valuable fruit tree crops that cannot be left unwatered or abandoned for a season, the campaign's choice to frame the extraction of groundwater as exportation proved savvy. By juxtaposing removal of water from the region for "unnecessary" bottling against the need to conserve increasingly scarce water for "important" uses—agriculture, local businesses, and homes—the campaign leveraged local norms and values to tap into a rich aquifer of support.

For this messaging to be effective, however, the campaigners had to convince voters that Nestlé's bottling twenty miles west in Cascade Locks posed a threat to their water. The argument that Nestlé would inevitably expand its water extraction to the Hood River Valley, impinging on farmers' access to water in the not-so-distant future, was central to their campaign. Finally, in a fairly conservative county with a strong libertarian streak, the campaigners found that opposition to control over local water by a global corporation united voters across the political spectrum. "At the end of the day," Perkins tells me, "I really did feel like it was a rejection of a particular corporation. And that it might've been different had it been a local company, doing things a little differently."

DEFENDING TREATY RIGHTS AND WATER

The other major element in this struggle was the crucial involvement of Indigenous activists and the four tribal councils. This played out in two distinct ways. First, independent tribal members were key protagonists in the anti-Nestlé coalition that emerged in 2015, originally to oppose the water rights exchange and then to support the ballot measure.

The independent activists spoke about the value of this collaboration. Kalama recalls that in rallying Warm Springs tribal members to travel repeatedly to Cascade Locks to oppose the bottling plan, "I was like . . . Come on, guys. . . . We've got to be seen. We have to be heard. We have a voice. We have a collective voice, of tribal and nontribal people working together.' And I can't say that I remember the last time that something like that happened—bringing nontribal and tribal people together." Westley addressed the timing of Native involvement in opposing the Nestlé proposal: "You know, the environmental groups fought for what, eight years, before we even knew about this? So that says a lot. And without them—Food and Water Watch, the Local Water Alliance, and Bark—I don't know if we would've been able to get anywhere as a tribe, once we stepped in, if it wasn't for them already doing all that work and laying the groundwork." She also says the Nestlé struggle marked a break from the historical pattern of relations between Indigenous and non-Native people in the Columbia Gorge: "During the fish wars in the '80s, everyone was against us. . . . We were alone. And in the Gorge, with some exceptions in Hood River, the racism and discrimination against Native people is generational and deeply institutionalized. [But] this time, it was different."

When the Columbia River tribes took formal, public stances on Nestlé's proposal, exercising their role as sovereign nations and framing the bottling plan as a threat to water, salmon, sovereignty, and (cultural) survival—and thus a violation of treaty rights—this clearly changed the course of events. Their strong opposition to the fast-track water rights exchange, partly a result of internal advocacy by Native activists, appears to have been the pivotal factor in derailing that process in 2015. And the tribes' credible threats to defend treaty rights in court against the state of Oregon were almost certainly one major factor—possibly the decisive one—in the decision by a governor concerned about her image and political future to finally pull the plug on the entire process.

During this same period, high-profile conflicts erupted elsewhere that underscored the vital role of Indigenous people and their allies as water defenders. "You should recall that at the same time there was the Standing Rock [protest] going on," Westley reflects. "And I think at least it let our state government know, Kate Brown . . . that we're not going to sit back and be docile. That we're going to say something. And that you are going to be up against us. That we are a sovereign nation, that we are to be respected. How many treaties have been broken?"

SHARING THE PLAYBOOK

Many participants in this campaign were also committed to sharing their experience with other communities facing similar proposals. "Goldendale was the next place to get hit," Larsen tells me in 2017. And they [were] emailing us, talking to us, and we [were] giving them information. . . . Sometimes I'll get an email from someone, 'Oh, we're trying to protect our water, and we want to hear your strategy.' . . . So we're sharing with everyone."[97] The requests have come from beyond the Northwest too. "I'd hear it over and over," Del Val says. ". . . People would [contact me and] say, 'You replace Cascade Locks with our town's name, and Nestlé with this company's name, it sounds like the exact thing.' . . . I'm really excited if our campaign and the learning from it can help other communities."

Among these lessons, participants stressed the urgency of responding immediately if community members learn that their local officials have been approached by a bottling firm. "You've got to keep them out," warns Busdieker, "because once they're in, if you try to make them stop or even slow down their pumping, they'll sue you into bankruptcy, and continue pumping all the time, because they have a contract." Westley draws a similar conclusion: "The key thing here with the Nestlé struggle is that the activism actually started before the papers were signed, sealed, delivered. And that was the only reason that we won, I believe. We would probably still be out there, protesting."[98]

Del Val stresses that the lessons of the campaign do not pertain only to Nestlé. "It's the industry. It's not one particular company. So that's why . . . we have to be really vigilant about protecting our water." She continues,

> The best way to be prepared is to actually talk to your neighbors. . . . I do think that corporations like Nestlé or the oil industry, fossil fuel industry—their playbook is always divide and conquer. And I think for communities, what you need to do is talk to each other, and that's how you can unite. . . . Instead of being characterized as "Oh, you hate our community and you don't want economic development," . . . we have to look at the risks . . . to our community and our environment if you let these folks in.

So many local environmental struggles have outcomes that are indeterminate, partial, or contingent or end in outright defeat. A definitive result like the battle over bottling in Cascade Locks raises useful questions about the significance of such a victory. Foster argues that the grassroots activism that began in the Gorge in 2015 marked a decisive turning point:

> This wasn't led by some big environmental group. You had environmental groups that had been raising money on Nestlé for five years, right? . . . That wasn't who did this. This was driven by people in Cascade Locks and kind of spread to Hood River, and people who saw that common theme of water. . . . Anyone who's sitting [at] their kitchen table, upset about a proposed bottled water plant, can look at this and say, "Aw, geez, these guys are beatable. We can do this." And not everybody has the ballot measure power, but even short of a ballot measure, government officials can be recalled. There's a lot of tools that are out there.

In late 2019, I speak with Westley one last time, and she recalls the moment when she heard that the governor had finally put an end to the Nestlé bottling proposal. "I couldn't believe it," she says. "My grandchildren were all dancing around the living room with me, and I was so happy that they could see that. And I told them, 'You can win, but you have to speak up. You have to organize; you have to do something. We can't sit back and watch this happen.' My granddaughters were at the rallies, and they saw what it takes."

Where do things stand today in Cascade Locks? After the governor's final decision to stop the bottling plan, I approached local officials once again, but they and the city council all declined to be interviewed. Del Val reflects on the lessons of the bottling plant struggle for her changing town and its leadership:

> Sometimes . . . people who are in office, or people who have been there for a long time, still don't have a broader vision, you know? And . . . they may not realize that we saved the town. . . . I think the folks who were in full support and remained in full support of the Nestlé plan . . . for them I think [the outcome] was hard and devastating. But [even] the city manager, who was the voice of the local town, he actually spoke up in some city council meetings [in 2016] and said, "You know, our town is going to be fine." . . . I actually think that there's a sense of "Okay, now we can move on."

The town's unemployment rate, over 18 percent when I first visited in 2010, is now just 5 percent. And Westley tells me that for the first time in its history, Cascade Locks elected not one but two Indigenous city council members. "It's hopeful," she says.

CHAPTER 6

Guelph and Elora: Watching Water, Broadening the Movement

The Go Bus from downtown Toronto spends what seems like forever traveling through older inner suburbs, then newer ones. Eventually we turn onto Highway 401, one of the widest freeways in North America, with sixteen lanes at one point. We pass through a sea of new development at the western edge of the ever-expanding Greater Toronto Area, whose population now surpasses seven million, on its way to a projected thirteen million by 2040. The leading crest of the sprawl consists of huge swaths of new houses spreading across the flat landscape in low-density, cookie-cutter style, consuming thousands of acres of farmland. After just a glimpse of countryside, we pass through Milton, marked by strip malls, a water tower, and a cluster of smokestacks. Suddenly the development disappears and I see a long green ridge running north and south, with a layer of white limestone cliffs. This is the Niagara Escarpment, a geological formation extending hundreds of miles from western New York, where it creates Niagara Falls, through southern Ontario, upper Michigan, and eastern Wisconsin. Here the escarpment also forms part of the Greenbelt, a massive two-million-acre (eight-hundred-thousand-hectare) buffer of rural land ringing the Toronto megalopolis that since 2004 has been largely protected from development.

We head uphill, pass through a notch in the escarpment, and enter a patchwork of fields and forest—the flat plains of the southwestern tip of Ontario, surrounded by three Great Lakes. The freeway is narrower now. Soon the bus turns off at a nondescript exit and heads north. Just two kilometers later we pass an enormous low building on the left, set back from the two-lane road, with a large sign proclaiming "Nestlé Waters Canada—Bottling Plant & Head Office." The bus heads on, eventually entering more development, makes a few disorienting turns, and finally drops us at a red-brick train station in the heart of a small city, surrounded by four-story nineteenth-century brick buildings, with the spires of a huge gothic-style stone church towering from a hill a few blocks away. This is Guelph, a university town of 130,000 and home to one of the most effective water advocacy organizations in Canada.

I am almost late for my meeting with the leaders of Wellington Water Watchers, but I quickly step into the fair-trade coffee shop at the end of the block to top up my mug. It has a bustling energy, with students and professionals busily chatting, reading newspapers, and working. Across the street is a modern city hall with an oval wading pool. It is warm and humid on this late summer day in 2015. I head for 10 Carden Street, a skinny Victorian row building squeezed between shops and restaurants, and enter a confusing space shared by several local nonprofit social and environmental groups. I am shown up a narrow staircase and into a room with butcher paper signs stored on the side, a big stuffed couch, and some mismatched chairs. Arlene Slocombe, the group's executive director, and Mike Nagy, the board chair, join me. Both look to be in their early forties; they have an obvious ease and camaraderie, and even finish each other's sentences occasionally.

We talk about the group's history, its work on water protection and water bottling, and the differences in water politics and policy between Canada and the United States. "The Canadian psyche is really the key thing," Nagy tells me. "We have this pioneer mentality that nothing will run out, including forests and cod, and water is always one. . . . We're water-rich—you hear this all the time; it is deeply ingrained in Canadian society. Like, everything's so abundant, right?" That myth of abundance is especially problematic in this region, he says. "Ontario has lots of water in the north, but we're in a water-starved area here. Guelph is the largest city in North America that is 100 percent reliant on groundwater." It is this same groundwater that has drawn the world's largest bottling firm to Wellington County. "So a big company comes along," Nagy continues, "and they're taking water that we all access, because it's from the same aquifer that they're tapping into. It's essentially the same water that we draw here, municipally, and people were starting to understand . . . that that water is mostly going out of our watershed."[1]

. . .

This chapter traces the story of Canada's biggest and longest-running struggle over groundwater extraction by the bottling industry. It involves the country's largest water bottling plant, and an ongoing campaign by two grassroots organizations and their allies to protect water in a region almost entirely reliant on groundwater. As in Cascade Locks, this conflict involves Nestlé Waters (and now its successor BlueTriton), but while there are parallels between the two cases, the differences are even more significant. In particular, this Canadian struggle revolves around the tension between commercial water extraction for bottling and the need to ensure drinking water for a growing population, a tension that is filtered through competing understandings of water scarcity.

While these water advocates' efforts have encountered major setbacks, they have also produced some significant victories. They have managed to complicate the bottling industry's extraction of groundwater in Ontario—raising its costs, curtailing its expansion, and shifting public attitudes toward bottled water in ways that diverge from the picture in the United States, while helping to erode consumer demand for this commodity. Their campaigning has also led to important changes to provincial policy on groundwater extraction by the industry and has halted a proposed new water bottling site entirely.

The chapter also asks how the sale of Nestlé Waters' North American bottled water business changes the game for the communities and organizations that have been challenging the corporation's water extraction in this region (and across the continent), and what it portends for their work in the future. More recently, these water activists have begun to forge alliances with neighboring First Nations and Indigenous activists building bridges between campaigns to defend groundwater, ensure the right to safe drinking water, and reclaim stolen land. In the process they are broadening and deepening this movement by embracing water protection as a vital issue of social justice.

FOLLOWING THE WATER

In a noisy downtown Guelph restaurant I meet Mark Goldberg, a neatly dressed man with a warm demeanor. Goldberg, an environmental consultant and a toxicologist, recounts the moment that led to the Water Watchers' creation:

> One night . . . I believe it was November 2006, and I was driving through Aberfoyle, where Nestlé's bottling plant is, and . . . I saw two large transport trailer trucks turn out of the driveway and head towards the 401, the main highway in Ontario. I thought to myself, "That's an awful lot of water leaving there." . . . So I decided to check on Nestlé's water bottling permit, to see how much they were allowed to take . . . and I was astounded to find that they could take 3.6 million liters a day.

As Goldberg dug further into the matter, he says,

> I realized that they had bought what used to be the Aberfoyle fish farm . . . for the [existing] water-taking permit. . . . I also noticed on this site that they were about to apply for an increase in their permit to take water. . . . And I began to talk to some friends, and just ask them, "Does this seem right to you?" . . . And most people I talked to were just astounded to learn this. So, we got together a group of people and we decided to form an organization we ended up calling Wellington Water Watchers . . . and it abbreviated to WWW.[2]

It turned out Nestlé also had a permit to extract 1.1 million liters per day from a satellite well in the town of Hillsburgh, which it trucked fifty kilometers to Aberfoyle. These two sites together constitute the epicenter of water bottling in Canada. The Aberfoyle plant is the nation's biggest water bottling plant, a huge facility that manufactures plastic bottles from dense PET forms, bottles and packages the water, and distributes up to fifty-six million cases—1.34 billion bottles—of Pure Life brand water nationwide every year.[3] Nestlé operated one other Canadian bottling plant, in Hope, British Columbia (both facilities were sold to BlueTriton in 2021).

The total of 4.7 million liters (1.24 million gallons) of water per day from these two Wellington County bottling sites is nearly two-thirds of the 7.6 million liters of groundwater permitted for extraction daily by all water bottlers in the entire province of Ontario.[4] Nestlé initially paid nothing for this water, except for a one-time permitting fee of $3,000.[5] In 2009, Ontario began charging a levy of $3.71 per million liters, which equates to $0.000014 (1.4 one-thousands of one Canadian cent) per gallon,[6] or roughly $17 per day for its entire authorized water taking. This represents a significant public subsidy to industry in the form of virtually free access to publicly owned groundwater.

According to Goldberg, WWW rapidly grew to about a thousand members and began informing local residents: "We took out a full-page ad in the local newspaper, had a graphic artist draw a picture of a plastic water bottle. And the caption on it was, 'What's Wrong With This Picture?' and below that were ten points about what was wrong with this picture. The fact that this is a public resource that has been turned into a commodity and sold for more than the price of gasoline; the fact that when you're done with the water you leave behind an empty plastic bottle, which is rarely recycled." The nascent organization held a number of public meetings, and interest grew rapidly. The first meeting brought together several seasoned local environmental and social justice activists, including Nagy, a former federal parliamentary candidate for the Green Party of Canada, and Slocombe, who had been active in environmental campaigns and became the group's first executive director.[7] An early statement described the group as dedicated to "the protection of local water and to educating the public about threats to the watershed."[8]

The newly minted WWW quickly turned its energies to Nestlé's upcoming application for a five-year renewal of its water extraction permit. Goldberg recalls, "We had over five thousand people say that they were opposed to Nestlé's permit to take water application being granted. And we delivered all of these [postcards] . . . to the Ministry of Environment's office and handed them [in]. At the time, that was the largest response to a posting on the Environmental Bulletin Board that had ever happened, by far." The postcards and public comments generated by WWW ground the bureaucratic gears to a halt for over a year. Despite the unprecedented opposition, the ministry eventually approved Nestlé's permit renewal, but for two years rather than five, and it added new conditions, including more rigorous groundwater monitoring. While the outcome disappointed some, the campaign helped establish WWW as a new political force at the regional and even provincial level.[9] The organization was formally incorporated as a not-for-profit group in 2008.

READING, WRITING, AND REFILLABLES

With the Nestlé permit settled for the moment, WWW moved in a new direction. The group launched programs to provide alternatives to plastic packaged water and educate youth and adults about water conservation and tap water quality. According to Goldberg, the WWW membership included a refillable metal water bottle:

> That was probably 1,500 people. But we decided that wasn't enough. So we got some funding to launch a campaign called Message in a Bottle. So we

> went into schools and we gave out a stainless-steel water bottle to twenty-five thousand students, grades 4 to 6. And . . . we delivered a water conservation message, and it was eye-opening to tell the kids where their water came from. They had no idea. They weren't even sure the tap water was safe to drink. . . . Our ideal goal was to try to change behavior by starting with people at a very young age.

He says this program "was kind of a controversial move, because we had to borrow money to buy these twenty-five thousand students steel bottles." Goldberg took a risk by guaranteeing the large loan personally, but ultimately it was paid off. "So as a result . . . there are probably about thirty thousand Wellington Water Watchers stainless-steel water bottles in Wellington County . . . almost a quarter of the population has these bottles. Our intention was to go to the point where everybody wandering down the street has a water bottle. And if they don't, other people would think they're uncool."

This effort to shift drinking water culture was especially important in the wake of the 2000 crisis of contaminated tap water in nearby Walkerton. "After the Walkerton tragedy," explains Goldberg, "people did not trust their tap water. . . . So, to change that behavior is very, very difficult."

Nagy adds that the youth program "was also a clever way of influencing parents at home, because you can't get to them. . . . Children met a lot of resistance at home. Because there were these little tiny [plastic] bottles that people would put in their lunches." The group also encountered resistance from another source. The program was partly funded by a grant from the provincial government, says Slocombe. "It was part of their climate change action program. . . . [But] they were getting trouble from Nestlé for funding us, and there were really threats of lawsuits. . . . Nestlé wrote letters—and I got a freedom of information [documentation]—they wrote letters to our funders, they wrote a letter to the government calling us a bad organization, they were sending emails to principals and teachers telling them not to listen to us. . . . They intimidated teachers not to even bring us in. That's how we knew we were having an impact."

LEAVING THE WATERSHED

The potential loss of control over fresh water supplies, including the prospect of bulk water exports, has been a sensitive political issue in Canada for decades, even surfacing as a theme in TV series and films.

However, the country's water is already being transported across large distances in the form of bottled water. Goldberg also sits on the water source protection committee for the nearby Lake Erie. He argues Ontario's environmental laws are antiquated when it comes to groundwater extraction. "Anybody can apply for a permit to take water, and once they take the water, they can move it anywhere they want to." He says the provincial Ministry of the Environment, which approves water-taking permits, has historically been unconcerned with the end use of the water or its destination: "I wouldn't mind if Nestlé took water and put it into amber-colored [glass] bottles, and those . . . bottles stayed within the watershed and were recycled and refilled. That'd be fine. But when you put it into single-use plastic containers and you transfer it out of the watershed, that's a net loss of water to the watershed, and it's not good for the ecosystem." I ask about the Great Lakes Compact, a U.S.-Canada treaty that bans large-scale transfers of water out of the Great Lakes basin. Goldberg tells me, "There's absolutely no way to prevent the bulk transfer of water, as long as it is in small containers." In fact, the "bottled water loophole" in the Compact is big enough to drive a beverage truck through: if the water is in containers of 21.5 liters (5.7 gallons) or less, there is no limit on how much of it can be removed from the basin.[10]

And Goldberg says Nestlé's Pure Life water from Aberfoyle was not just being removed from the watershed—it was being exported out of Canada. "Before they changed it to Nestlé's Pure Life brand," he continues, "they sold it as Aberfoyle Springs. People would report to me . . . from say, Sarasota, Florida, and they put bottled water on the table and it said Aberfoyle Springs. It's been seen in Rome, it's been seen in London, England. So it's all over the world." Nagy echoes this point: "Recently, [Nestlé has] been at the meetings up in Elora—they've been [saying], 'None of this is going to the United States,' and we've had truckers tell us, 'Of course it's going to the United States. I delivered it there myself!'"[11]

BOTTLING GROUNDWATER IN A DROUGHT

This inland part of southwestern Ontario is unusual in its almost complete dependence on groundwater: it is virtually the only source of water used by agriculture, industry, and municipal water utilities.[12] "Guelph is a very unique community," Goldberg says, "in that . . . our drinking water is taken from the deep aquifer, because it's really very, very clean, very, very good groundwater. . . . And of course that's why Nestlé's is here, because they're drawing from the same aquifer—it's the same very clean water that they bottle."

Yet this pristine groundwater is also vulnerable. The limestone rock layers that cover the aquifer attract the aggregate (gravel) mining industry, which is quite active here. Goldberg explains that one of the biggest local quarries, the Dolime Quarry, has excavated too much rock, puncturing the "aquitard" layer of hard rock that protects the deep aquifer, putting it at risk of contamination and potentially threatening the city of Guelph's drinking water supply. Opposing permits for expanded mining at Dolime was one of WWW's earliest campaigns and remains on its docket to this day.[13] I think about this a few days later as I swim in the deep, clear blue waters of the now-flooded Elora quarry, which draws hundreds of local people on a hot summer afternoon.

Despite its humid continental climate, this part of Ontario experiences recurring summer droughts, especially during the past two decades, possibly exacerbated by climate change. Because these aquifers are recharged by rainwater, even short-term drought puts great stress on municipal water supplies, leading to limitations on water use that would be more familiar to Californians. "The city has water restrictions that go into effect depending on how much rainfall we've had. . . . They will rate it as 0, 1, [or] 2," Goldberg explains, with level 2 meaning that "you can't wash your car, you can only water [vegetable gardens] on even-numbered days if you're an even-numbered house on the street, on odd days if you're an odd-numbered house. . . . We got to [level] 2 one year."

That year, 2012, there was an especially severe drought, which focused public attention—and anger—on water bottling. Goldberg recalls, "We were experiencing these drought conditions and having to comply with these regulations enforced by the city. But Nestlé was merely pumping as much water as they wanted to, and we thought, 'No, that's not right! There should be restrictions on water taking for commercial bottling as well as residential use.' It's only fair, it goes hand in hand."

That summer, Nestlé requested a ten-year extension of the permit for its satellite well at Hillsburgh in the township of Erin, northeast of Aberfoyle. Local residents became active, forming the ad hoc Friends of Hillsburgh Water to fight the permit renewal, supported by WWW. Nagy recalls, "When we were dealing with the Hillsburgh permit . . . the locals up there, who are affluent—there's poor farmers too, but horse people—all of a sudden, they are just enraged! We had . . . seniors knocking on each other's door and having meetings, and storming the council."

The Erin town council unanimously passed a resolution opposing the permit extension. Nonetheless, the ministry approved a five-year extension, with a key condition: that Nestlé reduce its water extraction during drought periods. The company appealed, saying the restrictions constituted an unfair burden, and settled with the ministry to remove the condition. WWW and the Council of Canadians filed a legal challenge to defend those limits. Finally, in late 2013 Nestlé cried uncle, withdrew its appeal, and agreed to accept the original drought restrictions. This small win for opponents applied only to the Hillsburgh Well but established a useful precedent.[14]

Like many local nonprofit groups, WWW has been largely volunteer-powered and operated on a shoestring budget for most of its existence. "Water Watchers hits way above its belt in weight size," Nagy says with a smile. "The problem is . . . everyone thinks we're this well-funded organism, but it's just a handful of us that are working all the time, and with that was this perception, 'Oh, they've got it under control.'"

TURNING UP THE VOLUME

Just a few months before I arrived in 2015, Nestlé announced it had secured an option to acquire yet another well—a defunct local bottling company in the town of Elora, north of Guelph, which had an existing permit to take water. It applied to the ministry for a permit to test the well and, if the results were satisfactory, to extract 1.6 million liters of groundwater a day and truck it back to Aberfoyle. If approved, the plan would increase Nestlé's total allowable water removal in the county to 6.3 million liters per day, or 2.3 billion liters (607 million gallons) per year. This was new terrain for the Water Watchers: a chance to stop a bottling site before it was ever approved, as opposed to the uphill climb of fighting one already in operation. The proposal also galvanized local opposition in Elora—a new grassroots group called Save Our Water.

. . .

Like the northeast Ohio countryside where I attended college, the terrain north of Guelph is mostly pancake-flat, with vertical relief coming in the form of river gorges cut down into the land. After twenty kilometers of straight driving through farms and forest fragments, I turn right, and suddenly the road curves down a steep hill, leading me across

FIGURE 25. Grand River in Elora Gorge, Ontario, August 2015. Photo: Author.

a bridge over a substantial river, then up into a picturesque main street lined with nineteenth-century limestone buildings: the historic village of Elora. Wandering on foot near the river, I come upon an enormous crumbling stone millhouse, under renovation to become a hotel and restaurant. Downstream, the water rushes around a massive rock outcropping. A few blocks later I'm at the edge of town, passing through a grassy park and then down a steep, forested trail into the protected Elora Gorge. Twenty-five meters below, I find myself at the bottom of a beautiful shaded limestone canyon, where the Grand River's green water is inviting in the late summer heat (figure 25). A narrow bridge spans the gorge overhead. I take off my shoes and wade to cool off.

Back up above and less than two kilometers downstream, near the opposite bank of the river, sits a seemingly abandoned, single-story industrial building with truck loading bays, surrounded by a chain-link fence with open fields on three sides. It could pass for a derelict post office. Inside this nondescript space is the Middlebrook Well, a 109-meter-deep hole sunk into an artesian aquifer. It is one of the least likely-looking flash points for a major regional water conflict you could imagine.

. . .

In a spacious older house on a leafy Elora side street, I sit around a large dining room table with three women, all cofounders of Save Our Water: Donna McCaw, Jan Beveridge, and Libby Carlaw. I ask them to characterize their community. "It's over 150 years old," Carlaw tells me, "and even from the start of its founding, it's had a diverse group of people who have lived here. . . . A lot of people came in during the 1970s. A lot of professionals, a lot of academics. . . . We've been known as an arts community since the late sixties, and that's brought in tourism and like-minded people as well. . . . People . . . will either go out east of Halifax, or Newfoundland, or they go out to British Columbia, Vancouver and different islands, or they're in Elora."[15]

The town has long been a haven for intellectuals and for artists drawn by the river and gorge scenery I saw earlier. Carlaw adds that Elora has a history of activism, often led by women, who have "always been feisty in fighting for what they wanted and what they believed in." This includes an earlier round of organizing around the Middlebrook Well, where a locally owned water company opened in 2002, at one point supplying bulk water for institutional uses but never extracting very large quantities. However, using an anonymous numbered company, the owner applied for a much bigger water-taking permit of 1.6 million liters daily. While many residents mobilized against the permit, it was eventually granted, making the site a highly attractive property for a large water bottling firm to purchase.[16]

Like many grassroots groups that form in response to resource extraction or industrial siting proposals, Save Our Water emerged quickly, its members had to learn skills and choose tactics on the fly, and its structure changed many times. The group formed in April 2015, just a few weeks after Nestlé's proposal became public. "We had an information meeting, we had a mailing list and a Facebook group," continues Carlaw. The emergent organization created several committees, "but the one group that kept meeting all the time were the nerds of the research committee [laughs]. And since we made a commitment to meet every Wednesday at the bungalow, we kind of ended up running things because we saw each other all the time."

One of Save Our Water's first public actions was a large water-themed, anti-Nestlé contingent in the town's annual parade on July 1, 2015, Dominion Day—the precursor to Canada Day. According to the group's website, "Participants carried bolts of blue and green cloth, and fanciful fish creating a fanciful river running down the main streets of Elora to remind onlookers of how important clean water is to our community and the need to protect it. This message was reinforced by signs that encouraged people to drink tap water. Friends chanted as they marched to drums, beat boxes and shakers, while onlookers joined to stomp, clap and hoot to the infectious rhythms."[17]

I get a glimpse of that artistic streak when I attend the Elora RiverFest, a music festival that draws crowds from Toronto and beyond. I find the Save Our Water group at a booth handing out literature, along with a striking blue four-meter-high water-themed puppet from the recent parade. Nearby is a large tent with Wellington Water Watchers volunteers, staffing a refill station with multiple spigots for filling water bottles and cups. On the initiative of WWW and Save Our Water, the entire festival is bottled water-free. The two groups have collaborated since Nestlé's proposal for Elora became public. "Water Watchers came and helped us right from the beginning," says Carlaw. "We would be in a lot bigger mess if we had not had our ties to Water Watchers and if they hadn't really mentored us a lot."

When I first visit, four months after its founding, Save Our Water is already engaged in a range of tactics to educate and mobilize community members, the local media, and elected officials. I ask the organizers about the level of local support versus opposition to Nestlé's bottling plans. In Elora, Carlaw says, "Farmers are against it, older people are against it, the young people are appalled by it. They're either against it or they're not quite informed . . . but they're very open to hearing about it." (Countywide, says Goldberg, opinions are "quite polarized. . . . [Some people are] very pro-Nestlé, and some are very, very anti-Nestlé. And the pro-Nestlé people of course work for Nestlé.")

Despite the strong opposition locally, from the outset Save Our Water confronted a political challenge: a permitting process that centralizes authority at the provincial level, giving communities virtually no say in decisions over the uses of local water. "The issue here is, how do we control our resources?," says McCaw. "Ontario has very, very poor legislation and policy concerning gravel and water. . . . It's a local issue, but really we have no voice in this local issue, except in a consultative process that is flawed. Seriously flawed."[18]

Both Elora and neighboring Fergus, formerly independent towns, now belong to a much larger township called Centre Wellington. Despite their township council's lack of authority, Save Our Water insisted that formal opposition from their local elected officials would make a difference. However, recalls Carlaw, "They said, 'Well, we can't say anything because we're not part of it. . . . We've got to be probusiness.'"

The group pushed ahead regardless. They hired a consultant to prepare a report for their local and provincial representatives about the potential impact of Nestlé's water extraction from the well on local aquifers. "Our councilors . . . know as much about the local skateboard park as they do about this," Beveridge says. "So we decided that we would do it one on one. . . . We prepared this [technical] document, we contacted each councilor, singly, and said, 'Would you like to come over, do you have time for a cup of coffee?' . . . And we just talked them through it . . . in a very positive way. Saying, 'We're all in this together, and we need them, and they need us,' and I think that really worked."[19]

. . .

The following day, I return to Elora to attend one of Nestlé Waters' regular public "community office hours." I climb the steps to the town's red-brick public library, then descend to a carpeted basement meeting room containing many empty chairs and just a few people, including a Nestlé Waters Canada staff member, a representative of a consulting firm hired by the company, and McCaw of Save Our Water. At the front of the room stand easels with slick posterboard displays. McCaw is questioning Andreanne Simard, Nestlé's natural resource manager and local press representative, about its proposed pumping test of the well. "There's [another] contaminated well within your drawdown zone. If your pump test causes contamination, who is liable?" she asks. The mood in the room is desultory, as if both sides have been through all this before but know they still need to be here.

I ask Simard a few questions. What is attractive to Nestlé about this particular well? "Our factory in Aberfoyle has about three hundred workers," she answers. "We can't shut it down. We need this well as a supplementary backup source for Aberfoyle—it's not necessarily about increasing demand." The Middlebrook Well, like Aberfoyle, would supply water for Nestlé's Pure Life brand. Unlike in the United States, where Pure Life is filtered public tap water, Simard says that in Canada

Pure Life consists of only spring water. I tell her I've heard that some of the water bottled in Aberfoyle is exported across the border. "None of that water should be going to the U.S.," she responds quickly. McCaw jumps on this—"But wait, John Challinor [former director of corporate affairs for Nestlé Waters Canada] said in a public meeting that you ship to Walmart in the U.S. all the time."

I also ask how public sentiment factors into Nestlé's siting decisions. If 90 percent of the local community were opposed to the company taking water from Middlebrook, for example, would they still push ahead? "Public opinion plays a role. Anywhere we go, we'll have opposition," Simard responds. "We've got nothing to hide. We know there's opposition; we want to work with the communities we're in."

. . .

By the time I return to Elora the following summer of 2016, the controversy over Nestlé's plan for the Middlebrook Well has erupted into a major regional and even national news story. As in Cascade Locks, the catalyst is another severe drought, which has highlighted the contradictions between restricted municipal water use for residents and the company's ongoing water bottling. A CBC story sums up the tensions: "What has some groups so upset is that Nestlé continued to pump millions of litres of water daily from local aquifers during one of the hottest, driest summers in the last decade. . . . In the nearby city of Guelph, Ont., homeowners faced a fine of $130 for watering their lawn. While crops wilted and lawns yellowed, Nestlé continued to bottle. . . . Several environmental groups, such as Wellington Water Watchers, the Council of Canadians and Save Our Water, claim Nestlé's Aberfoyle operation has led local aquifers to drop by as much as 1.5 metres."[20]

When I reconnect with Save Our Water, it is clear the scope and savvy of their campaign have taken a major leap forward. They have asked residents to show opposition to Nestlé by tying large blue ribbons to the street trees outside their homes. As I drive into Elora, the ribbons are impossible to miss. "We've basically got three arms that we're working with," McCaw updates me. "One is political engagement at the provincial level, because they're the decision makers. A little tiny bit at the county [level]. And also with our [township] mayor and councilors. . . . And then there's been . . . group meetings, large meetings. . . . We've been doing canvassing door to door. The blue ribbon campaign was part of that public engagement. It gets conversation going: 'What are those blue ribbons anyway?'"[21]

I ask which of the group's messages have resonated most with local residents. She answers, "Some people really respond to 'They're water mining, they're water extracting, they're stealing it, they're profiting from this, and it's just plain wrong,' very anti-Nestlé. Others just want to share their outrage that this kind of situation is going on, they just want to vent, talk about it. Other people are very strategic, very political—'Are you talking to the Wynne government, are you talking to the ministry?' . . . Others are just upset about plastic water bottles."

The activists tell me that perceptions in Centre Wellington have shifted substantially in one year. "Our local newspaper is a case in point," says McCaw. "Last year after the parade they put out a cartoon that made us look like thugs that had stepped into the parade. Unpatriotic thugs. This year there was an editorial talking about the importance of monitoring and protecting water." More importantly, adds Beveridge, the local officials are now firmly on side. "Our municipal councilors have done a complete shift. We had a few councilors on our side last year for sure; there were a lot of unknowns, and the mayor was not with us. Now we know that we have all the councilors and the mayor supporting us."[22]

What accounted for the change? While lobbying and pressure from Save Our Water played a key role, the pivotal issue turns out to be population growth. Under provincially and county-mandated growth plans, Elora and Fergus have been assigned a major share of the new residential development that is already starting to arrive. Since the Greenbelt is largely off-limits to development, Greater Toronto's growth will leapfrog over that buffer and into Centre Wellington, whose population is currently about thirty thousand.

I sit down with Don Fisher, one of the township's elected councilors, to discuss how this growth relates to the issue of water bottling and the fate of Middlebrook Well. Fisher tells me, "Just in the last year, new growth targets . . . came down from the province and the county. . . . Suddenly . . . we've gone from 'We grow incrementally, like everybody does,' to '50 percent of the entire growth in this county will be here—in this township, not in any of the others.' . . . When you ask any developer, this is where you want to be—because in part of our heritage, our history, our tourism, our economy, our location, our amenities, the way we run business here. We are it."[23] All of these new homes will need tap water. And the township, faced with a looming wave of growth, suddenly has been forced to reassess whether it actually has enough water to supply them all.

WHAT SORT OF SCARCITY?

This raises the crucial issue of water scarcity: what it means and who defines it.[24] The struggles over water in this corner of Ontario revolve around participants' understandings of scarcity, but often in ways that are implicit or unspoken. While the meaning of water scarcity might appear to be self-evident, it turns out to be a highly contested concept, with no commonly accepted academic definition. The term is used by a wide range of writers and institutions to describe both "permanent and temporary, [and] natural and human induced phenomena of low water availability," allowing for a great deal of ambiguity.[25] The most common framings of water scarcity that are used in public policy debates have been criticized for prioritizing purely quantitative metrics and ignoring vital issues of inequality and power.

More critical social science perspectives argue that water scarcity is not a physical fact but rather a socially constructed concept. Scholar-activist K. J. Joy and coauthors write that although water scarcity "is often presented as a natural problem rather than one of distribution or social relations of power," it is "always also deeply mediated by humans and codetermined by power relationships."[26] Geographer and political ecologist Alex Loftus emphasizes the need "to politicize understandings of the distribution of water," based on a recognition that "water and social power are . . . mutually constitutive"[27]—a version of the idea that "water flows uphill toward money" and power.[28] In other words, the question "Is there enough water?" must be followed by the accompanying questions "Enough for what, and for whom?" and "Who is able to access the water?"

Just as water scarcity is socially and politically constructed, so is the use of geographic scale in describing and contesting that scarcity. At what scale is water scarce? Water governance scholars Margreet Zwarteveen and Rutgerd Boelens write that the scale of scarcity is often disputed in struggles over what distribution of water is fair or equitable. They argue that "jumping" scales—such as placing the focus on a smaller or a larger watershed—can be an effective strategy to make water injustices disappear.[29]

How does this apply to struggles over bottling in Ontario? The recurring droughts in the region clearly dramatize the finite nature of groundwater. They highlight the conflicts between competing uses, such as commercial water bottling versus residential supply, as well as the implicit prioritization among them. But which scarcity is actually being discussed? Scarcity in the present moment, or potential future scarcity? Scarcity in biophysical or "volumetric" terms—the literal depletion of a particular aquifer, river, or wetland—or economic scarcity, in which water becomes prohibitively difficult or expensive to access? The conflict over Nestlé's water bottling in Ontario has involved or invoked all of these forms of scarcity at various moments.

For example, as mentioned earlier, Wellington Water Watchers used the contradiction between Nestlé's unrestricted pumping and the residential water use restrictions in its campaign and legal battle over Nestlé's permit renewal for its well in Hillsburgh. Even though the geographic scales of the drought (regional), the water use restrictions (both local municipalities and the Grand River watershed), and the groundwater pumping (aquifer-specific) did not coincide exactly, WWW saw the inequity in how residential and industrial water users were treated during droughts as a potent issue for mobilizing public anger. Save Our Water also used this perceived injustice in one of its early flyers, which featured two facing columns:

The Township of Centre Wellington is permitted to use 1.7 million litres per day.	A Big Water Company could take up to 1.6 million litres per day.
We pay $2,140 per million litres.	They pay $3.71 per million litres.
We are only allowed to water our lawns EVERY OTHER DAY 5am-7am and 7pm-10pm.	Big Water is NOT required to restrict use during a drought.

However, some people in Centre Wellington rejected the idea that the drought-related restrictions were draconian, or even that they signified longer-term water shortage or scarcity. Councilor Fisher told me that "those [residential use] restrictions . . . [are] minimal, not huge. . . . There's never an issue of we can't drink the water, or you can't have showers, or you can't run your taps . . . and there's no expectation that it would. But the Nestlé's thing brought that whole sort of concept to the fore."

Nonetheless, many water activists and local residents expressed a strong conviction that local (ground)water supplies are finite, and that by removing water from local aquifers or watersheds, commercial bottling causes long-term depletion and diminished water availability. Both WWW and Save Our Water have framed groundwater extraction for bottling as permanent removal, leveraging Canadians' long-standing concern about loss of control over freshwater supplies. Nagy told me the most potent message in the Elora campaign has been that "this water will leave forever, and we need it. . . . That one really speaks volumes to people, when they see that this water, for pennies, is going to be trucked away forever. . . . That tankers will show up, they'll pump it out of the ground, and they'll take it away. It's not going to be used locally, it's not creating jobs. . . . Some people are less concerned about plastic than others, but that one strikes at the heart of that community."

The choice to cast water bottling as "export" and permanent removal from the watershed—worsening local scarcity—has proven quite effective. Of course there are nested scales of watersheds, and a portion of the bottled water extracted from these wells in southwestern Ontario remains within the huge Great Lakes watershed. However, much of it leaves even that basin for other parts of Canada, and according to WWW some of it is literally exported across national borders.

Returning to Elora, some activists described a zero-sum conflict between groundwater pumping for bottling and the future water needs of a growing population. "A hydrologist was hired to review . . . the water system here in the township," McCaw told me. "He came up with some real concerns about . . . how basically there's not enough water to give the water to Nestlé and still have growth here in the township. . . . Really one of the only places we have to go for future water supply is near that well. So we have to keep that area available."[30]

Not everyone I spoke to accepted the argument that increased demand from population growth would overtax local groundwater supplies. Councilor Fisher argued that the township's water supplies will be sufficient. "As far as anybody knows, there's plenty of water for anybody who's here now and any reasonable projected use. . . . Droughts come and go, people get all excited about it, then it rains for a year, and people forget about it."

However, another town councilor, Kirk McElwain, was less sanguine. "A water bottling plant here . . . just doesn't make sense," he told me. "I don't know if I'd call myself an environmentalist, but I think it's the wrong thing to be doing with climate change, and not knowing where we're going, what's going to happen. We already had the knowledge from our previous water studies that said that this was an area where water could become a problem in the future. So I knew that already. So, having 1.6 million liters per day leave the aquifer makes no sense to me."[31]

It is worth noting the premise in this quote: that the extracted water will leave the aquifer permanently and reduce its volume long-term (by pumping faster than natural replenishment), putting current water bottling and future residential use in conflict.

THE VOLUME QUESTION

While these activists and some public officials embrace the argument that groundwater removal in the present is contributing to scarcity in the future, bottling firms tend to do the opposite, narrowing the frame to the present and stressing a narrative of water abundance. The bottled water industry frequently argues that its total water extraction is minuscule relative to other uses of water, and that it does not negatively affect aquifers or nearby water users. For example, in response to criticism of its actual and proposed bottling in Wellington County, Nestlé Waters Canada has insisted that its water extraction in the region is insignificant in volume terms. According to a company document, "The bottled water industry in Canada uses just 0.6% of the PTTWs [Permits to Take Water] in the Grand River watershed and just 0.2% of permitted water in Canada."[32] Similarly, in California during the 2015 drought, Nestlé responded to its critics with the argument that its total water withdrawals in the state only amounted to 705 million gallons per year, enough to water just two golf courses.[33]

However, while likely factually correct, such aggregate statistics are also irrelevant. Discussing the U.S. context, Peter Gleick, director emeritus of the Pacific Institute, tells me,

> The argument from the bottled water industry . . . [is] that bottled water is an infinitesimally small fraction of total groundwater withdrawals in the U.S. It's a complete bullshit argument. Because they're taking the total volume of bottled water [and] they're dividing it by total volume of water withdrawals in the U.S., or groundwater withdrawals in the U.S. You know, the denominator is huge. So analytically the number might be correct, but it's meaningless. What's really important is what fraction of a local stream's water resources a bottled water company takes, or local groundwater. And that's been the focus of some of these local activism efforts. They don't care what the total volume of groundwater is in the U.S.—they care about what these bottled water plants are doing to their local resources. . . . Ultimately these are local issues . . . and I think that's perfectly appropriate that these local communities band together to try and understand what the consequences are locally.[34]

In other words, such industry arguments represent the strategy of "jumping scales" mentioned earlier, to render the local impact of water extraction invisible.[35]

A crucial question is whether the volume of water extracted in a given place exceeds the recharge rate of local groundwater. According to Rod Whitlow, a water researcher from the nearby Six Nations of the Grand River Haudenosaunee First Nation, not all groundwater users are alike: "Water bottle companies like Nestlé go out and . . . they'll say, 'We're just taking this tiny little sliver of water, of all of the water that users are taking.' And they say the number one [user] is the municipalities. Well, that's a propaganda spin." He explains: "You're [Nestlé] taking it for extractive commercial purposes. You are taking it from a tertiary watershed faster than it can recharge, and you're exporting it. So it's not coming back into the watershed. [With] those municipal owners, for the most part, and farmer irrigations, it's staying within the tertiary watershed. It's for human drinking sources and for irrigation and farming. . . . So it's an interim net loss, but it's coming in a short turnaround back to the local watershed."[36]

So what are the actual effects of Nestlé's bottled water extraction on local water in Wellington County? Nagy tells me in 2016 that according to a hydrogeologist's scrutiny of data from the company's monitoring wells in Aberfoyle, "The [local] water table has been drawn down by 1.5 meters, and all they've done is increase pumping by 33 percent; they're not even at their maximum. . . . Aberfoyle is slowly running out of water, and eventually it's going to be contaminated, because pumps eventually bring contamination to themselves—that's the inherent nature of pumping."[37] However, Nestlé Waters Canada spokesperson Simard claimed a few months earlier that "we have 15 years' worth of data . . . that shows that the long-term sustainability of the aquifer is not impacted by our operation."[38] In such a battle of dueling scientific claims, grassroots groups are at a marked disadvantage. "We would never have enough money to win from a technical perspective," Nagy says, "because we're dealing with a $110 billion corporation. . . . That can just turn into a pissing match."

WHERE'S THE BENEFIT?

As the activists in Elora worked to mobilize local officials and the larger community around the Middlebrook Well, they faced important considerations about tactics and messaging. One theme they emphasized was the costs that commercial water bottling would impose on the local community, relative to the meager or nonexistent benefits. Councilor Fisher told me he believed Save Our Water had evolved in its approach. "Initially, they were kind of, 'We can't have these corporate, oppressive people come in; we don't want them coming here, this water's ours.'" He continued,

> Over time, they got much better . . . getting away from the emotional "How awful, how evil these people are," and getting down to the real issues. Eventually, they got professional advice . . . and they spread their base. So it wasn't just a few freaky people in Elora. It became all kinds of people. . . . They became more conventional. They started to raise different issues—more practical issues . . . not "Nestlé's a horrible corporation that exploits their workers in the Third World." [But] it's "What is Centre Wellington going to do about the impacts on infrastructure, the growth impact?" And the logical question from that is "Okay, where's any benefit to us from that?" They're not going to run a big plant there, they're not going to employ a lot of people, they're not going to pay a lot of taxes. They're just going to suck the water out, truck it away, and we'll never see it again. So that made sense to all kinds of people.

WWW's Nagy invoked scarcity as well: "You can have all the technical arguments you want, and I think it's always going to show in favor that this water is scarce. . . . Is this how we want to treat our water? . . . A lot of people have basically said no—this is not an acceptable use or permit for the water."[39] Others in Elora worried that corporate bottling of groundwater would lead to *economic* scarcity, making water either inaccessible or unaffordable for future public needs. According to councilor McElwain, "You can put a lot of conditions on what they [Nestlé] do to address a lot of the risks. The one risk you can't address is: there's a potential source of potable water for the future of this township that is now in private hands, that wasn't before. So therefore either it's not accessible, or only if you pay them a gazillion dollars to expropriate it." In other words, Nagy and McElwain suggest that letting the market control local water exacerbates scarcity because it forecloses access to finite groundwater for future use by local residents. They also imply that some test of social good should be required to justify water extraction.

Because the Centre Wellington Council runs the local drinking water utility, I asked Fisher whether he thought bottled water's growth also posed a threat to continued public support for adequately funding and maintaining public drinking water systems—the political will problem discussed in chapter 3. His response was unequivocal:

> Here people take it for granted; the natural assumption is municipal water. We don't have a history here of having real problems with municipal water. We love municipal water. . . . It's just that bottled water's a convenience where that's not available. . . . The default position will always be municipal water. We don't have things here like Flint. If I'm a U.S. citizen in Detroit, I'm beginning to wonder how can you deliver municipal water—if you're able to do it? Here we can do it. I don't think that mentality exists here. People expect their government to do it, and they're happy with how they do it. . . . We should be able to deliver it to our citizens, and we expect our government to do it. As an obligation.

However, as I describe below, the government has in fact failed to deliver that vital service to one important group of people in Canada.

"A REGULATORY ASSAULT"

Meanwhile, back in Guelph, the summer of 2016 had become a long waiting game for the Water Watchers. When I talked to Slocombe, she said the group had been preparing a major joint campaign with Save Our Water, to start as soon as the ministry approved Nestlé's pumping test for the Middlebrook Well:

> Time was ticking. . . . Nestlé's permit [at Aberfoyle] was expiring July 31, and we were expecting that they submitted their application bid in plenty of time for that whole process to happen. And it didn't happen. The ministry never posted. No posting, no comment period, no reason, no explanation why. And just a couple of days ago, their permit expired. [But] of course, they're still functioning. . . . We all have been floored by this. . . . Why has there been no transparency at all on this process? . . . And despite plenty of requests for information, there's been nothing—just silence. Mysterious silence.[40]

I asked Nagy about this bureaucratic delay. He speculated that fast-growing public concern about the bottled water issue might be a factor:

> At the provincial level, where maybe they thought this might be just another routine mouse-click renewal or application or other . . . I think they realize that's no longer the case. Whatever . . . they do, they're going to have to justify that process, that decision, a lot more than they had to in the past. Because the public eye is not just local, regionally, provincially, or nationally, it's internationally now on this. Because of many significant events since then—the Cascade Locks situation, which got a lot of airplay up here; the decision in Pennsylvania, a lot of online petitions. . . . Awareness of packaged water and its effects has increased dramatically. . . . And I think our campaign has helped a lot.[41]

Then, less than a week after I left Ontario in August 2016, Nestlé suddenly announced it had exercised its option to purchase the Middlebrook Well property outright, in order to stop an anonymous second bidder who had offered more money. Eventually, that bidder was revealed to be the Township of Centre Wellington, which by then had determined that maintaining access to the well was vital for expanding public water supplies for a growing population.[42] The local water advocates had managed to persuade their elected officials to act decisively, but it was too late.

Just a few weeks later, at the end of the hot, dry summer of 2016, the reason for the ministry's long silence finally became apparent. Opposition to groundwater extraction for bottling by the advocacy groups had become loud and sustained enough to capture the attention of the highest provincial officials, and WWW and Save Our Water were invited to Toronto to meet with senior Ministry of Environment staff. In September, Ontario's premier Kathleen Wynne declared that the province's groundwater permitting system, especially for water bottling, was out of date and needed to be changed. Speaking at a press briefing, she said:

> I believe that the rules should be changed around water taking for bottled water companies. I think that we have to look very closely at what those companies are paying, what they're allowed to take. . . . Of course, there are situations where bottled water is necessary. But I think we need to look at what our expectation is of those companies, and how we can put some different limits around it. It's not good enough, from my perspective, to say "well, there are lots of industries that need water." The water bottling is a different kind of industry, and we need to treat it differently.[43]

Wynne even questioned the necessity of the commodity itself. "I really think we need to look at the culture around bottled water," she told the *Toronto Star*. "Why are we all drinking water out of bottles when most of us don't need to?"[44] She cited climate change and a growing population as key reasons for the policy shift and directed her environment minister to "improve Ontarians' access to refillable water stations in public and private spaces, [and] increase awareness of the rigorous standards municipal water systems must meet to provide the tap water most Ontarians drink."[45] This stance—which directly embraced many of the critiques and demands of anti–bottled water campaigners—was unprecedented from a provincial leader.

Then in December, Wynne's government went even further, imposing a two-year moratorium on all new or expanded permits to take water for bottling in Ontario, in order to conduct a full review of the province's groundwater resources and the regulatory process for water taking.[46] This moratorium—a significant victory for local water activists—meant Nestlé could not proceed with its plans to test or pump water from the Middlebrook Well until January 2019 at the earliest. To the disappointment of WWW and other groups, however, the existing water extraction at Aberfoyle and Hillsburgh was allowed to proceed. Wynne also announced a major increase in extraction fees for water bottling, from $3.71 per million liters to $503.71, effective August 2017. This made Ontario's fees the highest in Canada, raising Nestlé's maximum payment for the water it extracted at Aberfoyle and Hillsburgh from $17 to $2,637 per day, or roughly one-half cent per liter.[47]

The bottled water industry's response to these changes was almost apoplectic. Elizabeth Griswold, executive director of the Canadian Bottled Water Association, wrote in an April 2017 op-ed, "Late in 2016, the Ontario government began a series of new initiatives which can only be described as a regulatory assault on the bottled water industry. . . . [W]e can only conclude that this has been driven by politics, in response to small but media savvy groups who oppose the industry. . . . The Wynne Liberals have abandoned the government's traditional role as arbiters and adjudicators and have entered the fray on the side of those who oppose the bottled water industry."[48] This remarkable statement suggested that the bottled water industry had been effectively shut out of provincial policy making around water extraction and that water and environmental advocates in Wellington County had been successful in forcing greater regulation of this industry across Canada's most populous province. Furthermore, as the quotes from Wynne indicate, leading politicians had actually adopted some of the discourses and frames around water scarcity that were developed by local water activists.

With Wynne's Liberal government facing elections in mid-2018, less than two years away, the drought-fueled conflict over bottled water extraction in Wellington County—and the efforts of WWW, Save Our Water, and allied groups—had apparently exerted a major influence on environmental politics at the provincial level. However, WWW and the Council of Canadians argued that the provincial government's policy changes were still insufficient to protect groundwater, and they demanded a complete phase-out of the bottled water industry in Ontario within a decade.[49]

GAINS AT RISK

Three years later, I meet Mike Balkwill in the same coffee shop I visited on my first day in Guelph. At 4 p.m. it's incredibly loud, but we manage to hear one another. Balkwill is Wellington Water Watchers' full-time campaign director, hired in 2016. A campaign consultant for nonprofits and labor unions for several decades, he speaks steadily and patiently, clearly accustomed to breaking down complex concepts for different audiences. "I didn't start off as an environmentalist at all," he tells me. "Climate change made me want to try and find a way into the environmental movement. And I actually found it quite hard." I ask him about the current status of WWW's campaigning and its effect on the bottled water industry. "It's hard to draw a direct straight line of cause and effect. But the moratorium that the Liberal government imposed three years ago is a direct result of all this campaigning. And we're not the only ones campaigning, don't get me wrong—there's many others. But we're leading the framing of the debate. Because we're on the ground. . . . We're not powerful enough yet to win what we want to win, but [we've] changed the conversation."[50]

The moratorium on new water bottling permits bought the organization some breathing room in its effort to stop the Middlebrook Well. In the meantime, Nestlé continued to pump water on its expired permits in Aberfoyle and Hillsburgh, while the ministry began a major review of the province's groundwater resources.

In December 2016, the water protection groups were alarmed to learn Nestlé was in talks with the Centre Wellington Council to share part of the water from the Middlebrook Well with the township, something critics alleged was a backdoor privatization deal. Nestlé's Simard told CBC the company wanted to discuss "potential revenue streams that contribute to shared and sustainable prosperity."[51] Balkwill recalls, "The mayor was initially prepared to engage in a private-public partnership with Nestlé. And we raised holy hell about it. And they backed off. And in fact the council became so pressured by that, that in the recent municipal election, every candidate for office, including the mayor, said, 'We oppose Nestlé.' And they passed a motion to that effect."

As the June 2018 provincial election approached, WWW cosponsored a public opinion poll on attitudes toward the bottled water industry. The results offered potent evidence of the effectiveness of the anti-Nestlé campaigns and the intensity of public support for the demands of

WWW and other water advocates. They showed that 64 percent of all Ontarians supported phasing out bottled water extraction in the province *completely* within ten years, and 52 percent favored doing so in only *two* years, while only 14 percent were against a phase-out.[52] Most relevant for the election, that sentiment crossed party lines, with support for a phase-out ranging from 60 percent among members of the right-wing Progressive Conservative Party to 77 percent of Green Party supporters. The argument that groundwater extraction for bottling had no place in Ontario—a radical proposition only a decade earlier—had become a majoritarian position.

Nonetheless, when the ballots were counted, the result was a shock for water protectors. Support for the incumbent Liberals crashed to only 19 percent, handing the election to the Progressive Conservatives (PC), led by the abrasive Doug Ford, brother of the highly controversial former Toronto mayor Rob Ford. On the results map, southern Ontario was a sea of PC blue, with only small outposts of Liberal support in Toronto, Guelph, and a few other cities. This came less than two years after the election of Donald Trump, and many observers found strong parallels between the two. I asked Balkwill what a year and a half of the Ford government had meant for WWW's work. "Well, they want to open the environment up for development. So they introduced an omnibus bill that's got all kinds of environmental deregulation in it. . . . It's interesting—Ford is a mini-Trump, and he got off on a really charging-around bad foot, and it hurt him." Ford has also made highly controversial moves to open protected land in the Greenbelt to new development and weaken land use planning laws.

The staff and volunteers of WWW and Save Our Water feared what would happen in early 2019 when the Liberal-enacted moratorium on new bottled water facilities was set to expire. The two groups held a string of protests, marches, rallies, and press events—some at the Middlebrook Well (figure 26), others at the gates of Nestlé's Aberfoyle headquarters—to maintain momentum, grab media attention, and increase the visibility of the issue. Save Our Water blanketed Elora with large posters with local residents posed in striking black-and-white photographs, under the caption "I Am a Water Protector" (figure 27). And WWW produced campaign flyers that focused on the provincial regulatory process (figure 28).

As the moratorium deadline approached, the Ministry of Environment was in a difficult position. According to *Sierra* magazine, "Denying Nestlé a permit would be inconsistent with the [ruling] party's

FIGURE 26. Save Our Water members at Middlebrook Well, Elora, November 2018. Image by Laura Amendola Photography.

FIGURE 27. Save Our Water contingent, Dominion Day Parade, Elora, July 2019. Photo: Paul Dimock/Artography Elora.

Four-point Program to Protect Public Water

Wellington Water Watchers calls on the Ontario government to protect drinking water and:

1. Phase out within five years all permits to bottle water in Ontario. (A majority of Ontario residents who responded to public opinion polls on World Water Day in 2017 and 2018 support this position.)
2. Adhere to the UN Declaration on the Rights of Indigenous Peoples (UNDRIP), especially regarding consent, in decisions on permits to take water.
3. Ensure public ownership and control of water management systems.
4. Support workers in the bottled water industry to transition to equivalent or better jobs.

Support our Say No to Nestlé campaign
WellingtonWaterWatchers.ca
info@wellingtonwaterwatchers.ca
thewellingtonwaterwatchers wellingtonwaterwatchers
@wwaterwatchers #SayNoToNestle #SayNoToBottledWater

Please donate to the campaign. Visit the website or send cheque* to:
Say No to Nestlé Campaign c/o Wellington Water Watchers
42 Carden St. Guelph, Ontario N1H 3A2
*Make cheque payable to Wellington Water Watchers

wellingtonwaterwatchers.ca

FIGURE 28. Wellington Water Watchers flyer, 2019. Design by Tony Biddle, Perfect World Design.

agenda. Approving it, on the other hand, could be political suicide."[53] Ultimately, the Ford government took the path of least resistance, extending the moratorium for another nine months while it carried out a major assessment of Ontario's groundwater resources involving academic experts, industry, water utilities, First Nations, and other parties.[54]

These developments indicate how sensitive and politically fraught the water issue had become. Nonetheless, the *Toronto Star* reported in late 2019 that "the betting money around Elora is on the Premier eventually allowing expansion" and permitting Nestlé to pump from the Middlebrook Well.[55]

"INSIDE THE BUBBLE"

With the action shifting to the realm of science and bureaucracy, the water protection groups faced big questions not only about their strategies and tactics but also about their overall vision for change. Challenging Nestlé's permit applications obligated WWW and Save Our Water to master the bureaucratic provincial regulatory process and to frame their official submissions at least partly around technical arguments about the impact of water extraction on local aquifers and municipal water supplies. Some members of Save Our Water became hydrological experts, able to educate their local officials.

Yet this focus on the technical realm has also been a subject of debate among the water protection groups. WWW's Slocombe tells me that if the group could prove Nestlé's well in Aberfoyle was causing the local water table to fall, "it would give us a really great argument when we need to respond on a technical, permit level." However, she adds, this approach has limits:

> One of the big pieces we learned in Elora, and we're continuing to learn, is that people want to believe—and I feel like I did at one point too—that the truth will set us free. You know, as soon as enough people know the information, that will be enough to change everything. . . . And that's great—it's helpful and worthwhile—[but] we have found that the most effective pieces come in this other venue, which is related to morals and values. And speaking to people at the door, if you have two minutes of their time, it's not going to be an effective use of our time to talk numbers and data.

The argument over prioritizing the technical and scientific realm goes beyond the question of whether it is an effective way to rally public support. According to Balkwill, it also poses a trap for activist groups:

"The Wellington Water Watchers, like lots of organizations, we're deeply invested in trying to dispute and challenge the well on technical grounds. . . . We use a metaphor about being in the bubble. The bubble is a decision-making process designed by governments to take you away from a political process, to defuse your energy, to deliberately frustrate you, to create the belief that this is a reasonable, fact-based, evidence-based process. And [that] if you just marshal enough evidence, you will win. And if you don't, you lose because your evidence wasn't good enough." Balkwill continues,

> And you know . . . this bubble is designed to make you feel the way you feel. It's designed to produce the results it does. And we have to get out of the bubble. And an important part of being in the bubble is it takes community leadership away from the community and puts them in a very time-consuming place, and keeps them busy with civil servants and politicians and endless conversations. And job one is to get out of the bubble. And it may mean you can't ignore the technical process. You have to participate. . . . But you have to put much more of your energy outside of the bubble and reframe the language.

Even some local elected officials agree. "If we hang our whole hat on the science," councilor McElwain tells me, "and they do this whole Tier 3 [groundwater] study, and it comes back and says, 'Oh, you've got lots of water,' then we don't have an argument left. That's why it's got to be a much broader umbrella."

Yet Beveridge, arguably Save Our Water's preeminent technical expert, views the distinction between these two arenas as a false dichotomy. "Even if the science had said something completely different," she tells me,

> the township is still going to be completely opposed to this for all kinds of value reasons. I mean, the community, the people who are out with their placards saying "I'm a Water Protector"—they're not up on the science. . . . They're opposed to this for all of the moral and ethical reasons that this is just wrong. And so our community as a whole is just so steeped in this. . . . People are totally opposed to this water taking on just simple ethical reasons, that the water should not be taken and put into bottles, and concern about the ecosystem and the watershed. So regardless of what the scientists said, our people would still be marching in the streets about this.[56]

In any case, action at the technical and regulatory level kept on coming. In November 2019, the ministry announced it had finished its Tier 3 study of the province's groundwater. The Save Our Water activists were happy to read its conclusion that Centre Wellington in particular faced a "significant risk" to its future water supply from population growth, possibly causing demand to exceed capacity as early as 2031. A separate engineering study done for the township also concluded that Nestlé's proposed pumping from the Middlebrook Well would lower surrounding wells by 2.5 to 3.5 meters, something the authors deemed "unacceptable."[57]

SIX NATIONS: LINKING BOTTLING TO WATER INJUSTICE

Meanwhile, WWW was starting to engage in a collaboration with nearby Indigenous communities and activists. The Six Nations of the Grand River Haudenosaunee is the most populous First Nation in Canada and has the largest land base. It is the only place where all six nations of the Haudenosaunee Confederacy reside.[58] Its reserve, sixty kilometers south of Aberfoyle, today constitutes only 5 percent of the original 950,000-acre, twelve-mile-wide swath along the Grand River—the Haldimand Tract—that was granted by the Crown under a 1784 treaty in exchange for the Mohawks and other Haudenosaunee Nations fighting on the British side in the American Revolution.[59] Six Nations has been pursuing land claims to reverse this dramatic dispossession since the 1970s,[60] and in recent years the vocal, treaty rights–based Land Back movement has increased pressure on the Canadian government to rectify the injustice, including demands for a development moratorium on the entire tract.[61]

However, it is not only land that is being contested here but water. The Haldimand Tract includes the Middlebrook Well in Elora, and Nestlé's (now BlueTriton's) Aberfoyle plant lies just outside it but within lands covered by the 1701 Nanfan Treaty and on which Six Nations claims traditional rights. Meanwhile, on the Six Nations reserve, fewer than one in ten inhabitants have access to safe piped drinking water in their homes. This is emblematic of a long-standing human rights crisis of unsafe water and lack of running water in Canadian First Nations, which is particularly egregious given Canada's great wealth in both water and material terms.[62]

This extreme juxtaposition—water injustice on Indigenous land, surrounded by near-universal clean water for the settler population—has drawn national and even international media attention. It has also set the stage for cooperation between WWW and activists, academics, and traditional leaders from Six Nations to oppose commercial water bottling.

According to Dawn Martin-Hill, an associate professor of anthropology at McMaster University who lives at Six Nations, "Ninety percent of our population does not have access to clean water, even though we're surrounded by Toronto and Hamilton, Brantford, Caledonia and Hagarsville. So if you live on this side of the reserve, you don't have any access to water. If you're over here and it's non-Native land, you have clean water. . . . And some of my students who have five small children . . . don't have running water. And there's a lot of reasons for that. . . . It's what colonization does. They deny you the basic human rights."[63]

What does this lack of safe water look like on the ground? Rod Whitlow, a former provincial and federal public servant who now does extensive research and advocacy around water issues at Six Nations, explains that there are four main ways people living on the reserve access water. The first is piped water from a $41 million treatment plant that was finished in 2013, which reaches only 9 percent of the residents, mainly in the town of Ohsweken.[64] "That water treatment plant," he tells me, ". . . theoretically could provide clean drinking water to a large portion of the population and the households. It's just that [the cost of] the infrastructure required to distribute that is too astronomical, and no government will touch it. So we have to do it on an incremental basis. And the projections are that it's going to be maybe another couple of generations before every single household will have access to piped clean drinking water."

Second, there are two types of household wells—some drilled deep into the aquifer, and more commonly shallow dug wells that are easily contaminated and are linked with health problems on the reserve, according to one study.[65] Third, he says, is trucked water:

> Part of the homes . . . are hooked up to [a] cistern, because we have an issue with delivery services for that water, and there's only a few local trucking companies, [so] often homes with many household members have a long wait. . . . And I grew up in one of those houses where you do have to ration the water, because . . . it could be one or two days, or sometimes three days, [if] there's a weather event. It could be longer where you don't have water. And so . . . some of them have actually connected their . . . downspouts to fill cisterns and dry wells [with rainwater]. They know that they can't drink that water. So what they do is at least there's enough water in the cistern, and the pump can bring it into the house to flush the toilet, but hopefully not have to wash dishes or cook or bathe with it without boiling it first.

Finally, packaged water in various forms fills the gap for at least 65 percent of Six Nations residents.[66] As Whitlow explains,

> Upwards of two-thirds . . . of the population still will rely on some sort of plastic bottle or plastic container to transport it to their homes. They don't trust the tap water in their wells. And every single household either has tons and tons of single-use plastic water bottles, or they have the 18.8-liter [plastic water jugs]. . . . And then we have a number of those [re]fill stations. They're just commercialized everywhere now. You can get them at the Home Depot, or you can get them at the carwash. . . . And it's self-serve and you put a dollar in, and you fill it up and then see how long that lasts. My house is one of those where we need bought water for cooking and drinking. . . . But those bottles are actually better than the single-use, because that's a reusable bottle—you rinse it out and chlorinate it and disinfect it.

This is the Canadian analogue to Flint and Warm Springs—thousands of people denied access to safe piped drinking water within one of the wealthiest nations on earth, where the vast majority take it for granted. In such a context of water injustice, explains Whitlow, rejecting plastic packaged water is simply not an option for many: "At Six Nations we're at a different spot on the spectrum. . . . Yes, we agree theoretically with this notion of eliminating all single-use plastic water bottles—we're just not there yet. Because how else are we going to transport the water, the drinking water, to our community? I had talked to some folks from Flint, Michigan, who are at a different spot on that spectrum of reaching water security. And when their water was contaminated with lead, they had no choice but to drink the Nestlé bottled water as well."

Since the election of Justin Trudeau in 2015, Canada's government has put additional resources into addressing the lack of safe drinking water on First Nations territory, promising to end all of the more than one hundred long-term boil-water advisories that were active on Indigenous reserves. As of late 2021, 68 percent of the advisories had been lifted, and nearly all those remaining were in Ontario. However, the official warnings don't capture the full extent of the problem. "Six Nations doesn't appear on those lists," says Whitlow, "and they probably never will. That's why, with the help of Wellington Water Watchers, we're trying to bring awareness of the fact that we have a First Nation, the largest one, and we continue to struggle with water security issues." In 2021, the federal government reached an $8 billion settlement with several First Nations to address the water crisis, but Six Nations is not a party to the agreement.[67] The estimated cost to connect all of the homes and business on the Six Nations reserve to the treated water system is $120 million.[68]

To highlight the link between commercial water bottling and the water crisis on the reserve, Whitlow calculated how much Nestlé was paying the provincial government for the water it extracted under the new fee of $503.71 per million liters, concluding that if it were to pump at full permitted volume, the fees would total nearly $1 million per year: "Where's that money going? It's coming from the Haldimand Tract. It's coming from a land claim area. It would be covered under UNDRIP Article 28, which is United Nations Declaration on the Rights of Indigenous Peoples . . . which basically says that First Nations, where you're exploiting and stealing [their] resources, they have to be compensated and there has to be redress. Well, they just ignore that." What could Six Nations accomplish with the money if Nestlé's water-taking fees were to go to the reserve as compensation, rather than to the provincial government? "[It] could go a long way," Whitlow responds. "If we're going to be stuck on wells for generations to come, then we need . . . to upgrade those wellheads. But we also need to have a dedicated water-hauling system that is affordable, and it's convenient. . . . Let's start trying to hook up a few more houses [to the piped water system]. Let's improve the wellheads. And let's set up a sanitation facility [for] people that are using their own makeshift water-hauling devices."

Six Nations has two sets of authorities—an elected band council officially recognized by the Canadian government that administers programs and services, and the traditional Haudenosaunee Confederacy Chiefs Council, which is chosen by the clan mothers. Makasa Looking Horse, a dynamic young activist at Six Nations who spoke at the UN youth climate summit, was instrumental in getting her traditional leadership to speak out publicly on the water bottling issue. She discussed the actions she helped organize at Nestlé's Aberfoyle bottling plant (figure 29): "We held multiple protests with the Wellington Water Watchers and made sure that [Nestlé] knew that we did not agree with this. And we want them to stop taking our water and get off of our land. That is our treaty land and our Indigenous land, and that's rightfully ours. But they need to pay in some shape, way, or form to the Six Nations community for all of the water that they have stolen—hand it back to the Six Nations territories. And we can decide for ourselves what we want to do with it."[69]

One of these protests involved delivering a formal declaration to the company from the Six Nations traditional leadership:

> The Haudenosaunee Confederacy issued Nestlé a cease-and-desist letter to stop immediately taking water from the aquifer . . . and it says, "Nestlé, this

FIGURE 29. March to Nestlé Aberfoyle plant led by Six Nations youth activists, November 2018. Photo by Federico Olivieri.

> is a message for you. We, the Haudenosaunee Confederacy, will continue to assert our rights as the Indigenous peoples of this territory. We will continue to assert our rights as those who have lived on our territories for countless generations. We will continue to rise up for the rights of our people, for the rights of life within our lands, territories and waters, and the coming faces of our future generations." . . . [We] handed this letter to the CEO of Nestlé . . . with the clan mothers.[70]

Looking Horse and supporters repeated the cease-and-desist action in 2021 after BlueTriton acquired Nestlé's bottling business (figure 30).

Whitlow tells me that so far "the elected council has not come forward with a position on Nestlé, whereas the hereditary chiefs and the Confederacy Chiefs and the clan mothers have." Why has the band council not taken a formal stand? "It's not just my First Nation," he says:

> It's kind of a Catch-22. You've got to play nice with the federal government if you want something from them—including honoring the treaties. To me, and this is my own personal perspective, [the band council] are not pushing the agenda because we rely on revenue from the federal government for our day-to-day operational services. We don't have a tax base and we are forced to rely on capital support from them. So they're not going to rock the boat too hard, and some First Nation political leaders might be reluctant to stick

FIGURE 30. Six Nations community members deliver cease-and-desist letter from Haudenosaunee Confederacy Council, BlueTriton headquarters, Aberfoyle, September 2021. From left: Colin Martin, Colleen Whitlow, and Makasa Looking Horse. Photo: Troy Bridgeman, Multimedia Journalist.

> their necks out, and they have to remain vigilant every year to make sure that we get more money to widen the [water] distribution system.

Nonetheless, this Indigenous-settler alliance around water appears to be bearing fruit. "Solidarity with like-minded people and the Wellington Water Watchers is the most fabulous thing that we could do," Whitlow says. "They supported me with a community petition . . . for a strategy to deal with the water security, and on giving notice to Nestlé once and for all, and all water bottle companies, that we don't want you extracting water from our watershed." Slocombe tells me that these collaborations have also helped to shift the emphasis of WWW's work: "We've been working to support the voices at Six Nations that have been calling for the return of the Aberfoyle Well, and the justice issue around there being a multinational corporation present, extracting water for profit from treaty lands where the treaty people, many First Nations communities, still do not have access to clean potable drinking water. . . . That's the work that our more or less middle- to upper-class White movements need to just do our homework and get behind in supporting and lifting those voices up."

As we finish our conversation, Whitlow says that although the provincial and federal governments have not yet felt sufficiently pressured by water activists and traditional Six Nations leadership to take more action,

> we have to keep the momentum going. It's going to take academic researchers, it's going to take the women, the traditional role of the women [to protect water], and the warriors. . . . In my Mohawk language, the word that we use for warrior is *Roti-sken-hrakete,* which means he who carries the burden of peace. We want a peaceful solution to this. There's this negative connotation—oh, you're going to protest and [do] civil disobedience and you're warriors, and you're all radical, and whatnot. But the men in my community really do need to step up. The youth are doing their share, and the women, but the men really need to step up. And that's not coming from me. That's coming from a lot of people in the community.

NEW EXTENSION, NEW REGULATIONS, NEW CONCESSIONS

Meanwhile, things were moving again on the provincial front. In October 2020, the Ministry of the Environment announced it would extend the moratorium on new bottling permits yet a third time, for six more months, to consider the thousands of comments on its proposed changes based on the massive groundwater review it conducted. Finally, in March 2021, the ministry announced the review results and the new regulations. One of the top-line conclusions was good news for Nestlé: it declared that "groundwater takings for water bottling are being managed sustainably in the province."[71] However, most of the changes were better reading for the water protection groups and their allies. One reform, approved by the provincial parliament, imposed a new official hierarchy of water uses—with drinking water, the environment, and farm animal production given highest priority, general agricultural use in the second tier, and industrial and commercial uses (including water bottling) in third place. This has important implications for future decisions over competing uses in areas with limited groundwater like the Grand River watershed.

Most important for WWW and Save Our Water, the province announced a new "municipal veto" over large permits to take water specifically for commercial bottling, whether new or expanded. From now on, water bottling companies will need the support of the local government body for any permit to extract more than 379,000 liters (100,000 gallons) per day.

This is a major change—local authorities previously had no say over water extraction whatsoever. It also effectively rules out a permit for the Middlebrook Well in Elora, since the Centre Wellington Council had already declared its opposition. The volume threshold for the veto is less than 25 percent of the amount Nestlé was seeking to extract at Middlebrook.

What do the water activists think of the changes? McCaw of Save Our Water says that "this municipal veto is a great boost and a profound change. However, a company could still seek and be granted a permit without council permission for the lesser amount. And then we're back to fighting a permit again."[72] Slocombe is less sanguine in her assessment:

> I think the wins were fairly small, quite honestly. . . . I think it's a bit of that "throw a few bones out." There were some significant losses in the new regulations too. . . . The win around a municipal veto is really great, but . . . most of these places where these wells are put in are within or adjacent to multiple municipalities or townships. So in Centre Wellington, only Centre Wellington has the ability [to veto water taking] . . . [but for] nearby Guelph, or nearby Puslinch township . . . which are fairly adjacent and also accessing the same wellheads—there's no say.

Whitlow argues there is one glaring omission in the province's new regulations. While now requiring permission from local authorities, "they left out the First Nations, and the treaty right holders, and the original people, the Indigenous peoples whose traditional territory they're taking the water from. . . . It's not even consistent federally with what Canada has agreed to internationally."

All told, what does this long, drawn-out provincial process signify politically? "The moratorium in Ontario lasted for four years," Balkwill tells me in 2021, "and the Conservative government, which was elected two years after the moratorium was implemented, extended the moratorium three times. And . . . they were much more vulnerable and much more attentive to public opinion on this than we ever really expected. So it's a soft spot. That's what I take away from it. . . . [They] are trying to push through highways. They're overriding planning principles to approve housing subdivisions and industrial developments. Like, they don't give a fuck. But on this they did. And that's interesting."[73] It appears that the combined efforts of these water advocates and their allied groups have succeeded in shifting public opinion—and hence the politics—around commercial water bottling so substantially province-wide that even a conservative government with a deregulatory agenda was forced to pay attention, take water protection seriously, and enact meaningful, albeit limited, reforms.

FALLING FORTUNES

In addition to the opinion polls showing widespread public support for ending commercial water extraction for bottling in Ontario, bottled water itself is suffering from slowing demand across Canada. In contrast with its southern neighbor, per-person consumption of bottled water in Canada is projected to remain flat in future years at just under 73 liters (19.3 gallons) annually—less than half the amount consumed in the United States.[74] More dramatic is the share of Canadians who report that they consumed bottled water during the past day: after soaring from 29 percent of households in 2012 to 53 percent in 2018, this figure fell for two straight years to only 44 percent in 2020.[75] The commodity of bottled water has become increasingly stigmatized in Canada, at least partly because of the well-publicized battles over groundwater mining for bottling in Ontario (and also British Columbia).

There have been indicators of this downturn at Nestlé's facilities. In 2018, the local newspaper reported that the company had laid off a substantial number of workers at its Aberfoyle bottling plant and headquarters. Nestlé's corporate affairs director said that "we have faced some business challenges over the past year . . . including unprecedented cost increases. As a result, we have had to make the decision to make a permanent reduction in our workforce."[76] The "cost increases" may refer to the major 2017 hike in water extraction fees.

According to Nestlé's monitoring data, the amount of water the company extracted from its two wells has fallen noticeably. "At the Hillsburgh Well, the water they're taking there is at the lowest level they've ever taken it," says Balkwill. In 2020 the company extracted only 6 percent of its permitted water volume from the Hillsburgh Well, and just 35 percent of its total allowed volume between the two wells—an overall decline of roughly 30 percent since 2016.[77]

NESTLÉ WALKS AWAY

In July 2020, Nestlé made a surprise announcement: it intended to sell its entire Canadian bottled-water operations to the far smaller Ontario-based water bottler Ice River Springs.[78] Slocombe called the planned sale a "strategic retreat"—a product of the "momentum we together leveraged on this issue . . . which disrupted Nestlé's expansion plans at the Middlebrook Well," as well as a boycott of Nestlé products promoted by the Council of Canadians and Six Nations activists.[79] Within months, however, the sale was blocked by the Canadian Competition Bureau on antitrust grounds.

Nonetheless, the corporation clearly wanted out of large parts of its bottled water business and announced it was seeking buyers for its water brands in Canada and the United States. But with the Covid pandemic throwing global commerce into chaos, several months passed without further developments.

Finally, in February 2021, Nestlé made global headlines with the revelation that it would sell its entire North American bottled water business—including its twenty-seven bottling facilities, forty-two springs, six spring water brands, and the Nestlé Pure Life brand in the United States and Canada—to a consortium of the private equity firms One-Rock Capital Partners and Metropolous & Co., for US$4.3 billion, effective March 31. The buyers announced that the new entity's name would be BlueTriton Brands. Dean Metropolous, a billionaire investor known for buying and "flipping" struggling corporations, including Bumble Bee and Hostess, would become the new company's board chair.[80]

At the same time, Nestlé held onto "a slimmed down group of upscale and trendy brands" of water, including Perrier, San Pellegrino, and Acqua Panna.[81] It also retained its bottled water brands and bottling facilities in the rest of the world, including in Europe and in the global South, where sales continue to grow. Nestlé remains one of the Big Four bottled water firms, but it has gone from first to third place globally.[82] It is still the world's largest food and beverage corporation, and it continues at the helm of the 2030 Water Resources Group.

What caused Nestlé to sell? Only a few years earlier, profits in its water division were high and sales strong. Yet a *New York Times* article chalked the company's decision up to "tepid sales and criticism from environmental groups," describing an interview with the firm's CEO, Mark Schneider: "Mr. Schneider said Nestlé had decided to consider exiting the U.S. water brands in part because they were not selling as well as the company would like. American consumers are less willing to pay for bottled water than Europeans are. Mr. Schneider acknowledged that environmental concerns had hurt sales. Those concerns are easier to address with imported brands that command a higher price, he said."[83] Other reporting added further context. According to Swiss television channel SRF, Nestlé's "water division is weakening, and massively. Sales fell by more than 10 percent in the first half of the year—more than in any other division. . . . Water packaged in plastic is therefore increasingly becoming a reputational risk for manufacturers."[84]

Michael O'Heaney, director of the Story of Stuff Project, tells me that public scrutiny and threats to the company's reputation were also fundamental: "I think there was public brand pressure. . . . Ultimately the plastic thing probably is a not insignificant part of what drove Nestlé [to sell], because in order to meet the commitments that they were publicly talking about in terms of plastics reduction . . . they just got rid of one of the biggest plastic-intensive parts of their portfolio. . . . The brands they kept [are] more likely to be in aluminum and glass. So Aqua Panna, Pellegrino—they kept the premium brands, and then got rid of this cheaper set of brands." O'Heaney also credits the impact of site fights—the numerous local struggles to stop the company from accessing new water sources or pumping more water at existing ones:

> Nestlé's crummy reputation goes back decades. . . . The sort of reputational damage, the frequency with which they were in local, regional, and national press for screwing communities, couldn't have been good. I also think we made it more difficult for them to get access to water. Part of the site-fight decision was, Why are we fighting this when it's already in the bottles and it's on the store shelves? Why aren't we fighting this at the source? Let's go make it more difficult for them to get access to water in the first place. That's their bread and butter, is finding new water sources. So when you have Cascade Locks and then their difficulty in finding new water sources in the Pacific Northwest, I think that makes it really hard for them.

After Nestlé announced the sale, but before BlueTriton assumed full control, water activists began assessing what the new ownership meant for their own local struggles. Unlike Nestlé, BlueTriton is a private equity venture whose multi-billion-dollar purchase was financed largely through $3.2 billion in "junk bonds." Typically this strategy creates a major incentive to aggressively slash costs and/or strip a firm's assets, then resell it within a short time frame at a large profit.[85] An alliance of anti-Nestlé community groups in the United States and Canada, including WWW, calling itself the "Troubled Waters" coalition, raised concerns about the sale in an open letter: "This sale . . . would be a massive private transfer of 'water wealth,' an especially ominous development in light of Wall Street's accelerating interest in 'water futures' trading. . . . The combination of decades of pollution, depletion of aquifers and effects of climate change make clean water scarcer. . . . It is also clear that a private equity firm, freed of Nestlé's reputational responsibilities, will seek to cut expenses at the cost of the limited promises its predecessor made regarding environmental sustainability and community benefit."[86]

Thus the groups were in the ironic position of arguing "Better the devil you know"—acknowledging that Nestlé's sensitivity around its brand image had at least given them some leverage. The question of whether the BlueTriton acquisition truly represents a move in the direction of water trading—long-term speculation on fresh water as a bulk commodity as it becomes more scarce and valuable—is impossible to answer at present. However, water activists are worried. "Nestlé was at least a *bottling* company," Balkwill tells me: "They were at least in the water bottling business, right? This [BlueTriton] is a straight profit exploitation. And [we're] using the climate crisis as a background to say it's even more important that water not be in private hands and that water bottling can even be understood as an interim use, until you can really figure out what you're going to do. . . . It's an interim use for fifty years until the crisis gets here, and then you can say we've got water and we're going to sell it to the highest bidder."[87]

A POST-NESTLÉ WORLD

When the sale was announced, some water protector groups expressed hope that as a new company, BlueTriton would need to reapply for permits to extract water from the wells it had acquired. However, it has received permission to continue operating on Nestlé's existing water permits in every case to date—including Aberfoyle and Hillsburgh, where in 2021 BlueTriton's water taking was renewed for another five years.[88]

From Florida to British Columbia and from California to Ontario, the communities where Nestlé for one or two decades had been *the* face of the bottled water issue—the target of repeated protests, constant media attention, and lawsuits—suddenly found these hotly contested water sources in the hands of an unknown entity. What does a post-Nestlé world signify for these grassroots water movements? What does it mean to contest water extraction by the bottling industry without Nestlé as an utterly familiar and widely reviled target? "It's got a big impact for sure," Slocombe admits. "People came to us because they love to hate Nestlé. And so there is an effect of taking the wind out of the sails a bit now. And I think that was so strategic. That's exactly what they were after."[89] O'Heaney likewise acknowledges that for the community groups in the Troubled Waters coalition, "It's a not insignificant challenge. And you even see that in the conversations with the local partners. . . . Their local constituents, journalists, elected leaders—people know Nestlé, they know the name, they understand that company. BlueTriton doesn't mean anything to anyone. . . . But I don't think it's an insurmountable challenge. Ultimately it's the same company—it's the same people, except for the top ranks of leadership."

While the loss of a widely known target with a long and checkered corporate history complicates the situation for activists, they are still confronting a single common corporate owner across North America—which for campaigners is a preferable situation to Nestlé having broken up its assets among multiple buyers. On the other hand, the sale may limit the North American groups' ability to coordinate efforts with organizations elsewhere in the world that are still intensely focused on Nestlé. Whether the sale will reduce public scrutiny of Nestlé's overall water practices remains an open question.

THE STATE OF PLAY

In the wake of the Nestlé sale, it is worth gauging what has been achieved during this fifteen-year campaign. WWW and Save Our Water have managed to substantially slow and likely halt the expansion of the bottled water industry in Wellington County, and thus in Ontario. Assessing the impact of these campaigns on Nestlé and its successor firm, WWW's Balkwill sees a cascading effect on their fortunes in the region:

> Well, the positive spin is we slowed them down. Right? We put things on pause, we've stopped the expansion of Nestlé in our community. . . . Because the veto has been put in place, BlueTriton's not asking for a permit [for Middlebrook Well]. At least under the current conditions, they would be told no if they did that. . . . Our understanding all along has been that for their business model to work, they needed that well . . . [that] if they expanded and got the third well, we would never stop them. . . . So that's the first victory.[90]

The water advocates also appear to have changed the political calculus on this issue province-wide, not only for the previous Liberal government, but for the current Conservative one, which eventually enacted water policy reforms that ran counter to its political ideology. "The second victory," Balkwill tells me, "[is] that one government introduced a moratorium and a second one continued it." He continues, "The third thing is, we have four elected parties in Ontario, and we've moved two of them to commit to phasing out permits to bottle water in Ontario. The NDP and the Greens. And we're in discussion with the Liberals. . . . We have moved a social movement demand into a political platform demand. So that's progress."

In Elora, the Save Our Water members are savoring at least a provisional win. According to McCaw, "We know we've made a positive difference for provincial regulations for water taking in Ontario." However, she continues, "Centre Wellington water is not safe yet. Middlebrook is still owned by two American companies. . . . We know that no company has made a penny taking groundwater at Middlebrook for the last six years. And now, BlueTriton needs to get that message that it will never make a penny here. . . . We're not done yet."[91]

WWW and Save Our Water have also been effective in stigmatizing the commodity of bottled water (at least for settings where clean tap water is readily available) and have helped build a majority public base across Ontario that is receptive not only to a damning critique of plastic packaged water and its environmental and hydrological impacts but to ending the industry's extraction of groundwater entirely.

The involvement of activists and traditional leadership from Six Nations in this alliance has been crucial in shifting participants' understanding of the scope of the issue. The lack of safe drinking water for most people at Six Nations and dozens of other First Nations across Canada, juxtaposed against the ongoing, high-volume extraction of pristine aquifer water from Six Nations' unceded treaty lands, has moved commercial water bottling from being viewed as primarily an environmental problem to being understood as a social justice issue for the first time. Slocombe believes it may have also been decisive in changing provincial water policy:

> It's my suspicion that the involvement of Six Nations here is what pushed it over the top, and is what will push it over the top across many issues. That's what's happening with pipelines, that's what's happening with major infrastructure developments—that Indigenous land title treaty rights are the thing that makes it completely unpalatable for those multinational corporations. . . . I think we are going to have way more power when we get behind and lift up those [Indigenous] voices. I think it terrifies corporations. . . . Here, the jurisdiction is provincial around [water] permits, but it starts to bleed into national jurisdiction with treaty rights . . . and I think they don't want to deal with that too much.

CROSS-NATIONAL CONTRASTS

It is useful to step back and consider the main parallels and divergences between the two detailed stories portrayed here and in the previous chapter: two cases of community members and activists who found themselves challenging the same global bottled water corporation in two nations and two different political, cultural, and social settings. The first major distinction is that while the campaign in Cascade Locks aimed to stop a proposed new water bottling facility, Nestlé was already well established in Wellington County before the Water Watchers ever formed. For WWW's first eight years, its work focused largely on slowing or stopping the firm's existing, permitted water extraction. In fact, it is worth noting that those efforts to stop or shrink Nestlé's water taking at its existing wells have been only marginally successful, limited to winning shorter permitting periods, reduced pumping levels during droughts, better monitoring, and other minor concessions. This was largely a defensive effort. It was only when Nestlé proposed a new well in Elora that WWW—along with the newly established Save Our Water—was able to shift into offensive mode.

The provincial policy changes relating to the Middlebrook Well—the increased extraction fees, the unexpectedly long moratorium on new water taking, and now the municipal veto on large-scale water permits for the bottling industry alone—can be read as direct wins by the water advocacy groups and their allies. The most immediate practical result of these developments? The Middlebrook Well is unlikely to be developed for commercial water bottling, although whether it will ever be sold to local government for use as a municipal water source is another matter. Blocking the Middlebrook Well from operating after a six-year battle is the clearest victory for these Ontario water activists. BlueTriton continues to extract water in Ontario, but its footprint appears far less likely to expand.

As some of the Oregon activists quoted in the previous chapter warned, based on the experience of other U.S. communities, it is notoriously difficult to dislodge a corporate water bottler once it has become established, built industrial facilities, and acquired rights or access to spring water or groundwater. In both Oregon and Ontario it is the efforts to halt *proposed* facilities that have been successful. Blocking the industry's expansion turns out to be easier than loosening its hold on existing water sources.

The differences between these two cases also include questions of strategy and tactics. Unlike in Cascade Locks, where the Local Water

Alliance and allied groups pursued a high-stakes ballot measure campaign to change local law and ban water bottling entirely, the political structures in Ontario and Canada did not allow WWW and Save Our Water an opportunity for the same kind of game-winning "big move." Instead, the Ontario groups' efforts were directed at the provincially controlled realm of water policy, necessitating engagement with "the bubble" at key moments, but also obliging them to focus on longer-term community education, campaigning, and the building of alliances and political power.

These differences had implications for the nature and longevity of the activist organizations themselves. In Oregon, despite what became a decade-long campaign involving a coalition of existing environmental, labor, and civil-society groups, the key organization behind the ballot measure campaign (LWA) was formed for a specific purpose—defeating Nestlé's bottling plant proposal—and eventually became inactive after that goal was won, while the preexisting groups in the coalition pivoted to other issues. In Ontario, by contrast, WWW has built a far more permanent and broad-based apparatus. While its primary focus has been commercial water bottling, it is increasingly mobilized around multiple threats to water in the region and province, including aggregate mining and urban development in protected watershed areas including the Greenbelt, as well as newer work to support Indigenous land and water sovereignty.

On one issue—the framing and messaging used in these two campaigns—there is both overlap and divergence. In each case, activists used narratives of scarcity to mobilize local opposition, and both were aided by the timing of severe droughts that dramatized water scarcity more vividly than words ever could. Those droughts amplified public concern about loss of control over local water and were transformative in the outcomes in both nations: in Ontario, the bottling moratorium and increased fees announced in 2016, and the municipal veto, lower volume limit, and hierarchy of water uses enacted in 2021; and in Oregon, the resounding "yes" vote on the ballot measure and the governor's eventual decision to terminate the water exchange and halt Nestlé's bottling plans.

However, this broad frame of scarcity and threats to water availability posed by commercial bottling played out differently in each region. In Oregon, it was the potential risk to the Hood River Valley's high-value irrigated agriculture that was the winning frame for the Local Water Alliance. In Wellington County, it was the threat to municipal drinking water supply in the near future—the prospect that bottling would impede local government's ability to expand public water supplies to meet the needs of a growing population—that was the decisive frame in blocking Nestlé from using the Middlebrook Well. In a region almost totally dependent on finite groundwater, those messages were especially salient.

The parallels between the two conflicts are instructive as well. In both the Oregon and Ontario contexts, activists successfully portrayed water extraction for bottling by a global corporation as a threat to sufficient future water supplies. That framing—extraction as permanent removal of limited water from the watershed for profit—is a key similarity between the two campaigns. Both effectively used the notion of water "export" to dramatize the finiteness of water and advocate for its protection.

Second, both the Oregon and Ontario campaigns prove that water protection can transcend political boundaries if handled properly. The Local Water Alliance and the water defenders in Wellington County both succeeded in uniting multiple constituencies across traditional divides—rural/urban, progressive/conservative/libertarian—to build overwhelming majorities in favor of protecting local water from extraction and removal by global corporations that produce a commodity widely viewed as unnecessary. Many interviewees in both sites mentioned that there is something unique and powerful about water that unites people. With 69 percent of the voters supporting the ballot measure to end water bottling in Hood River County, and 64 percent of Ontarians supporting an end to all bottled water extraction in the province, that is clearly not hyperbole. In an era of climate emergency and extreme political polarization, these examples offer a glimpse of one effective model for local defense of the natural world.

Finally, in both cases the key protagonist groups would not have been successful without forging strategic alliances with a range of local, regional, and national groups. Most significantly, in both settings the core organizations entered into vital collaborations with independent Indigenous activists and leaders of First Nations that hold and claim rights to land and water on these same territories—linkages that in Oregon, and possibly in Ontario as well, were pivotal in the outcomes. These two conflicts illustrate the power of current and potential moves by Indigenous nations to exercise or expand their treaty rights to land and water. In both the United States and Canada, asserting treaty rights can involve those Indigenous nations "jumping scales" over subnational authorities to engage directly with federal governments and courts.

Even the threat of such action can influence political outcomes at the state or provincial level. Many current battles over planned fossil fuel infrastructure across North America offer further evidence that Indigenous and non-Native water protectors have the potential to become vital strategic allies.

LESSONS FROM ONTARIO

Just as the story of Cascade Locks highlighted the critical role of local community groups collaborating with regional and statewide actors and Indigenous communities, and embracing key local meanings of water in their campaign messaging and framing, the ongoing struggle in Ontario also offers valuable lessons for other communities facing similar proposed or continuing water extraction by the bottling industry. While the trajectories of both Wellington Water Watchers and Save Our Water are multilayered, a few core lessons stand out.

The first involves the question of how substantially water protection movements should engage with the scientific and technical realm that underpins the public policy process. Balkwill recalls his first meeting with the staff and leaders of WWW and Save Our Water:

> I said to them, the science doesn't matter. . . . We're never going to win on science. Which was hard . . . because they really wanted to believe that it was a reasonable process. . . . So job one was to say, you've got to get out of the bubble. We have to reframe the debate. Why do you really oppose this? You really oppose it because of the values; you believe water should not be a commodity. That's job number one. That's the high moral ground. . . . Don't try and fight them on the science. Do enough. You need a technical report to counter their technical report so then you can have a political fight, right? . . . And you've got to find and create an arena in which you can have that fight.[92]

A second lesson regards the importance of campaign strategy. Like the LWA in Oregon, which worked closely with seasoned political organizers in crafting and waging the ballot measure campaign, WWW chose to hire an organizer. Slocombe says this decision was transformative: "We started off like a lot of grassroots groups that are in the environmental movement . . . middle- to upper-class White settler folks, worried about something in our own backyard, and activating around it, and learning along the way the effective ways to organize. And Balkwill came and joined us, who has experience as a campaign organizer. And that really elevated our ability to have impact in terms of campaigning. So first of all, I'd say, connect with a campaign organizer."

A third lesson from WWW's water fights in Ontario is the value of embracing water protection as a justice issue. Through its alliance with Six Nations leaders and activists, WWW has come to embrace an organizing model more focused on social justice and forging connections with other movements, beyond water alone. "If I were to imagine [our] trajectory," Slocombe tells me, ". . . it is about seeing the intersectional nature of the issue that we're addressing. And not just staying siloed within that issue, but seeing how it aligns with so many other justice issues—climate justice issues, and building coalitions and alliances around those pieces that can really bring on major pressure."

A fourth lesson is the notion of escaping from the tyranny of the possible—daring to envision the world that participants truly hope to achieve, and working toward that vision. Balkwill continues recounting his early effort to encourage the water protection groups to think in such terms:

> When I [first] met these people [WWW board and staff, and Save Our Water leaders] . . . I said, what do you want? They said, well, we want to have more conditions on Nestlé's well . . . a shorter time permit, et cetera. I said, is that what you *really* want? And they went, well, actually, no. We don't want them here at all! I said, well, let's make that our goal. And it was a moment of deep release for them. . . . They'd just been conditioned into a place where incremental change was the most they could hope for. And it just released all kinds of things. And we said, no—we want to kick them out. Well, great. Let's make that our demand.

Finally, these water protectors and their allies have managed to partially decommodify water in Ontario in at least four ways: blocked expansion, increased costs, reduced operating freedom, and decreased demand. Nestlé's efforts to expand to the Middlebrook Well—vital to its business plans—were halted by the new groundwater regulations and local veto. The water advocates also managed to increase the company's costs in several ways: the drought-related pumping restrictions and the new monitoring requirements WWW won in its early campaigns; the higher royalties enacted by Wynne's government; and the long provincial moratorium, which obligated the firm to expend resources while sitting on a newly purchased but unusable well site. BlueTriton has no public plans to develop new sites in Ontario, and the local veto on new large water permits greatly limits its ability (and that of other bottlers) to do so. Finally, by eroding public support for the industry, the activists have helped reduce consumer demand for bottled water, particularly the spring water brands sold by Nestlé and now BlueTriton. All of these factors translate into lower profits, which played a key role in Nestlé's decision to sell.

Yet beyond these specifics, for many of the participants in this linked set of campaigns to protect water in Canada, what motivates their dedication to this work is simply a fundamental opposition to the commodification of water, and to profiting from its enclosure. As we end our conversation, Balkwill says that ultimately the fight to defend water boils down to "a struggle with the capitalists, a struggle with the private interests, over the source of life."

CHAPTER 7

Empty Bottles: Water Justice and the Right to Drink

I believe that it is totally possible, and that it's the only choice, for us to have fully funded public water systems that are democratically accountable to people, that provide water to everyone equitably at rates they can afford, and that the water is safe and clean.

—Lauren DeRusha, Corporate Accountability

When I return to Guelph in mid-November 2019, an unseasonably early cold spell has brought several inches of snow and subfreezing temperatures. The packed snow sparkles and crunches under my feet as I cross the crowded church parking lot in a residential neighborhood. Despite the cold, the church hall is packed with at least two hundred people who are here to attend an unusual forum titled "All Eyes on Nestlé," the last of four such events held around the region. A row of tables along one side of the undecorated hall is covered with coffee urns and plates of cake. At the back of the room sit more tables with a crowded set of displays, flyers, and maps staffed by a range of local groups, including Save Our Water, youth from Six Nations, and more. By the time the bugs in the sound system are worked out and the event starts behind schedule, almost every seat is full.

Mike Balkwill takes the microphone. He tells the crowd,

> In 2018, we started to reach out asking if there were communities around the world fighting Nestlé to see what their experience was, what were they dealing with, what could we learn from that, et cetera. And we were introduced a year ago to Franklin Frederick . . . from Brazil, where he was part of the successful campaign to stop Nestlé, and a part of the campaign he was involved in meant going to Switzerland [and] animating a relationship with the Swiss churches, putting pressure on Nestlé there—that was important to

> their victory. . . . And Franklin said, I think it'd be important if . . . someone from Wellington Water Watchers went to a meeting in Vittel, France, in February, which we did.

Earlier that day, I had met those international activists in the spacious home of one of WWW's board members, down a snowy cul-de-sac. Frederick, an energetic Brazilian writer in his early fifties, described his role leading a movement opposing Nestlé's water extraction in the spa town of Sao Lourenço in the Brazilian state of Minas Gerais: "Perrier, the French group, they were the owners of the water park, and they were bottling one [spring] and selling it as bottled water, but there was no conflict with the population. . . . Perrier was only taking the water that was naturally coming [up in the spring]. They never pumped water [out]—and better, they were even having profits." Frederick recounted the transformation that occurred once Nestlé acquired the springs:

> Nestlé bought Perrier in the '90s, and the profits . . . were just not enough for Nestlé. So they decided to enlarge the bottling facility inside the water park to produce, besides the mineral water, this new water that they created called Pure Life. It's not mineral water. . . . It's always the same [taste] . . . so it was like a Coca-Cola of the waters. And . . . shortly after Nestlé started the production of Pure Life . . . the amount of water that was surfacing in the other [springs] diminished, and the taste changed. Why? Because mineral water, it needs time inside the earth.

Faced with this situation, "the people from Sao Lourenço, they reached out to other people asking for help. I proposed to them to bring the issue to Switzerland. It's . . . a Swiss company. Let's try to have Swiss public opinion on our side." Frederick traveled to Switzerland to make connections with civil society groups opposing the company there for other reasons. The various opposition efforts, he says, eventually bore fruit back in Brazil. "In early 2006, finally, Nestlé did make a deal with our public ministry. . . . It's a victory, because they stopped pumping water and they moved away from Sao Lourenço. Pure Life is not produced anymore in Sao Lourenço. So we won."[1]

In the process, however, Frederick said he was a target of corporate espionage. According to the Swiss daily *Le Courrier,* between 2003 and 2004 (and allegedly longer), Nestlé hired the security firm Securitas AG to monitor a local branch of the activist group ATTAC, with whom Frederick was collaborating and which was working on a book about Nestlé's corporate practices.[2] In 2008 the TV program *Temps Present* reported that a Securitas employee had infiltrated the group under a false name, reporting her findings to Nestlé security and communications staff.[3] While criminal charges were ultimately dismissed, ATTAC filed suit against Nestlé and Securitas on the basis of these revelations, and in 2013 it won a civil judgment in a Swiss court against both companies, which were convicted of "unlawful violation of the rights of the person" for illegal infiltration and were ordered to pay 3,000 Swiss francs compensation to each plaintiff.[4]

I also met Bernard Schmitt and Reneé-Lise Rothiot of the grassroots group Collectif Eau 88, from the mountain town of Vittel in the Vosges region of northeastern France. The town is home to Nestlé's Vittel-Contrex-Hépar bottling plant, where the company employs nearly one thousand people and has enjoyed a good deal of public support due to the jobs and what Schmitt calls its paternalistic management style. The husband and wife duo recounted the story of their campaign to hold the corporation responsible for the drawdown of one of the main aquifers beneath the town, in a context of increasing drought. The French government eventually built a pipeline from a nearby town to supply Vittel's municipal drinking water system because of the falling water table, instead of restricting Nestlé's water extraction.[5]

Schmitt discussed the impact that connecting with the Canadian water activists had on the French group's efforts. One thing that "helped a lot was the internationalization of the whole story," he told me:

> It was at the end of 2018 that I made contact with [Wellington Water Watchers]. I proposed to them to invite the Canadians to go there [Vittel] in February. So this year we all went there. The press was very much interested in these issues, so the visit by the Canadians and Franklin appeared a lot in the French press. And we discovered a new discourse: plastic bottles; ecocide. . . . The main thing was that it not only changed our minds but the local people's minds [in Vittel]. Because before, they [accepted] this discourse that the problem was only there—that they could find a common solution with the [French] authorities—like, trust us. And when they started to see that in Aberfoyle exactly the same thing is happening—exactly the same—this helped them to see our authorities with different eyes.[6]

When the members of WWW visited Vittel, they too identified many similarities between the company's practices in Canada and France, including extracting the water for very little cost while local residents paid far more for tap water, and high-volume pumping that activists claimed was steadily lowering the water table. The Canadian group used these parallels in its campaign communications. "If Nestlé will behave this way in Vittel," read one WWW blog posting, ". . . how can we trust what they propose for Middlebrook well in Elora? How can we trust that the Ontario government will protect water—rather than protect the business interests of corporations like Nestlé?"[7]

. . .

These kinds of cross-national connections between communities affected by the bottled water industry—sharing lessons learned, identifying parallels, advising on campaign strategy, carrying out reciprocal visits—have become increasingly extensive, and they also include links between activists from the global North and South. Can the fledgling transnational alliances and the various local campaigns profiled in this book be understood not merely as a network but as parts of a larger social movement opposing packaged water? And if so, what is the relationship of that movement to the more established global water justice movement?

BUILDING A MOVEMENT ACROSS BORDERS

Back at the public forum in the Guelph church hall that evening, Gina Luster from Flint Rising speaks, her dynamic presence holding the audience's attention. She describes the latest developments in the water struggles in Flint and in Michigan and then recounts her recent trip to Lagos, Nigeria, to speak at a summit organized by an international coalition of water justice groups, including Corporate Accountability:[8]

> It was in January 2019, and wow. This was the largest convening of water warriors from all over the world. . . . We couldn't even understand each other's language. But Nestlé was the one common thing in a room of thousands of people. So we decided while we were there, let's go pay Nestlé a visit. So . . . we went over to this Nestlé plant that sits in the middle of a village, with kids . . . like you see on TV with the malnourished bellies, no clothes, living in huts, no water. And right walking distance from them is the Nestlé plant that is pumping the water out of the river of that village. . . . The women you see with the buckets, they're walking five to ten miles [for water] because Nestlé sold this village a dream and said, we're going to build you some wells. And some fountains. Well, when we got here, the fountains were built, and they were dry as a bone.[9]

This gathering in Guelph also grew out of efforts to forge new connections within North America. WWW reached out to grassroots groups across the U.S., starting with Maine, where Nestlé's Poland

Springs brand had been extracting large volumes of water for over two decades and the group Community Water Justice had fought an uphill legal and regulatory battle against the company. Balkwill recalls,

> I called up Nickie Sekera in Maine to talk about our researcher coming there. . . . The story was pumping out of her like an artesian well. . . . So we went to Maine to spend a weekend with Nickie and a bunch of people down there. And then we said to ourselves, you know what? Why don't we spend our money to bring everybody together in one place, rather than us send some researcher off [to] one [community] at a time, because we'll generate all kinds of analysis together, and in fact it'll create buy-in [from] everybody to this project going forward.

By the time the November 2019 event in Guelph came together, it included not only the activists from France and Brazil but leaders of grassroots groups in three U.S. states where Nestlé has bottling operations—Our Santa Fe River in Florida; Michigan Citizens for Water Conservation; and Community Water Justice in Maine. The Canadian presence included representatives from Six Nations, the Council of Canadians, faith groups and environmental NGOs, researchers, students, and a large contingent from WWW and Save Our Water. Some of these groups had already met two years earlier in Flint at the gathering that generated the Water Is Life Alliance.

The day after the packed public forum, the representatives from these groups met on the top floor of WWW's narrow office for a full-day session to share experiences and strategies from their local campaigns (figure 31). Frederick told me his hopes for the gathering: "If we start exchanging our experiences, we'll realize all the common points. . . . And once people realize all these common points they see, then it's not a local issue. It's a pattern. And if we have a pattern, we have to fight the pattern. And then the second movement, I hope would be, after we are more connected, [to] then connect with the people fighting in defense of public water services."

After this gathering, the organizers intended to escalate their international efforts further, planning to meet in Switzerland on World Water Day in March 2020 to highlight links between Nestlé and the Swiss international development agency, but the Covid pandemic ended those plans. Yet unlike with many such one-off meetings, the momentum from Guelph did not dissipate. The international gathering had offered participants a taste of the possibility of building a broader North American network to confront commercial bottled water extraction. A chance to crystallize that vision arrived in 2020 when Nestlé announced

FIGURE 31. Reneé-Lise Rothiot and Bernard Schmitt, Collectif Eau 88, November 2019. Photo: Author.

it was seeking a buyer for its North American bottled water operations. As Michael O'Heaney of Story of Stuff recalls,

> Nestlé said basically they were looking for someone to buy their North American water business. And we thought, okay, here's the opportunity to actually bring these groups sort of more tightly together into a network. I think one of the things that we had always seen was fantastic local activists who, like local activists on any number of issues, sometimes felt a bit disconnected from either allies in other parts of the country, or national allies, or even international allies. And so when the sale was announced, we brought the groups together and said, "Hey, why don't we ask the company to basically give back these most troubled water sources?"—which they of course declined to do. But we were able to generate some additional attention

> around both the sale, and then sort of welcome BlueTriton ultimately to the arena.[10]

This was the genesis of the "Troubled Waters" network and campaign. According to Balkwill, the connections WWW had built for the November 2019 event put it in an ideal position to coordinate such an effort. "There's a stronger informal network now than there was when people came together for our gathering in Guelph. . . . The outreach we did to create that meeting was good. I think it created a basic set of relationships that were respectful and trustworthy, then Story of Stuff came along . . . and I think we took it further."[11]

The Troubled Waters effort has grown beyond the groups that met in Guelph to include activists opposing Nestlé/BlueTriton's water extraction in Southern California and in Chaffee County, Colorado. In this process, WWW has served as a bridging organization, reaching beyond its own campaigns in Ontario and helping to convene a North American network of communities fighting commercial bottled water extraction. (In 2022, WWW formally changed its name to simply Water Watchers, to reflect its national and international reach.) O'Heaney elaborates on how these grassroots groups use the network to coordinate and share lessons between their far-flung local efforts:

> These [local] groups and then outside allies: Sum of Us, Courage Campaign, us [Story of Stuff], Food & Water Watch, Corporate Accountability . . . have really opened up opportunities for collaboration and sharing. So the group that we've been supporting . . . in Colorado—the [Chaffee] County commissioners just voted two to one to approve [BlueTriton's] permit. . . . [The local activists] were considering legal action and asked, "How did you all figure out whether or not to use the law as a tool?" And to have folks in San Bernardino, to have folks in Florida, to have folks in Michigan, all kind of weigh in and say, here's the pluses and minuses, here's the things that you need to be thinking about, that kind of thing.

The network and Story of Stuff achieved a tangible win in early 2021 when the State of California ordered Nestlé/BlueTriton to stop extracting water from the San Bernardino National Forest, which it had continued to do despite a long-expired contract.[12]

FROM NETWORKS TO MOVEMENT

This kind of communication between local bottled water fights long predates the Troubled Waters effort. "There were always informal networks where these groups were finding each other," says O'Heaney. "The folks in Cascade Locks in particular. . . . When they passed that county ordinance, they were quite generous for a year with, I will talk to this person, can we introduce you to so and so—they really thought of themselves as organizers." Similarly, the connections forged between urban activists fighting toxic water and shutoffs in Flint and Detroit and groups opposing commercial bottled water extraction in rural western Michigan have strengthened both of those constituencies. That network, part of the People's Water Board Coalition (and the former Water Is Life Alliance), partially overlaps with the binational Troubled Waters campaign, suggesting the outlines of a larger North American movement that links opposition to commercial water extraction and bottling with activism for water justice. "I think there is a movement," O'Heaney tells me. "There is a movement that has taken a couple of different forms over the last decade and a half. You see it pop up in Massachusetts towns banning bottled water sales. You see it in the campus stuff. You see it in the city purchasing agreements that Corporate Accountability and others really pushed. . . . And I also think that movement has contributed to the overall fight for water justice. So maybe it's a vein in the water justice movement. But I think it's an important vein."

Yet to what extent are these movements contesting packaged water coordinated, cohesive, or complementary? Examining the entire sweep of efforts described in these pages reveals an interesting pattern. While many local campaigns opposing groundwater extraction in rural communities, or pushing to remove bottled water from city property or university campuses, have emerged in spontaneous grassroots fashion from a core of highly motivated residents or students, some degree of coordination and networking among these efforts has also been present at least as far back as the early 2000s. That was the post-Seattle moment when the Polaris Institute pulled together a network of Canadian municipalities and universities aiming to ban bottled water, and when CAI and later Food and Water Watch launched campaigns to support similar efforts in the U.S. By the end of that decade, Council of Canadians had begun its Blue Communities campaign, and Story of Stuff emerged on the scene with a viral video critique of bottled water, eventually expanding its work to support local fights against the industry's commercial water extraction.

The centrality of these five North American organizations and (with one exception) their continued work on the packaged water issue to this day attest to the durability and longevity of what can be called a bottled water movement coalition. The key role of these groups indicates that the two main facets of this activism—the site fights at the water extraction end and the efforts to reclaim the tap and halt single-use plastics at the consumption (and disposal) end—are complementary parts of the same broad movement, which is challenging the industry all along the bottled water commodity chain.

At least four of these key coordinating groups came to their work on bottled water from earlier anticorporate and global justice activism focused on the impact of neoliberal policies and international institutions on the environment, human rights, and democracy. To varying degrees all of them are explicit about how the bottled water industry relates to the worldwide picture of water commodification, and how contesting bottled water connects to efforts to fight water privatization and defend public water systems. Finally, campaigns opposing bottled and packaged water (especially from the consumption side) are now also increasingly linked to, or an outgrowth of, the global climate justice movement and the international movement against single-use plastics.

Even in local struggles over groundwater extraction for bottling, the increasing threats to fresh water availability from climate change and other causes have become a powerful driver of activism. As the stories of Cascade Locks, Oregon, and Wellington County, Ontario, make clear, water scarcity is a common thread that powerfully mobilizes local concerns about the removal of fresh water supplies that are seen as increasingly finite, precious, and needed for local uses and ecosystems. Drought intensifies these concerns, dramatizing scarcity in ways that are difficult to ignore. Yet ultimately, while quite real in its effects, water scarcity is socially and politically created, not merely a natural fact. This suggests for social movements the importance of stressing "how resource shortages and ecological degradation are primarily a result of the uneven social measures that manufacture scarcity all over the world for the economic and political gain of powerful interests."[13]

These two cases—in two countries and regions that differ culturally, economically, and politically—also highlight important parallels. One is the way that proposed water extraction by a global beverage firm has united local residents across lines of race/ethnicity and class, linking rural and urban dwellers and bridging entrenched political divides. Both of these stories also embody what geographer Zoltán Grossman calls "unlikely alliances" that link Indigenous and non-Indigenous activists, illustrating the power of yoking water protection campaigns to efforts to honor treaty rights and restore stolen land.[14] Water is different. Everyone is connected to water, and threats to it appear to have particular power to transcend the lines that separate people.

RETURNING TO THE SOURCE

In order to pull together the wide range of issues and movements covered throughout this book, it is useful to return to the start, connecting the overarching questions and themes described in the Introduction with the concrete cases and stories described in these pages. In the process, I make five major arguments.

The first regards the particular ways that bottled water commodifies water. To begin with, packaged water poses fewer obstacles to private profit than does private management of complex municipal tap water networks, which partly explains how it has become so ubiquitous within only a few decades. Bottled water constitutes a more perfect commodity, because it is not hindered by many of the obligations, tethers, and barriers to capital accumulation that have slowed the privatization of tap water systems globally. Packaged water's key traits—its far greater portability and profitability—make it distinct from tap water. The technological development of PET plastic, along with cultural and political-economic changes wrought by neoliberalism, have facilitated its rapid spread, in societies both with and without widespread access to safe drinking water from the faucet. Bottled water bypasses the huge infrastructural networks of water treatment and piped delivery systems built by governments over the past century or more and rushes to fill the gaps where such networks do not exist. It skips over those infrastructures to deliver private drinking water in small volumes, but at a high cost. Compared to privatized tap water, it represents a far more complete severing of drinking water from any shared societal endeavor.

Bottled water also epitomizes Karl Polanyi's incisive arguments regarding the perils of the commodity fiction. Water—an inseparable part of nature—is not literally produced for sale in the marketplace, yet it is indeed sold as a commodity for profit on a huge and growing scale. However, treating it as a genuine commodity generates highly destructive human and ecological effects. Nonetheless, although bottled water represents a prime example of commodification, and a clear case of accumulation by dispossession, in most cases it does not strictly constitute privatization, because it does not involve the appropriation or acquisition of publicly owned goods or assets by private capital. David Harvey argues that privatization represents the leading edge of accumulation by dispossession, but outright privatization is only one route by which capital can enclose water. Yet it is also important to recall that for many water activists the word *privatization* serves as a blanket term encompassing all forms of commodification.

Even within the realm of packaged water, some segments are more amenable to commodification than others. Compared with the commercial extraction of springs and groundwater in Michigan, Ontario, and elsewhere, the bottling of already-treated public tap water represents an especially convoluted example of accumulation by dispossession, a dynamic that Harvey describes as involving "cannibalistic, as well as predatory, practices."[15] By piggybacking onto municipal water systems, bottled water firms are parasitizing and profiting from more than a century of taxpayer and ratepayer investments in tap water infrastructure. In serving up the very same substance in bottles for thousands of times the cost while it simultaneously "delegitimizes public water,"[16] the industry contributes to further hollowing out the public sphere.

However, it is bottled spring water, not refiltered tap water, that sells for higher prices and offers greater profits. And rural communities with important attachments to local springs and groundwater have generated the strongest opposition to the industry, both in North America and worldwide. As Peter Gleick told me in 2019, "It must be incredibly frustrating for Nestlé to believe internally that 'Look, this isn't a lot of water we're taking; we're bringing jobs to a community' . . . true or not—and then to be stymied after ten years of effort. But it's hard to feel much sympathy for them. They've decided they want to do spring water." He continued,

> Now, Coke and Pepsi don't have these problems. This is because Nestlé continues to insist on looking for spring water sources. So it must be economically worth it to them. . . . These people aren't in it for anything other than the money; there's no other reason. . . . And so the ability to say this is spring water from some pristine—you've seen the branding, you've seen the advertisements—that's the advantage. And so it sells at a premium compared to the cases of municipal water that they can sell for a lot cheaper. So there's some money—follow the money.

Two years after I spoke with Gleick, Nestlé, whose bottled water sales had been stagnating or falling in recent years, threw in the towel and sold its U.S. and Canadian spring water operations and brands to BlueTriton, retaining (in North America) only a few imported water brands. Those falling sales were due in part to the backlash against plastic bottled water, as well as consumers switching to cheaper private-label brands, and likely also in part to the results of multiple site fights. That pincer effect—serious challenges at both the extraction and consumption ends—appears to have played a key role in Nestlé's decision to sell off a large portion of its global water business.

One lesson from the outcomes in both Oregon and Ontario is that battling with rural communities for years over establishing new spring water extraction sites may be too costly and brand-tarnishing, and thus a threat to profitability, for an image-sensitive global beverage firm. Opposition movements have made expanding to new sites more difficult, but it is easier for companies to hold onto existing permitted sites—likely the route BlueTriton will take.

Interestingly, in the U.S. (in contrast with Canada), those battles over spring water bottling pertain to only about one-third of the total bottled water market, leaving out the majority, which comes from public tap water supplies. That refiltered tap water segment has gone almost entirely uncontested at the point of extraction. While so-called purified tap water sells for less than spring water, companies acquire the prime ingredient for very low cost, often at discounted rates. Access to this cheap municipal water represents a major public subsidy to capital. It also entails fewer costs and less effort on the front end—there are typically no legal battles over siting, little or no public process is required, extraction takes place at existing beverage bottling plants, and the water is distributed via bottlers' established supply chains. In sum, bottled municipal water is far less visible and harder to challenge at the point of extraction than spring water, yet it contributes equally to the overall loss of public trust in tap water and to the negative effects of this commodity on social inequality.

However, bottled municipal water is not going entirely uncontested. The wide range of local policies, bottled water bans, and efforts to revalorize and increase access to public tap water that constitute the "reclaim the tap" segment of this movement challenge *all* forms of packaged water, usually without distinction by source or brand. By problematizing the generic commodity of bottled water, they are taking on 100 percent of the market, not just a portion.

Ultimately, whether it comes from rural springs or already-treated municipal supplies, bottled and packaged water accelerates the process by which fresh water as a whole is being commodified on a global scale. Maude Barlow argues that bottled water "allows people to view water as a commodity and sets the stage—one bottle at a time—for the complete corporate takeover of water."[17]

A THREAT TO PUBLIC WATER

This leads to my second argument. Given the large proportion of the public who already turn to bottled and packaged water for at least part of their drinking water supply, the continued growth of this commodity represents a significant threat to the future of universal safe public drinking water provision—possibly even more serious than that posed by the privatization of tap water utilities.[18] Packaged water poses this threat mainly by weakening the political consensus needed to build, expand, maintain, and adequately fund public water infrastructure. This is true both in places that currently have the privilege of reliable clean tap water and in those that do not, including large parts of the global South. Either way, these dynamics have opened up space for private capital to appropriate the state's role and to benefit from the scarcity or unavailability of clean water. "The bottled water industry," writes Gleick, "is successfully capitalizing on, and profiting from, the decay of our comprehensive safe drinking water systems, or, in the poorer countries of the world, their complete absence."[19]

This same pattern also pertains to "structurally adjusted" places within the global North where governments have either abandoned or never fulfilled their role to guarantee a safe supply of clean water to drink. These communities underscore the interaction between (often racialized) austerity and water injustice.

In contrast, for the large majority of the population in wealthy nations, who are still able to access abundant supplies of clean tap water without interruption, the process by which widespread adoption of bottled water weakens public water systems is different: a spiral of growing distrust, inverted quarantine, and loss of public support.[20] "Protecting drinking water isn't just a matter of money," writes Elizabeth Royte. "It takes political will to allocate and spend it. But the more people . . . opt out of drinking tap water, the less political support there may be for taking care of public water supplies. . . . Committed bottled-water drinkers will have little incentive to support bond issues and other methods, including rate increases, of upgrading municipal water treatment. . . . It's a self-fulfilling prophecy: the fewer who drink from public supplies, the worse the water will get, and the more bottled water we'll need."[21]

However, the nature of bottled water's threat to public water is actually a bit more nuanced and complicated. Drinking and cooking together typically account for only a minuscule proportion of total household water consumption. Except in severe cases of toxic water, most people will not stop using tap water for taking showers, washing cars or clothes or dishes, watering gardens, or even, in many cases, cooking. Public water utilities have not experienced a meaningful *economic* threat from the wholesale shift by households to bottled water.[22] Thus most utilities have not fought back—at least so far.

Yet what is jeopardized by the inverted quarantine response of switching to bottled water, repeated across millions of households, is the political will for continuing to ensure that tap water is treated to the extremely high standards required for it to be safe to drink. "Water utilities are well aware that residents drink or cook with only 1 to 2 percent of the water that enters their home," explains Royte. "Why spend millions to bring water up to high standards . . . if so little is actually consumed?"[23] In other words, for most people in wealthy nations the risk of allowing disinvestment to continue is not that public tap water will someday cease to flow but rather that what comes out of the tap might no longer be reliably safe to drink for a growing share of the population. Tony Clarke, summing up the pioneering 1999 NRDC report on bottled water, wrote: "If society shifted to bottled water as the main source of drinking water, and public systems were therefore allowed to deteriorate, people would then be exposed to contaminants when they used tap water for showering, bathing, washing dishes, and cooking. In other words, tap water must never be allowed to drop below drinking-level quality or the public will be at risk."[24] To read these arguments—written a decade and a half ago as a dystopian vision—in the post-Flint present is unsettling to say the least.

BOTTLED WATER AND INEQUALITY

The third argument involves the relationship of bottled and packaged water to social inequality. There are two major aspects to this issue. As the stories of Flint, Six Nations, and Warm Springs attest, widespread dependence on bottled water is a key *marker* of water injustice. It serves as an indicator that the human right to water is being abridged or violated. In these settings, bottled water often comes to serve not as a short-term emergency fix in the immediate wake of crises and disasters, but rather as an indefinite, even permanent replacement for tap water. As multiple activists in Flint have stated, bottled water may act as a Band-Aid to patch the wound of unsafe water, but it will never be a cure. "The answer," O'Heaney argues,

> is investment in public drinking water, and access to safe, clean water. And the idea that any of these bottled water companies are going to reliably or sustainably fill the void in a community that doesn't have access to public water is laughable—it's off the wall. And if you ask people in Flint, "Would you rather have clean tap water or would you rather go and drive up once a week to a parking lot where they throw four cases of bottled water into your trunk, and you use that for cooking and bathing and all of these other kinds of things?" People would tell you they want safe, public drinking water access—not bottled water.

The other facet is how bottled water consumption *exacerbates* existing class, racial, and ethnic inequalities. In 2008, writing about the U.S., Royte asked, "Is a two-tiered [water] system—bottled for the rich, bilge for the poor—that far-fetched?"[25] However, since then her prediction has ironically been inverted. Today it is middle- and upper-income people who can typically count on safe tap water and are increasingly rejecting bottled water, while (as a result of austerity, disinvestment, systemic racism, environmental injustice, and/or predatory marketing) it is low-income people, communities of color, and recent immigrants who disproportionately distrust their tap water and spend the most on bottled water—both in absolute terms and as a proportion of their income. They must also continue to pay their tap water bills, which across the U.S. are rising sharply and already unaffordable for many.

This double truth—that dependence on packaged water both illuminates and intensifies social injustice—is what links communities such as Jackson, Mississippi,[26] East Porterville, California,[27] Six Nations, and Flint to parts of the global South where reliable access to safe tap water is similarly absent. In all these places, people haul heavy bottles and jugs from stores or refilling stations to their homes, while many also pay for the undrinkable water that (at least sometimes) flows from their taps. These settings also share other traits: government either has abandoned or is unable to fulfill its obligation to deliver consistent supplies of safe water, forcing people to depend on packaged water for drinking and cooking, and their higher spending on packaged water can represent a major economic burden. For the poorest people, however, even cheap refill water may be unaffordable.

Frederick insists that this packaged water paradigm is not sustainable anywhere: "The fact [is] that actually poor people cannot afford bottled water. They say it's a solution. No, it's not a solution. These guys cannot afford it, period. . . . The money that people are spending on bottled water, this amount of money would be enough to repair a public water system practically anywhere in the world. So we even have the finances to do that."

However, that very paradigm is being actively pursued for the global South by international institutions and governments. As I described in chapter 2, the definition of an "improved water source" for the purpose of meeting the UN Sustainable Development Goals was recently revised to include packaged water for the first time. "Qualifying access to bottled water as access to an improved water source has changed the rules of the game," writes Joshua Greene. "The model provides clear advantages to policymakers who are looking to achieve the SDGs without increasing national expenditures on water infrastructure. . . . The bottled water paradigm offers low-cost solutions for providing water to the poor . . . while the better-off receive superior service."[28]

Returning to North America, the connections between bottled water, social inequality, and water injustice so visible in Flint, Six Nations, and elsewhere are increasingly influencing the analysis and strategies of bottled water movements. Efforts like the Troubled Waters network and the Michigan coalition described in chapter 3 are making connections between racialized austerity, unsafe tap water, privatization, and groundwater extraction for bottling, leading some local groups fighting the bottled water industry to shift their analysis and strategies to embrace a broader social justice frame. Balkwill tells me that "frankly, part of the problem of the bottled water struggle in the past is it was largely a White middle-class environmental commodity kind of thing. It wasn't really a justice thing." However, he says, these struggles are now becoming "an essential fight that is rooted in justice, and that has the capacity to bring together lots and lots of people across demographics."

A MOVEMENT FOR DECOMMODIFICATION

My fourth argument relates to the nature of the challenge that bottled water movements pose to the packaged water industry. The range of (often seemingly disparate) countermovements that are opposing packaged water at both the water extraction and consumption ends of the commodity chain—much like the local and international movements opposing water utility privatization—collectively constitute an increasingly coherent and effective movement for water decommodification.

Several observers have used the concept of decommodification to characterize such challenges, expanding on sociologist Gosta Esping-Andersen's original use of the term as a response to the commodification of labor power.[29] John Vail describes decommodification as a wide range of processes that "insulate non-market spheres from market encroachments; increase the provision of public goods and expand social protection . . . and undermine market hegemony by revealing the market's true social costs and consequences."[30]

The broad sweep of movements challenging packaged water—from city-specific campaigns reclaiming the tap to rural-urban alliances linking toxic tap water to commercial groundwater extraction—has been partly successful in decommodifying water and obligating the industry to alter its accumulation strategies. This partial decommodification has happened through several routes: (1) reducing demand for this product by stigmatizing bottled water, expanding access to tap water in public places, and promoting refilling as an alternative; (2) blocking the industry's expansion to new water extraction sites, as in the cases of McCloud, Cascade Locks, and Elora; (3) raising the industry's costs and reducing its room to maneuver by waging years-long struggles over proposed new sites (Oregon, California, Ontario), or winning stricter regulations on water extraction, as in Ontario; and (4) inflicting reputational damage on valuable global brands. All of these dynamics can threaten profits and compel firms to shift strategies. The clearest example of such a shift is Nestlé's retreat from the North American bottled water market, with the CEO acknowledging that falling profits and environmental opposition were largely behind the move.

Polanyi, writing in the mid-twentieth century, likely could not have foreseen the conversion of water into a mass-market commodity through bottling, but his incisive arguments for rolling back market power in relation to the fictitious commodities are clearly relevant to this context. He vividly described how the harmful social and environmental repercussions of commodifying nature and labor inevitably generated powerful "movements of self-protection," which he also termed "countermovements" against commodification.[31] The constellation of local and supralocal movements that are contesting bottled water exemplify such countermovements for decommodification in the challenges and obstacles that they are posing to the expansion of the fictitious commodity of water.

STRONGER TOGETHER

My final argument relates to the importance of bottled and packaged water in the analysis and strategies of movements against water privatization. The growing centrality of packaged water in the process of water commodification on a global level means that this industry and this commodity should not be relegated to a subordinate role for water justice movements. As a soon-to-be $500 billion global industry, led by four of the largest multinational food and beverage corporations, bottled and packaged water can no longer be viewed as tangential or incidental to a "main story" of water utility privatization. The rapid growth of packaged water around the world is clearly hastening the overall conversion of water into a commodity, accustoming entire populations to accept that safe drinking water comes in plastic containers at high prices. Precisely because of its higher portability and profitability, this global market is growing not only faster than anticipated but faster than the private water utility industry, which it will almost certainly surpass in size within one to two decades.

Given those facts, why have the packaged water industry and the commodity of bottled water so far not been central to the demands and platforms of the global water justice movement, which has primarily focused on opposing water utility privatization? Speaking to several longtime water activists, I hear a range of answers. Richard Girard of the Polaris Institute says, "I've witnessed that, working in water justice movements around the world . . . the bottled water issue never really got in there as a key issue. . . . But I think the keys are the commodification, privatization, and also the fact that the biggest corporations in the world are making a huge amount of money off of selling this stuff."[32] Frederick argues that these two movements have not historically embraced one another's core issues because of their different constituencies: "The defense of the public water services, at least in the Southern countries, has been done more by trade unions, left-wing political parties, and so on. . . . But the bottled water fights have been done by very small local groups and mostly NGOs. And these two worlds don't go very well together. And we have to bridge these two movements, you know?"[33] Finally, I pose the question to Maude Barlow, who spearheaded the successful fight for the UN declaration of the human right to water. She responds, "It is very important for the water justice movement to incorporate bottled water into its platform, and [that] is why it is a key component of our Blue Communities campaign. Many in the movement just don't understand the connection between the denial of safe public water to billions and how the 'water hunters' come in with their answer of bottled water."[34]

There are already partial linkages between anti–bottled water movements and the broader global water justice movement, as demonstrated by the role played over the last two decades by several key organizations—especially Council of Canadians, Food and Water Watch, and Corporate Accountability—in campaigns focused both on the packaged water industry and on fighting water privatization internationally. Yet these connections have not been fully integrated into other international and domestic campaigns to oppose water utility privatization and defend the human right to water. This can be partly explained by the fact that the private water utility sector represents a different target: it is dominated by a separate group of multinational firms; fights over privatization are typically focused on specific cities and nations; and that industry's agenda of "public-private partnerships" has been implemented through different venues, including the World Bank and the World Water Forums. Yet as the global packaged water market has grown dramatically and as efforts by the Big Four to manage water globally have increased, these distinctions have become blurred. Frederick notes that the 2018 Forum in Brasilia "was the first World Water Forum that was fully sponsored by the bottled water industry—AmBev, Coca-Cola, Nestlé."

The linkages between these two movement coalitions can be greatly strengthened, and they can more explicitly embrace one another's issues. They have the potential to unite around a shared critique of the unique threat that bottled and packaged water poses to water justice, to public water systems, and to realization of the human right to water. The global water justice movement, writ large, should acknowledge bottled and packaged water as central to the global process of water commodification and as a key impediment to realizing the human right to water, not to mention a major environmental threat. In turn, movements challenging bottled water, whether focused on the consumption or extraction end, should reassess their own goals and vision. Local groups opposing water extraction by the bottling industry ought to embrace an understanding of their work as part of a larger international movement for water justice that intersects with other racial and economic justice struggles led by communities of color and low-income people, in both the global North and the South. They need to forge links with antiprivatization and water justice movements and their constituencies (public sector unions, urban water movements, Indigenous communities, and others) and consider how best to become allies to those sectors. Thus it is vital for these two movements against water commodification to embrace one another's critiques and forge firmer bonds. Their work will be stronger together.

Conclusion

> This [is] an issue for everyone. It is not a niche issue. Bottled water uses fossil fuels. Bottled water creates an enormous amount of plastic on our fragile planet. Bottled water privatizes water and disconnects people from nature. Bottled water is offered as a substitute for safe tap water, thus denying billions their human right to water.
>
> —Maude Barlow

Let us take a parting glance at the relationship between the fairly new commodity of bottled water and the much older (and still largely public) good of tap water. The fates of these two forms of drinking water provision are closely intertwined. The growing embrace of packaged water as an acceptable, or even a preferred, mode of drinking water has troubling implications, not only for the century-old project of universal public provision of clean drinking water in many countries, but also for the possibility of guaranteeing the human right to safe water worldwide. Because bottled water's growth is both an effect and a further cause of disinvestment in our tap water infrastructures, this commodity poses a unique challenge to public water. Packaged water, unlike all other beverages, advances the private enclosure of water globally because it delivers the very same substance—plain water—in small volumes, at much higher prices and with far greater environmental impact, and in the process also helps weaken the political will needed to sustain public provision of high-quality drinking water.

While the majority of citizen-consumers in the global North at present drink *both* bottled water and tap water, our choice to do the former is contributing to the decline of the latter. Similarly, on a global level, disinvestment or underinvestment in safe public drinking water supplies renders water more amenable to commodification by the packaged water industry but less accessible to the majority of humanity, to whom it is essential for life. However, the inverse is true as well. "If everyone on the planet had access to affordable safe tap water," argues Peter Gleick, "bottled water use would be seen as unnecessary."[1]

RECOMMENDATIONS FOR CHANGE

So what should be done? I now turn from the analytical realm to the pragmatic, to outline a series of recommendations for curtailing the negative social, economic, and environmental impacts of bottled and packaged water and for countering the threat its growth poses to public water systems. They range from the reformist to the radical, from low-hanging fruit to highly aspirational, and from the local to the national and even international levels. Some of these suggestions are appropriate only in settings where the great majority of people have access to clean tap water, while others apply more broadly. Many of them have been generated and promoted by local activists or consumer and environmental groups, and many are already being implemented in communities around the world. Rather than presenting a single comprehensive, internally coherent platform, my aim here is to outline a broad menu of potential changes—to provoke thought and debate among consumers, activists, academics, local officials, policy makers, and others about the range of actions that are needed to loosen the grip of bottled water on society.

Reclaiming and Expanding the Tap

The first set of recommendations builds on the dynamic efforts by local governments, nonprofits, and even private actors to reclaim the tap that are described in chapter 4. Many of these models have already been widely implemented and are easily adaptable, making this the lowest-hanging fruit of all. Most of these initiatives are applicable in places where safe tap water is already widely available.

First and foremost, local and state/provincial governments (as well as public, publicly funded, and nonprofit institutions, including universities) should eliminate—or expand existing bans on—the purchase and sale of single-use bottled and packaged water in their facilities. Using public property to sell a product that undermines trust in tap water and public water utilities should not be permitted. While many bottle bans in the past were the result of one-off local efforts, this work is increasingly being coordinated by international efforts—notably the Blue

Communities campaign—that provide a template for localities to enact new bans. But there is room for more. Following the example of towns in Massachusetts and a few other places, local governments should also consider taking the next step and prohibit sales of single-use plastic packaged water in all private and public settings. The inevitable legal challenges will provide a valuable platform for debating the necessity of packaged water where clean tap water is available. Beyond this, existing and new governmental bans on single-use plastic products should be expanded to include water bottles (and all beverage containers), which constitute a major portion of the global plastic pollution problem. We will never solve the single-use plastics crisis without eliminating single-use plastic water and beverage bottles. Political resistance from industry cannot be allowed to eviscerate this crucial reform.[2]

The counterpart to restricting bottled water sales in public places, of course, is to expand the availability of free drinking water in those same places. The innovative work by San Francisco, New York, Sydney, Berlin, Paris, London, and other cities to increase access to tap water offers a promising set of blueprints. But such initiatives must become mainstream and ubiquitous. "If we're committing to public health," says Gay Hawkins, "what states and governments should be doing is intensifying people's access to free water in public. We should see more water fountains everywhere, and they should be clean and readily available all over urban space."[3] Forward-thinking municipal, state, provincial, and federal governments need to encode this expansion of public water access into legislation and policy, and support it through taxes.

More broadly, governments must mandate the right to access free, filtered drinking water in public spaces. The European Union's new Drinking Water Directive, which requires member states to do exactly that, is a promising development. However, the directive does not extend to crucial privately owned spaces used by the public, including restaurants. In some countries, diners can typically expect to be served free tap water, but that is not the case everywhere. At the World Water Forum in Brazil, Catarina de Albuquerque, former UN Special Rapporteur on the Human Right to Water and Sanitation, told an audience:

> Each time I go to a restaurant in a place where the tap water is of good quality, I say that I want tap water. I was in Geneva, Switzerland, two weeks ago . . . and I asked for a glass of tap water. And the waiter refused. And he said the legislation in Switzerland allows for the restaurant owner to refuse to give you tap water and force you to drink bottled water. And I said, "But look—sell me the tap water. I understand that it costs you money to have

> somebody washing the glass, you have to pay your water bill, da da da—let me pay for the tap water." And he said, "No, the law allows me to only allow *you* to drink bottled water."

Despite the urgings of water campaigners, the final EU directive does not require restaurants to provide tap water "because the traditions in the Member States vary widely"[4]—a major missed opportunity. Mandating such a change is essential, because this is a key place where norms of consumption can be shifted.

An even more crucial site for shaping social norms is schools. After the widely successful campaigns that removed sugary soft drinks from most public schools in the U.S. and other nations, the beverage industry's profits—and its efforts to "brand" young consumers—have been partly sustained by replacing soda with bottled water in school vending machines.[5] In cities where bottled water has filled the gap created by lead-laden pipes and fiscal austerity, school budgets (and taxes) are being used to supply bottled water to children year after year, while shut-off water fountains sit rusting. It is unacceptable for schools to be providing children and youth with plastic single-use bottled water—effectively teaching them that water comes from a bottle—where the tap water is potable or can be rendered safe to drink by replacing internal piping and installing lead-removing filters. Thus the next logical fight must be to restore safe public drinking water to *all* schools and then to remove bottled water from schools. Families should take the lead in demanding that governments prioritize schools in allocating funding to fix drinking water infrastructure.

In addition to the repair of pipes in old buildings, new ones must not be built without full access to free tap water. It is critical that governments reform building codes to mandate abundant water fountains and refilling stations in all new and existing buildings—not just stadiums, airports, train stations, libraries, schools, and universities, but also offices, stores, and apartment buildings. Codes that fail to ensure free drinking water access are hard-wiring dependency on bottled water into our built environment, a state of affairs extremely difficult and expensive to reverse.

Next, public water utilities, and the local governments they are part of, should launch or expand major campaigns to promote the quality of their cities' tap water, encourage refilling, and discourage single-use plastic water. These efforts need to use the tools employed by capitalist firms—advertising (social marketing), public relations, branding such as city-specific logos on free refillable bottles—and spend what it takes to do that well. Cities should also make free or subsidized point-of-use water filters available to all residents to address concerns with tap water taste and quality. In short, local leaders and water utilities need to stand up and push back against the largely one-sided war on tap water, acknowledge the stakes in this fight, and summon the political will to counter it. They should also partner with nonprofit environmental, public health, and community organizations who have a stake in the outcome of these struggles.

Linked to this are crucial reforms to guarantee the right to water. Shutoffs of household water service must be outlawed across the board—they are an egregious violation of the human right to water, disproportionately affect communities of color, and tragically increase bottled water consumption, as people without running water have few other options. An adequate minimum volume of water—sufficient for drinking, cooking, bathing, and washing—must be guaranteed to everyone regardless of ability to pay. Income-based water and sewer rates (like those in Philadelphia and Baltimore) should be adopted by all water utilities and mandated nationally, so that combined water and sewer bills remain below the affordability threshold of 4.5 percent of income for all households.[6] Low-income residents' water bill debts must be canceled, and all currently shut-off water service restored. Quite simply, access to clean drinking water must never be dependent on the ability to pay.

Finding the Funds, Fixing the Pipes

Most of the above recommendations require funding. After decades of neoliberal austerity, where are fiscally stretched local governments to find the money? The U.S. Congress in 2021 did pass historic investments in water infrastructure, some of which will reach local and state governments, but water justice advocates insist this does not begin to match the full scale of the problem. Many groups express a vision of a permanent fund to pay for direly needed water infrastructure improvements. Mary Grant of Food and Water Watch tells me that "at the federal level, our big vision is to set up a trust fund for water systems across the country, water and sewer systems and household wells and septic systems. Really fully funding our water infrastructure nationally, restoring that federal government commitment to safe water for all."[7] National governments must fully fund the cost of local water infrastructure improvements and deferred maintenance, including removal of all lead pipes and service lines, and also subsidize local water utilities' operating costs so they are not obliged to raise water rates to backfill for missing federal funds. In the United States, Congress must at least restore the peak federal funding levels of the late 1970s (adjusted for inflation and population growth) and provide the great majority of money to localities in the form of grants, not loans. In allocating these funds, the highest priority should go to restoring water systems in structurally disadvantaged communities, particularly those with predominantly Indigenous, Black, and Latino/a residents, as well as small water systems, all of which have a disproportionate share of water quality problems. These investments will make water systems far less prone to breakdown and more resilient to extreme weather, significantly reducing unsafe water events as well as dependence on bottled water. Passage of the WATER Act or similar legislation would be a major step toward realizing this broad vision.

There are other revenue options not dependent on federal spending. Local, state, and provincial governments should impose taxes on single-use bottled and packaged water in order to restore public water systems, address the product's negative externalities, and disincentivize consumption. In the U.S., the clearest model is Chicago's five-cent tax on bottled water, enacted in 2008. In its first five years the tax raised $36 million,[8] a figure sure to grow after its recent expansion to include online sales. On a larger scale, Washington State in 2010 "imposed" a sales tax on bottled water in 2010 (by removing its exemption from the state's sales tax). Although it was rescinded after only six months, in that time the tax caused bottled water sales to drop by more than 6 percent.[9] Akin to local or state taxes on cigarettes or sugary soft drinks that fund programs to mitigate the health harms those products cause, bottled water tax revenue could be used to expand and repair drinking water infrastructure including fountains and refill stations in schools and parks, and/or to compensate local governments for the waste collection and recycling costs imposed by billions of single-use plastic water bottles. Journalist Laura Bliss writes that we need to "tax bottled water, use the money to upgrade our water systems, and change the country's mindset."[10] Whatever the end use, the principle is clear: tax what we need less of and use the proceeds to fund what we need more of.

Stopping the Plastic Tsunami

Next come policies to tackle the devastating effects of plastic bottle pollution on water, air, climate, ecosystems, and human health, and on local governments' efforts to manage waste. Many of these ideas dovetail with larger initiatives to reduce or halt the use of all single-use beverage containers, of which water bottles represent the largest share. Some of these approaches fall under the rubric of "extended producer responsibility"—policies that obligate corporations to bear the burden of collecting and recycling their products after use, creating incentives to change product design and shift toward reuse.

I somewhat reluctantly include here legislation to dramatically increase both bottle recycling and recycled content, such as California's mandate of 50 percent recycled plastic content by 2030 and the EU requirement for a 90 percent recycling rate by 2029. Despite repeated promises, the beverage industry has long resisted such changes. While forcing industry to create bottles from recycled rather than new material will reduce carbon emissions a sliver, save landfill space, and ramp up recycling programs, it would also extend the shelf life of the destructive single-use plastic paradigm. If enacted more broadly, these policies must only be a stopgap on the path to far more ambitious changes.

In tandem with the above, governments must enact or substantially expand container deposit laws or bottle bills, which are the most effective way to increase recycling and recovery rates. Bottle bills must include all types of bottled and packaged water (and all beverage containers), and deposit fees must be set at levels that achieve very high container recovery rates like those in Germany, Oregon, and Michigan. The entrenched industry opposition to new and expanded bottle bills must be broken. According to Story of Stuff's Michael O'Heaney, "The bottle bill thing is really important. The reason that companies hate it, is it sends a negative price signal. It adds a nickel or a dime to each bottle of water. And so the companies don't like that—even though the consumers get it back, you've still got to lay down the money for it."[11] Bottle bills also place the onus on individuals to collect and return single-use containers and on cities to recycle them, while producers should at least partly share the burden. So this approach, while vital, is also a stopgap.

For beverages as a whole, the true solution is to phase out single-use plastic beverage containers entirely—to end the "no deposit, no return" era forever—as part of a comprehensive movement to eliminate all single-use plastics. One crucial step toward this goal is legislation mandating beverage makers to shift to 100 percent reusable, refillable containers, with clear interim targets for the transition process. For example, Germany's 2017 packaging law includes a 70 percent quota for refillable beverage and food containers by 2022.[12] Beverage return and refilling systems with refundable deposits are not a pipe dream—they were omnipresent just a few decades ago, and they still account for almost a fourth of beverage containers worldwide, although they are shrinking or nonexistent in many countries.[13] As an incentive, refillables could be exempted from taxes or fees on plastics and on packaged water. However, the role of bottled water (as opposed to other beverages) in the refilling picture is complicated. In much of the global South, twenty-liter jugs and microtreatment plants already represent deposit-based refilling systems, but refillable packaged water must be viewed only as an interim measure while water systems are fixed or expanded—never as a permanent solution. The same applies to parts of the North where the tap water is unsafe because of pollution or disinvestment. By contrast, in places where safe tap water is widespread, the investment and energy should be focused on expanding access to drinking water in public places and keeping water quality high—not on converting packaged water to refillables. In those settings, having the industry switch to refillable containers for plain bottled water would be essentially a form of greenwashing.

Reining in the Industry

The practices of the bottled water industry must also be far more strictly scrutinized and regulated. First, any advertising for bottled or packaged water that makes claims implying or stating that it is safer than tap water, or disparaging the quality of public tap water, must be prohibited. Similarly, any product claims regarding health or fitness benefits, including those related to hydration, must be outlawed. The European Union since 2011 has banned bottled water firms from making any hydration-related claims.[14]

Second, the regulatory approach to bottled water safety must be radically overhauled. At a minimum, the playing field must be leveled between the typically weaker regulations and scrutiny required of bottled water firms and the more stringent ones that apply to tap water utilities. In the U.S., several specific reforms are direly needed. The loophole exempting bottled water that does not cross state lines from federal regulations must be promptly closed. Authority for regulating bottled water should be transferred from the FDA to the EPA, which regulates tap water, and staff levels increased to be commensurate with drinking water utility programs. The frequency and rigor of water testing and inspection must be increased to match that of tap water utilities, and the results of those tests should be quickly made public. Product recalls for contamination must be automatic, swift, and publicized, with repeat violators fined or stripped of their licenses. Bottled water firms, like water utilities, must be required to publish frequent online reports detailing contaminant levels (including microplastics) and any regulatory actions including recalls or fines. If bottled water does continue to exist, it should not be allowed to undermine public tap water by being subjected to weaker regulation and flying under the radar when it comes to safety.

Third, the water extraction practices of the industry must be tightly regulated and reformed. In the U.S. and Canada, regulatory frameworks vary widely between states and provinces. In many cases bottlers can extract water for free or nearly free, existing wells or springs can be transferred from agriculture or small businesses to high-volume commercial water bottling with little scrutiny, and new extraction permits can be approved with little regard to their potential impacts on aquifers, aquatic ecosystems, fisheries, or nearby residential and agricultural users. All water extraction proposals—whether for new wells or boreholes, or existing ones regardless of their current use—must be subject to robust environmental impact assessments. Permit lengths must be short (three to five years maximum), strict monitoring and maximum volume levels should be imposed, and permit renewals must take into account changing climate and hydrological conditions as well as permittees' compliance records. Groundwater regulations must treat consumptive commercial water extraction—water "mining" that permanently removes water from local or regional watersheds—differently from other uses that return it to the same basin. Exemptions allowing large-scale export or removal of packaged water from regulated watersheds—such as the bottled water loophole in the Great Lakes Compact—must be eliminated.

Where commercial groundwater bottling is permitted to continue, steep royalties should be imposed on such water taking to defray its social and ecological costs, not merely the cost of administering regulations. Peggy Case of Michigan Citizens for Water Conservation points to a bill introduced in 2020 by Democratic state legislators in her state: "This model is first of all saying that companies like Nestlé that want to extract water should pay a royalty on it. . . . [It's] the same thing that the oil and gas industry does—you get a royalty. And that money has to go into a public trust that is focused on infrastructure and equality."[15]

Ultimately, however, national, state/provincial, and even local governments should phase out all existing commercial bottling of spring water or groundwater and prohibit all new such extraction, as the 2020 Washington State legislation would have done.

That leaves the issue of companies that bottle already-treated municipal water, which in the U.S. constitutes the majority of bottled water sales. Some city water utilities—particularly in deindustrialized cities and wetter regions—have excess water treatment capacity, and some have come to rely on the rates paid by bottlers to balance their budgets or reduce rate hikes. "A lot of cities are actually facing a crisis," says Grant, "where . . . the water efficiency standards have decreased the per capita usage of water, which is a great thing. But for utilities that means declining revenue, which makes them even more dependent on these big bulk buyers like the bottled water industry." However, allowing corporations to bottle public tap water and sell it as still bottled water undermines those same water agencies by communicating to the public that packaged water is trustworthy while tap water is not. Local governments should take several steps in response. Taxes on bottled water sales should be instituted and used at least partly to replace the revenues cities currently raise from selling tap water to water bottlers, allowing them to end such contracts. Bottled water firms should also no longer enjoy bargain-basement deals on their prime raw material. In fact, cities should set water rates for commercial water bottling higher than standard residential and business rates to reflect the high social and environmental costs of this product. Gleick tells me that "if I were a water utility and I had a bottled water plant sucking down hundreds of thousands or millions of gallons of water, I would charge them a special rate, a higher rate, because they're driving demand for water."[16] Since refiltered tap water is a low-cost market segment, this step alone may well deter some bottlers. Moreover, packaged water firms should not be allowed to bottle municipal water in arid and water-stressed regions.

But in the end, the bottling of already-treated municipal water for profit must come to a halt. If some cities are determined to be in the water bottling business, it should be limited to sales or distribution by public agencies and nonprofit groups only, in refillable containers. Critics will respond that beer, soda, and juice makers also use reprocessed tap water to make their products. It is true that firms like PepsiCo and Coca-Cola use municipal water to make a wide range of drink products, not just Dasani and Aquafina, although bottled water sales now far surpass those of soft drinks. However, as I have argued above, water is different. Plain bottled water purports to be a replacement for tap water—the provision of which is a fundamental task of local government. It implicitly persuades us that our tap water is unsafe or untrustworthy. Beverage firms should not be allowed to use public water supplies to undermine public water itself.

Ending Bottled Water's Emergency Clause

Ultimately, there is a compelling case for eliminating all forms of plain bottled and packaged water sold for profit—ending its manufacture and sale entirely in places that have widespread safe tap water, and phasing it out while repairing and extending public water systems in places that currently do not. Typically, the critics of bottled water who make this same argument go on to add a proviso: "except in cases of emergency." Yet while natural disasters may be unpredictable, water crises such as Flint are socially and politically created (although their human effects are quite real), and states of "emergency" rarely have clear endings. There are alternative ways to ensure access to safe water even in disasters that render tap water unsafe or damage supply networks. "Even in emergencies—bring in water tankers," argues O'Heaney. "It irritates me to no end when you see these companies bring in their truck with the branding on it, and they hand out cases of bottled water. You're telling me FEMA doesn't have water tankers?" Yet this infrastructure is currently far from universal. National governments should mandate the use of bulk water tankers and water stations in disaster response plans, allocating the funds needed for states/provinces and localities to build up this vital public sector capability. Without the ability to lean on the crutch of plastic water, the logistical capacity for bulk water delivery would already have been well developed. Where safe bulk water sources are available, single-use bottled water should be the last recourse in emergencies, not the first.

Removing Packaged Water from Global Goals

Finally, packaged water must not be counted by international institutions as an improved drinking water source. It is true that in many parts of the world, significant numbers of people currently have few or no sources of safe drinking water other than packaged water. Yet acknowledging this reality is utterly different from validating it. Widespread dependency on packaged water is a signal of global water injustice—a damning indictment of the failure of states and suprastate institutions to provide everyone with access to clean public drinking water. Including this commodity in the UN Sustainable Development Goal 6 is not merely a surrender but a ratification of growing corporate control over fresh water. According to Joshua Greene, "The [packaged water] industry is in a phase of expansion and the new discourse by international agencies to include bottled water as a category of [improved water] access is recognition and de facto endorsement of this expansion."[17] Yet when a vital human need is commodified, access to it is based on the ability to pay, meaning millions of people will be forced to go without. Including packaged water in global goals gives governments a permission slip to abandon or ignore their vital obligation to provide this human right to everyone, and it must be reversed.

. . .

In the memorable book *Hope in the Dark,* Rebecca Solnit writes that "what we dream of is already present in the world."[18] While the task of loosening the grasp of bottled water on our societies and planet is a daunting one, the early blueprints for doing so are already available. Almost all of the ideas listed above are currently being implemented in communities and nations around the world, providing a fertile seedbed for experimentation. These models are being shared, adapted, and adopted, and residents and activists can pressure public officials to enact them far more widely. Taken together, they suggest an outline that may enable us to imagine a different, more sustainable, and more just drinking water future.

RESTORING A VISION OF PUBLIC WATER

At this point, even some sympathetic critics may object that the genie cannot be put back into the bottle—that it is unrealistic to advocate a return to a world where billions of people do not depend on packaged water to drink, or a repeat of the era of massive state spending on water infrastructure that characterized the postwar global Keynesian project. In other words, they might continue, governments simply cannot or will not allocate the trillions of dollars needed to fix our decaying drinking water systems, let alone build them out where they don't yet exist. Or they may argue that if such a prospect is at least imaginable in wealthy nations, it is naive to envision most Southern governments constructing massive new centralized piped water systems or even restoring unsafe tap water to potability. Like it or not, they may conclude, packaged water is a necessary evil, providing an imperfect solution to a sorry state of affairs.

To these critics, I would ask: On what basis should access to water be triaged? On which moral or ethical grounds can a large portion of the world's people plausibly be excluded from the enormous benefits of access to safe and reliable tap water that are enjoyed by the large majority in wealthy nations—consigned to live permanently in a costly packaged water world?

It is true that the project of rehabilitating and extending large-scale public drinking water systems is a massive and costly undertaking. Nor is it necessary to romanticize public or state management of drinking water, which in some settings can be plagued by inadequate investment, regressive rate structures, inconsistent service, unresponsive or corrupt management, and/or poor water quality. Where community-managed (as opposed to public sector) water systems exist and function—many of them in rural communities or neighborhoods unserved by public water networks—they should be firmly defended and supported. More decentralized approaches, including rainwater harvesting, should also be encouraged and expanded. Nonetheless, for the great majority of people living in urban areas worldwide, large-scale, well-maintained piped water utilities are far and away the most efficient, economical, and sustainable way to deliver clean water to large numbers of people. "That's why we *have* water systems," says Grant, "because that's a cost-effective way of making sure people have access to safe water."

The primary causes of our collective inability to imagine restoring and extending such a massive public endeavor are not new: four decades of neoliberalism and austerity in both North and South; stratospheric global inequality; the ongoing extraction of enormous wealth from the South through debt, unequal exchange, and other mechanisms; brazen tax avoidance by the rich; corporate capture of governments; and the relegation of swaths of the rich nations (primarily poor and/or non-White) to second-class status—populations who are not entitled to expect reliable, safe drinking water to flow when they turn on the tap.

And the possible solutions to this state of affairs are not new either. They include many of the same demands that movements for global economic justice have been making for decades. For the global South, this includes forgiveness of external debt without conditions, an end to the systematic theft of wealth through unfair terms of trade and unjust investment rules, reparations for the damage wrought by structural adjustment, and climate debt payments for greenhouse gas emissions that have enriched wealthy nations while devastating the poorest. For all countries, it includes more heavily taxing wealth, corporate profits, financial transactions, and ecologically damaging industries and goods. The wealth exists. And among the many direly needed uses for this money is a renewed commitment by governments (pushed and scrutinized by social movements) to provide clean, affordable, publicly delivered drinking water for all people, fully realizing the fundamental right declared by the UN in 2010. The alternative—increasing dependence on costly private packaged water by a growing portion of the world's people—would be a denial of that right. According to Lauren DeRusha of Corporate Accountability,

> We have to invest in our water systems, and . . . we have to start speaking up. Because the other choice is to throw our hands up in the air and rely on these abusive corporations to provide water for us for the rest of our lives. And we know that these corporations, they'll provide water to people who can pay the most. . . . I think it requires us all to [have] a vision, and not to fall prey to the narrative that the bottled water industry would like us to believe, that we're living the dystopian future and we have no other choice.

. . .

The commodification of water is an essential part of the historical and ongoing efforts by capital to commodify nature—indeed, to commodify virtually everything. Bottled and packaged water is central to the process of private appropriation of water by powerful interests; it constitutes a major element of the accelerating global fresh water grab. Packaged water is not a parenthetical side story to this enclosure of the world's increasingly scarce and valuable fresh water—on the contrary, it represents a more complete form of enclosure that is rapidly growing. Similarly, the movements opposing water extraction by the bottling industry, and contesting the consumption of packaged water, form part of the long tradition of resistance movements against the enclosure of the commons by the wealthy and powerful reaching at least as far back as sixteenth-century England.[19] They also embody the vibrant resistance to accumulation by dispossession, especially to the grabs of public goods and nature that have proliferated worldwide in recent decades.

Thus the two facets of the movements opposing water commodification—those fighting to halt or reverse privatization and deterioration of public tap water systems, and those challenging the bottled water industry at both the extraction and consumption ends—

are not merely distant cousins but siblings. They are talking about and working for the same thing. Their differences are minor, and their work will be stronger together.

Although capital has asserted substantial if highly uneven control over water worldwide, the hope for achieving genuine water justice ultimately rests in the hands of states, communities, civil society, and social movements. The unique nature of water—unsubstitutable and essential for life—explains the great intensity and vitality of the movements for water justice that are opposing private control of municipal water and contesting the growth of packaged water worldwide. These movements insist that fresh water must remain part of the commons, that it is far too precious to entrust to the market. They have proven partially successful at slowing, preventing, and reversing both the privatization of public tap water and the commodification of water through bottling. They constitute a collective force that is rendering the fictitious, yet quite real, commodity of bottled water somewhat less perfect.

Finally, let us return to the question that I posed at the outset. Do we agree that affordable access to clean drinking water for all people should be a fundamental human right, an essential part of the social contract, or do we not? If the answer is yes, this right to drink will never be realized or delivered through the market, and it certainly will not come in a plastic bottle.

Acknowledgments

So many people contributed to making this book possible over many years that it is impossible to thank them all individually. I am greatly indebted to Jane Collins and Gay Seidman, who read the entire manuscript as well as early chapter drafts, and to Lucy Jarosz, who read and commented on several chapters; all of them offered invaluable suggestions for strengthening the book. My thanks to the talented Lisa Colleen Moore for helping me to shorten and improve the text substantially. I am also grateful to Robert Case, Benjamin Pauli, and Michael O'Heaney for their valuable comments on individual chapters.

I want to express my deep gratitude to all of the community members, volunteers, grassroots activists, local officials, and many others who generously shared their experiences and helped to facilitate this project in myriad ways. In Ontario, sincere thanks to the staff, board, and members of Wellington Water Watchers and Save Our Water, especially Arlene Slocombe, Mike Nagy, Mike Balkwill, Donna McCaw, and Jan Beveridge, and also to Rod Whitlow. In Oregon, my huge appreciation to the volunteers and leaders of the Local Water Alliance, as well as to Klairice Westley, Whitney Kalama, and Julia DeGraw. A big thank-you also to the photographers and designers who granted permission to publish their fine work: Blue Ackerman, Will Doolittle, Aurora del Val, Laura Amendola, Troy Bridgeman, Paul Dimock, Federico Olivieri, KatsanDesign, and Tony Biddle, and to Glynnis Koike for the cover illustration.

I am grateful to everyone at the Centre for Sustainability at the University of Otago, New Zealand, for generously providing a great community during a sabbatical visit that was unexpectedly prolonged by the pandemic, and office space to write much of the book in. Special thanks to Hugh Campbell, Marion Familton, and Karly Burch for their sharp insights, warm welcome, and camaraderie.

I want to acknowledge the following publishers for permission to use short sections from my previous work (much of it in revised form), and thank my coauthors. Portions of "A Bottle Half Empty: Bottled Water, Commodification, and Contestation," published in *Organization and Environment* (© 2013 Sage), and "A More Perfect Commodity: Bottled Water, Global Accumulation, and Local Contestation," published in *Rural Sociology* (© 2013 John Wiley and Sons), both coauthored with Soren Newman, appear in chapters 1 and 5. Extracts from "Draining Us Dry: Scarcity Discourses in Contention over Bottled Water Extraction," coauthored with Robert A. Case, published in *Local Environment* (© 2018 Taylor & Francis), appear in chapter 6. Segments of "Enclosing Water: Privatization, Commodification, and Access," published in *The Cambridge Handbook of Environmental Sociology* (© 2020 Cambridge University Press), appear in the Introduction, chapter 1, and the Conclusion. Full references for these publications are in the References section.

The research for this book was partially funded by two Faculty Development Grants from Portland State University.

My sincere thanks to the wonderful staff at UC Press, especially Aline Dolinh, Summer Farah, Emily Grandstaff, LeKeisha Hughes, Teresa Iafolla, Katryce Lassle, and Emily Park, and also to copyeditor Elisabeth Magnus and indexer Cynthia Savage. Naomi Schneider, my editor, deserves the biggest thanks of all for her unflagging belief in this project from the start, despite numerous delays. I count myself extremely fortunate to have benefited from her support and guidance for two books and almost two decades. A writer could not hope for a better editor and advocate.

To my sister Lisa Jaffee, my heartfelt thanks for her unconditional support and hilarity over all these years. Dawn Nafus, my partner, first reader, and best critic, read through drafts of every chapter with surprisingly good cheer and made amazing suggestions—and she even came up with the main title. I am grateful for her always. This book is dedicated to my mother, Iris Jaffee, a microbiologist who worked for many years with the Seattle Water Department. I wish that she were still here to share it with. Her concern for social justice and her belief in the vital role of safe public water helped to shape my values and these pages.

Notes

PREFACE

1. Gina Luster, interview, November 2019.

2. Ellison 2019.

3. WEYI TV 2022. In 2021, Nestlé sold its North American bottled water brands and operations to a private equity consortium that operates under the name BlueTriton Brands.

4. MST 2018; de Mora e Souza 2018.

5. Frederick 2018; Dawson 2018.

6. Watts 2018.

7. Browdie 2018.

8. Kretzman and Joseph 2020.

9. Oaten and Patidar 2020; Sunder 2021; R. Brown 2022.

10. IBWA 2020b; Shoup 2017b; Felton 2019b.

11. Statista 2022b.

12. Statista 2022b; Grand View Research 2022.

13. Szasz 2007; Gleick 2010.

14. Food and Water Watch 2018a.

15. Laville and Taylor 2017.

16. Laville 2020; PET Planet 2022.

17. The concept of *water justice* has been defined in a wide range of ways by different organizations, researchers, and other actors. Water governance scholars Margreet Zwarteveen and Rutgerd Boelens (2014: 154) argue for a definition of water justice as a relational concept that incorporates not only material and economic dimensions but also cultural and political ones.

INTRODUCTION

The chapter's epigraph is quoted in Wagenhofer (2005).

1. IBWA 2022.

2. Drewnowski, Rehm, and Constant 2013.

3. Statista 2022b; IBWA 2022.

4. Rodwan 2019, 2021; Statista 2022b.

5. Euromonitor International 2022b. In Europe and Canada, however, the vast majority of bottled water comes from springs and groundwater.

6. Some bottling firms employ water hunters, typically hydrologists or hydrogeologists, who scour the planet in search of clean, accessible water sources (Conlin 2008b).

7. Barlow 2019; 2030 Water Resources Group n.d.

8. See, e.g., Rodwan 2019.

9. The term *Majority World,* connoting the formerly colonized nations that are home to the majority of humanity, was coined by Bangladeshi photographer Shahidul Alam, with *Minority World* equating to the wealthy nations of the former First World (*Masalai Blog* 2009). In this book I utilize *global South* and *global North,* which, although imperfect, align with the terminology used by the great majority of sources quoted and cited here.

10. Hawkins 2017.

11. Girard 2009; J. Greene 2018; Prasetiawan, Nastiti, and Muntalif 2017.

12. Gleick 2010; Parag and Roberts 2009.

13. A bacterial outbreak in the Walkerton water system in May 2000, due to inadequate treatment after heavy rains, contaminated a supply well with bacteria from cattle manure, causing six deaths and sickening over two thousand people.

14. Readfearn 2018; Gleick 2010; Felton 2019b; Tyree and Morrison 2019.

15. As of 2017, 62 percent of U.S. bottled water came from municipal supplies (Antea Group 2018).

16. Gleick and Cooley 2009; Laville and Taylor 2017.

17. Schnell 2012; Prasetiawan, Nastiti, and Muntalif 2017; Raman 2010.

18. Scholars disagree whether drinking water should be considered a public good. Bakker (2010) argues that water is neither a public good nor a private good, calling it a common-pool resource. In contrast, Vail (2010) calls municipal water supplies a clear example of a public good. Both these positions have salience. Although tap water does not meet economists' formal definition of a public good—which includes the criteria of nonrivalry (consumption by one individual does not detract from use by others) and nonexcludability (prohibiting use is difficult or impossible)—universal access to clean tap water is clearly a common good with great societal benefit.

CHAPTER 1

1. Barlow and Clarke 2002; Abrams 2014.

2. Fergusson 2015.

3. Zwarteveen and Boelens 2014.
4. Goff and Crow 2014.
5. The UN SDGs define an improved water source as "water accessible on premises, available when needed and free from contamination" (United Nations 2018). However, this is not synonymous with a safe drinking water supply; approximately one billion people have access to an "improved" water source but not to safe water (Walter, Kooy, and Prabaharyaka 2017: 642).
6. WHO and UNICEF 2021; UN News 2020.
7. UNDP 2006.
8. Zwarteveen and Boelens 2014: 143.
9. Mehta 2016.
10. Bakker 2010.
11. Malkin 2012.
12. Narain 2014.
13. Jasechko and Perrone 2021.
14. UN Convention to Combat Desertification 2022.
15. Bond 2008.
16. Subramaniam and Williford 2012.
17. Bakker 2005.
18. United Nations 1992.
19. Goldman 2007.
20. Castro 2007.
21. As of 2010, the global private water market had annual revenues of US $507.5 billion, of which $395.6 billion represented private water utilities—the statistic most relevant to this discussion (Arup 2015). In 2018, the private water market was estimated at $770 billion, but that figure includes sewage utilities and spending by industry on water and wastewater treatment. As of this writing, no reliable current estimates for the size of the worldwide private water utility market alone are publicly available, but a figure of approximately $500–600 billion is plausible.
22. Arup 2015.
23. Harris 2013.
24. Goldman 2007; Conca 2008.
25. Bakker 2005. The water utilities in Scotland and Northern Ireland were not privatized.
26. Goldman 2005.
27. Bakker 2014: 474–75.
28. Arup 2015.
29. McDonald 2016: 109.
30. Bakker 2014; Castro 2007.
31. Bakker 2014.
32. Bakker 2005: 559.
33. Harvey 2003.
34. Jaffee and Newman 2013b.
35. Claudia Campero Arena, water campaigner, Blue Planet Project, interview, January 2016, Mexico City.

36. Bakker 2007.
37. Castro 2008; Swyngedouw 2005.
38. World Bank 2005.
39. Ruiters 2007; Whiteford and Whiteford 2005.
40. Kurland and Zell 2011: 329.
41. O'Reilly 2011.
42. Ahlers 2005: 57.
43. Thara 2017: 265.
44. Castro 2008; Spronk 2015.
45. Olivera and Lewis 2004; Driessen 2008; Spronk and Webber 2007.
46. Spronk 2015.
47. Barlow 2014; Lobina, Terhorst, and Popov 2011; Nelson 2017.
48. Snitow, Kaufman, and Fox 2007; Barlow 2007.
49. United Nations 1948.
50. Barlow 2019: 48.
51. Barlow 2019; Lederer 2010.
52. United Nations 2010.
53. Bakker 2010; Sultana and Loftus 2012.
54. UN-Water 2018; Karunananthan and Tellatin 2016.
55. Hall, Lobina, and Corral 2011; Vidal 2006.
56. Esterl 2006.
57. Loftus 2009.
58. Loftus 2009: 957.
59. Bakker 2010.
60. Arup 2015: 5.
61. Kishimoto, Steinfort, and Petrovic 2020.
62. Hall, Lobina, and De la Motte 2009; Harris 2013.
63. Driessen 2008; Spronk 2015.
64. Food and Water Watch 2012b.
65. Snitow, Kaufman, and Fox 2007: 5.
66. Cutler and Miller 2005: 3.
67. Varghese 2007.
68. Esterl 2006; Snitow, Kaufman, and Fox 2007; J. Robinson 2013.
69. American Water Works Association 2012; Food and Water Watch 2022.
70. Barlow 2019: 75.
71. Peck 2015.
72. Sloan 2016; Gurley 2021.
73. Food and Water Watch 2012a. *Private equity* refers to investment firms that acquire or invest in publicly traded or private companies, often using leveraged debt financing, typically with the goal of taking them private, cutting costs, and reselling within a short time frame for high returns.
74. Klein 2007.
75. Food and Water Watch 2016c, 2017; Wait and Petrie 2017.
76. Mack and Wrase 2017.
77. Lakhani 2020.
78. Teodoro and Saywitz 2020.
79. Kornberg 2016; Rushe 2014; Food and Water Watch 2016a.

80. Zhang et al. 2022. Low-income households served by private utilities spent 4.39 percent of annual income on water bills (not including sewer), versus 2.84 percent for public utilities.

81. See, e.g., Pempetzoglou and Patergiannaki 2017; Mercille and Murphy 2015.

82. Melo Zurita et al. 2015.

83. Arup 2015.

84. Exceptions to this trend include Mexico, Colombia, and Brazil.

85. Pierce 2015.

86. Bakker 2013: 257.

87. Arup 2015.

88. Polanyi 1944.

89. Polanyi 1944: 72.

90. Polanyi 1944: 73.

91. Polanyi 1944: 76.

92. Laxer and Soron 2006.

93. Marx 1867; Roberts 2008; Glassman 2006.

94. Luxemburg [1913] 2003.

95. Harvey 2003: 149.

96. Harvey 2003: 149.

97. Harvey 2003: 148.

98. Harvey 2003: 166.

99. See, e.g., Ahlers 2010; Spronk and Webber 2007.

100. Swyngedouw 2005: 87.

101. S. Ball 2004; Gilbert 2008.

102. Vail 2010: 313.

103. Laxer and Soron 2006: 28.

104. See, e.g., Bond 2005.

105. Kloppenburg 2010.

106. Mascarenhas and Busch 2006: 125.

107. Kloppenburg 2010: 370.

108. Bonny 2017.

109. Pechlaner and Otero 2010: 185.

110. See, e.g., Kinchy, Kleinman, and Autry 2008.

111. Shiva 2016; Bezner Kerr 2013; Kloppenburg 2010.

112. Snitow, Kaufman, and Fox 2007: 197.

113. Bakker 2005: 559.

114. Jaffee and Newman 2013a.

115. J. Greene 2018: 11.

116. Boreal Water News 2010.

117. Salzman 2012: 177.

118. J. Greene 2018.

119. Felton 2020b.

120. Bakker 2010: 200.

121. Barlow 2007: 84.

122. Clarke 2007; NRDC 1999; Castro 2007; Spronk and Webber 2007; Bond 2005.

123. Bakker 2010; Driessen 2008.
124. Girard 2009.
125. Arup 2015; Bond 2005; Bakker 2010; Kishimoto, Steinfort, and Petrovic 2020.
126. Harvey 2003.

CHAPTER 2

The chapter's epigraph is from Shoup (2017a).
1. Salzman 2012.
2. Chapelle 2005: 15.
3. Chapelle 2005: 3.
4. Salzman 2012: 174.
5. Holt 2012: 245.
6. Szasz 2007.
7. Nestlé 2020.
8. Szasz 2007: 129. Since 2002 Coca-Cola has distributed Danone's water brands in the U.S. and Canada. Danone is the top-selling bottled water company in Mexico.
9. Holt 2012: 248.
10. Nestlé Waters Canada 2017.
11. CBS News 2007.
12. Beverage Marketing Corporation 2010: 253.
13. Felton 2020b.
14. Holt 2012: 245.
15. Overstreet 2021. Metal containers accounted for 41 percent of sales, and glass 10 percent.
16. Hemphill 2018.
17. Rodwan 2011.
18. Hawkins 2017: 5; Race 2012.
19. Jaeger 2018.
20. Loria 2020.
21. Felton 2019b.
22. Royte 2008: 167.
23. Royte 2008: 169.
24. Szasz 2007: 127.
25. Race 2012: 86.
26. Holt 2012: 247.
27. Race 2012: 72.
28. Race 2012: 85.
29. Race 2012; Szasz 2007: 127.
30. Sparrow 2015.
31. Harvey 2005.
32. Hawkins 2017: 5.
33. Szasz 2007: 126.
34. Hawkins 2017: 3.
35. Holt 2012: 251.

36. Gleick 2010.
37. Radonic and Jacob 2021.
38. See, e.g., Gleick 2010: 2–3.
39. Girard and Shaker 2008.
40. Pierre-Louis 2015.
41. Sparrow 2015.
42. Pierre-Louis 2015.
43. Rodwan 2019.
44. Girard 2009, n.p.
45. Rodwan 2021.
46. Clarke 2007; Girard 2009.
47. WHO and UNICEF 2017.
48. J. Greene 2018. Another variant is "water ATMs," automated vending stations or standpipes that dispense water with the use of prepaid cards.
49. Prasetiawan, Nastiti, and Muntalif 2017: 1.
50. Prasetiawan, Nastiti, and Muntalif 2017: 2.
51. Prasetiawan, Nastiti, and Muntalif 2017: 4.
52. Bakker 2010: 108–9.
53. Walter, Kooy, and Prabaharyaka 2017: 643.
54. Euromonitor International 2020a; Prasetiawan, Nastiti, and Muntalif 2017.
55. Walter, Kooy, and Prabaharyaka 2017: 646.
56. Heller 2017.
57. Montero Contreras 2019; IDB 2011.
58. J. Greene 2018.
59. Montero Contreras 2019.
60. A Coca-Cola subsidiary that includes Mexico and much of Latin America.
61. Euromonitor International 2020b; Mestre Rodríguez 2019.
62. Lemus 2019: 260.
63. Malkin 2012.
64. Heller 2017.
65. J. Greene 2018.
66. J. Greene 2018: 8.
67. Montero Contreras 2019.
68. Montero Contreras 2016.
69. Heller 2017: 11.
70. Montero Contreras 2016.
71. J. Greene 2018: 11.
72. Pacheco-Vega 2019.
73. Agua Para Todos n.d.
74. Elena Burns, interview, November 2019.
75. WHO and UNICEF 2017: 12.
76. WHO and UNICEF 2017: 12.
77. J. Greene 2018: 11.
78. Heller 2017: 9.
79. IBWA 2022a.

80. Euromonitor International 2019, 2022a.
81. Euromonitor International 2022a.
82. Mintel 2021.
83. Rodwan 2019; Beverage Marketing Corporation 2020.
84. Statista 2021b.
85. Oreskes and Conway 2010.
86. See, e.g., Gleick 2010.
87. Gleick 2010: 6.
88. Gleick 2010: 6.
89. Primo Water n.d.
90. Opel 1999: 68.
91. Bottled water firm representative, interview, May 2010.
92. Gleick 2010: 8.
93. Rodwan 2019: 13.
94. Mintel 2021: 62.
95. J. Kaplan 2016.
96. Pierre-Louis 2015.
97. Szasz 2007: 127; Gallup 2017.
98. Pierre-Louis 2015.
99. Holt 2012: 246.
100. See, e.g., McSpirit and Reid 2009.
101. Javidi and Pierce 2018: 6109.
102. Szasz 2007: 127–8.
103. Szasz 2007: 202.
104. Jaffee and Newman 2013b.
105. This dynamic also exemplifies the framework used by sociologist Albert O. Hirschman in the book *Exit, Voice, and Loyalty* (Hirschman 1970). Faced with the declining quality of a service, individuals can either "exit" by switching to a competing service, or stay and exercise their "voice"—engage in political action or protest to improve the quality of the service. This framework has been applied to services such as schools and transit, where exit entails shifting to a higher-cost private alternative, incentivizing further disinvestment in the public service.
106. Rodwan 2019: 13.
107. Statista 2021a.
108. Beverage Marketing Corporation 2021.
109. Mintel 2020.
110. IBWA 2020c.
111. Gleick 2010: 12–13.
112. Hemphill 2018.
113. Quoted in Gleick 2010.
114. Mintel 2020, 2018.
115. IBWA 2022b, 2020c.
116. Rosinger et al. 2018.
117. Hawkins 2017: 5.
118. Hawkins 2017: 8.

119. Quoted in Erbentraut 2016.
120. Snitow, Kaufman, and Fox 2007: 14. While this quote refers to privately owned water utility firms, the observation applies equally to the mismatch between private bottled water companies and public water utilities.
121. Gleick 2010: 160.
122. Food and Water Watch 2011.
123. Elliott 2003.
124. IBWA 2001.
125. Pierce and Gonzalez 2017: 9.
126. Salzman 2012: 183.
127. Cournoyer 2012.
128. Royte 2008: 143–144.
129. Felton 2019a.
130. Felton 2019a.
131. NRDC 1999: vii.
132. NRDC 1999.
133. Stephenson 2009.
134. Felton 2019a.
135. Felton 2020a.
136. IBWA 2020a.
137. Royte 2008: 145.
138. Salzman 2012: 185.
139. Gleick 2010: 91.
140. Montuori et al. 2008.
141. Westerhoff et al. 2008.
142. Sax 2010.
143. Royte 2008: 151.
144. Bittner, Yang, and Stoner 2014: 1.
145. Mason, Welch, and Neratko 2018.
146. WHO 2019.
147. Heid 2019.
148. Readfearn 2018.
149. Cox et al. 2019.
150. Sherri Mason, quoted in Heid 2019.
151. The American Water Works Association cites this figure as the average consumer cost of tap water.
152. Food and Water Watch 2018a: 8.
153. Szasz 2007: 131.
154. Ross 2020: 17.
155. Ross 2020: 9.
156. Felton 2019b.
157. Gorelick et al. 2011.
158. Rosinger et al. 2018.
159. Vieux et al. 2020: 11.
160. Javidi and Pierce 2018: 6108.
161. Rehm et al. 2020.

162. Rosinger et al. 2018.
163. Mintel 2020: 43.
164. Food and Water Watch 2018: 5–6.
165. McSpirit and Reid 2009; Food and Water Watch 2018b.
166. Javidi and Pierce 2018: 6108.
167. Teodoro and Saywitz 2020.
168. Felton 2019b; Pulido 2016.
169. Felton 2020b.
170. Gleick and Cooley 2009.
171. Villanueva et al. 2021.
172. Antea Group 2018.
173. Royte 2008: 140.
174. Gleick 2010; Barlow 2014.
175. Geyer, Jambeck, and Law 2017.
176. Schroeer, Littlejohn, and Wilts 2020; Pew Charitable Trusts and Systemiq Ltd. 2020.
177. S. Kaplan 2016.
178. Parker 2014; Five Gyres 2017.
179. Laville and Taylor 2017.
180. Young 2019.
181. Laville and Taylor 2017.
182. Rodwan 2021.
183. PET Planet 2022.
184. Geyer, Jambeck, and Law 2017; Volcovici 2022.
185. Schlosberg 2019.
186. Parker 2019.
187. NCSL 2020.
188. TOMRA 2021.
189. Wilkins 2018.
190. Royte 2008.
191. Chaudhuri 2020.
192. Wong 2017.
193. Schroeer, Littlejohn, and Wilts 2020. However, this level of material and climate impact is still far higher than that of tap water.
194. Cabernard et al. 2022.
195. Carpenter 2019.
196. Laville and Taylor 2017; American Chemistry Council 2018; World Economic Forum 2016.
197. CIEL 2019.
198. Mintel 2022: 23.
199. Marini Higgs 2019.
200. Mosbergen 2019.
201. Mosbergen 2019.
202. Schlosberg 2019.
203. Laville 2020.
204. Leonard 2020.
205. Rankin 2019.

206. California's landmark 2022 plastics reduction law, AB54, requires increased plastic recycling and reductions in single-use packaging, but its recycling mandate does not apply to beverage bottles.
207. Bloomberg News 2020.
208. Parker 2019.
209. Loria 2019.
210. D. Thomas 2020.
211. Peter Gleick, interview, December 2019.
212. Royte 2008: 165–6.
213. Mintel 2019.
214. Mintel 2020.
215. Mintel 2020.
216. Koltrowitz and Thomassen 2020.
217. Nestlé 2020.
218. Statista 2020.
219. Mintel 2019.
220. Rodwan 2021.
221. Euromonitor International 2019: 1.
222. Mintel 2020:27.
223. Mintel 2020: 18.
224. Maloney 2019.
225. Mintel 2020: 34.

CHAPTER 3

The chapter's epigraph is quoted in Felton (2019b).
1. Gina Luster, interview, November 2019.
2. Quoted in Felton 2019b.
3. There are many different ways to approach the story of Flint and what it signifies about U.S. society. A large volume of excellent work has been published about the Flint crisis, some of which is cited here. This chapter focuses primarily on the role played by bottled water in the disaster.
4. Hawkins 2017.
5. U.S. Census Bureau n.d.
6. Pauli 2020a: 3.
7. Rector 2016; Klein 2007.
8. Fasenfest 2016.
9. Fonger 2015.
10. Earley later claimed this decision had been approved by the Flint City Council, but no record exists of such a decision, and city officials reject Earley's claim.
11. Pauli 2020a.
12. Lynch 2016.
13. L. Butler, Scammell, and Benson 2016; Ingraham 2016.
14. Pauli 2019, 2020a.
15. Davey 2015.
16. Fonger 2015; Pauli 2020a, 2019.

17. Hanna-Attisha et al. 2016; Pauli 2020a.
18. Pauli 2020a: 3.
19. Council on Environmental Health 2016.
20. Bellinger 2016; Pauli 2020a.
21. Pulido 2016; Ranganathan 2014.
22. Radonic and Jacob 2021.
23. Zdanowicz 2016.
24. Story of Stuff Project 2018.
25. Benjamin Pauli, interview, December 2019.
26. Peggy Case, interview, December 2019.
27. WEYI TV 2022.
28. Pauli 2019: 23.
29. The lawsuit was filed by Concerned Pastors for Social Action, Natural Resources Defense Council, the ACLU of Michigan, and Melissa Mays.
30. However, comprehensively repairing the city's water infrastructure would cost over $1 billion. Benjamin Pauli, personal communication, October 2021.
31. City of Flint, Michigan 2020.
32. Quoted in Chariton and Dize 2021.
33. Robertson 2020.
34. Chariton and Dize 2021; Fonger 2022.
35. Fonger 2021; Murdock, Murray, and Simpson-Mersha 2020.
36. Pauli 2020b.
37. Rushe 2014; Rector 2016.
38. Associated Press 2020.
39. UN Human Rights Office of the High Commissioner 2014.
40. Lauren DeRusha, interview, December 2019.
41. Food and Water Watch 2016a.
42. Quoted in Felton 2020b.
43. Food and Water Watch 2016a.
44. Colton 2020; Lakhani 2020.
45. Pauli 2020b.
46. Pauli 2019.
47. Gina Luster, presentation at forum, "All Eyes on Nestlé," Guelph, Ontario, November 14, 2019.
48. And later also by For Love of Water (FLOW), led by attorney Jim Olson.
49. Mary Grant, interview, December 2019.
50. Nestlé Waters North America 2018.
51. Matheny and Egan 2016.
52. Ellison 2020.
53. Rouda and Tlaib 2020.
54. Ellison 2020.
55. The member groups of the Alliance as of 2021 were Aamjiwnaang First Nation, Coalition to Oppose the Expansion of "US Ecology," Corporate Accountability, Council of Canadians, Flint Democracy Defense League, Flint Rising, Food and Water Watch, For Love of Water (FLOW), Grand Rapids

Water Protectors, Great Lakes Commons, Michigan Citizens for Water Conservation, Michigan Welfare Rights Organization, People's Water Board, Water You Fighting For, and Wellington Water Watchers (Water Is Life Alliance n.d.).

56. Peggy Case, personal communication, December 2022.

57. American Society of Civil Engineers 2020.

58. Felton 2019b.

59. Facundo 2022.

60. Izundu, Madi, and Bailey 2022.

61. Milman 2016.

62. Rose 2019; London et al. 2018. As of this writing, the FDA and EPA have not set standards for allowable levels of per- and polyfluoroalkyl substances (PFAS) in bottled water or tap water. PFAS, a class of highly toxic chemicals found in products including firefighting foams, waterproof clothing, and nonstick coatings, have been discovered in groundwater, some drinking water systems, and human bodies, raising alarm. Some scientists argue levels above one part per trillion (ppt) for all PFAS combined pose a health risk. A peer-reviewed study found PFAS in 39 out of 101 bottled water brands tested, 19 of them above 1 ppt, with the highest levels in spring water (Felton 2021).

63. Peter Gleick, interview, December 2019.

64. Maude Barlow, interview, December 2019.

65. Gallup 2017.

66. Holden et al. 2021.

67. Switzer and Teodoro 2017.

68. NRDC 2019.

69. Viscusi, Huber, and Bel 2015: 450.

70. Pierce and Gonzalez 2017: 9.

71. Felton 2019b.

72. Peck 2015: 6–7, 2.

73. Meehan et al. 2020: 28700.

74. Pulido 2016: 1.

75. London et al. 2018.

76. Szasz 2007.

77. Pulido 2016: 11.

78. Vieux et al. 2020; Mintel 2021.

79. See, e.g., Swanson 2016.

80. J. Kaplan 2016.

81. J. Kaplan 2016.

82. Jeffrey 2009.

83. Hawkins 2017: 6–7.

84. Glenza and Milman 2019. Portland voters in 2017 approved a school bond measure including $150 million to replace the contaminated pipes, add lead-removing filters, and restore water fountains in all of the city's schools.

85. Gleick 2010: 177.

86. Story of Stuff Project 2018.

87. Pacheco-Vega 2019: 2.

88. Herr 2021.
89. Snider 2021; Grant 2022.
90. Lakhani 2021.
91. Zhang and Warner 2021.
92. Pauli 2020b: 324; 311.
93. IBWA 2020d.
94. Felton 2020b.
95. Mintel 2021.
96. Royte 2008: 140.
97. Maude Barlow, interview, December 2019.
98. Swanson 2016.
99. Pulido 2016: 4–5.
100. Fedinick, Taylor, and Roberts 2019.

CHAPTER 4

The chapter epigraph is from a presentation by John Caturano on the "Life after National Sword" panel at the 2019 Plastics Recycling Conference and Trade Show, March 11–13, 2019 (quoted in Joyce 2019).

1. Richard Girard, interview, August 2015, Ottawa.
2. Clarke 2007. A second edition was published in 2007.
3. Hawkins, Potter, and Race 2015: 155.
4. Hawkins Potter, and Race 2015: 159.
5. Clarke 2008.
6. NRDC 1999.
7. Holt 2012: 249.
8. See, e.g., Ferrier 2001.
9. Polaris Institute 2014.
10. This agreement, enacted in 1988, was superseded by the North American Free Trade Agreement in 1994.
11. Council of Canadians n.d.-b.
12. Maude Barlow, presentation at World Water Day panel, "Whose Water Is It, Anyway? Taking Water Protection into Public Hands," March 21, 2021.
13. Barlow 2019: 76.
14. As of this writing, the list of Blue Communities includes seventy-seven municipalities in eight countries, plus universities, school districts, churches, and other institutions (Council of Canadians 2022).
15. Maude Barlow, interview, December 2019.
16. Van Esterik 2013: 513.
17. Penny Van Esterik, presentation to "All Eyes on Nestlé" forum, Guelph, Ontario, November 14, 2019.
18. Lee 2008.
19. CNN 2007.
20. Lauren DeRusha, interview, December 2019.
21. Corporate Accountability n.d.
22. Food and Water Watch 2009, 2010, 2011, 2018a.
23. Mary Grant, interview, December 2019.

24. Fox 2007.
25. Fox 2010.
26. Michael O'Heaney, interview, August 2021.
27. Hawkins 2017: 7.
28. Gleick 2010.
29. Gleick 2010: 151.
30. DeCalma 2020.
31. Davis 2017.
32. Pierre-Louis 2015.
33. NYC Mayor's Office of Sustainability 2020.
34. Davis 2018.
35. Sydney Water n.d.
36. The Greens/EFA in the European Parliament 2019.
37. Staff member, San Francisco Department of Environment, interview, December 2019.
38. Quoted in Lagos 2014.
39. Levin 2017.
40. G. Thomas and Medina 2019.
41. Will Reisman, San Francisco Public Utility Commission (PUC), personal communication, June 2021.
42. John Scarpulla, San Francisco PUC, interview, December 2019.
43. Foley 2009.
44. Sustainable Practices n.d.
45. Cain Miller 2016.
46. FindTap.com n.d.
47. Council of Canadians n.d.-a.
48. This actually understates the case, since it leaves out provincial, regional, and county governments that have also declared such policies. Those bans encompass 2.7 million people, but they overlap with several local bans, resulting in some double-counting.
49. Upadhyay 2019.
50. Scroll.in 2017; DNA India 2017.
51. Haydock 2009.
52. Food and Water Watch 2016b.
53. Vasquez, Carter, and Valko 2015.
54. Kingkade 2015.
55. UCLA Sustainability n.d; California State University 2022; UCLA 2020.
56. Industry Week 2013.
57. Toner 2021: 52.
58. A. Peters 2015.
59. IBWA 2015.
60. IBWA 2020c: 5.
61. Milman 2022.
62. Refill n.d.
63. Evan Pilkington, interview, August 2015, Guelph.
64. O'Mahoney 2020: 50–52.
65. Mintel 2019.

66. Holt 2012: 249.
67. Mintel 2021; Ross 2020.
68. Hawkins, Potter, and Race 2015: 146.
69. Catarina de Albuquerque, presentation at World Water Forum Panel, "Public or Private Provision: Human Rights and Social Risks," Brasilia, March 22, 2018.
70. Johnson 2019.
71. Barlow 2019.
72. DNA India 2017.

CHAPTER 5

1. Anna Mae Leonard [Klairice Westley], interview, August 2017, Cascade Locks.
2. Conlin 2008a.
3. Conlin 2008b: n.p.
4. Conlin 2008a.
5. McCloud resident, interview, June 2011, McCloud.
6. Martin 2008.
7. Member of McCloud Watershed Council, interview, June 2011, McCloud.
8. Conlin 2008a.
9. Conlin 2008b: n.p.
10. California Trout staff member, interview, June 2011, McCloud.
11. Nestlé Waters North America 2009.
12. Dan Bacher, interview, September 2015, Sacramento.
13. Hecht 2015.
14. Hecht 2015.
15. Associated Press 2015a.
16. James 2015.
17. Dangelantonio 2015.
18. Crunch Nestlé Alliance member, interview, September 2015, Sacramento.
19. Powers 2015.
20. Gleick 2010.
21. Confluence Project n.d.
22. Mapes 2020.
23. D. Ball 2010.
24. The interviews in 2010 were jointly conducted with Soren Newman. I gratefully acknowledge her contribution to this early research that forms part of the chapter.
25. Not his actual name. Several interviewees in this chapter have been given pseudonyms (as indicated below in notes).
26. Jerry Rogers [pseudonym], interview, April 2010, Cascade Locks.
27. Emily Caples [pseudonym], interview, April 2010, Cascade Locks.
28. Deanna Busdieker, interview, August 2017, Cascade Locks.
29. Robert Hanford [pseudonym], interview, March 2010, Cascade Locks.
30. Caleb Townsend [pseudonym], interview, April 2010, Cascade Locks.

31. Les Perkins, interview, August 2019, Hood River.

32. Nancy Marquardt [pseudonym], interview, August 2017, Cascade Locks.

33. Brent Foster, interview, May 2018, Hood River.

34. Julia DeGraw, interview, May 2010, Portland.

35. Bottled water company staff member, interview, July 2010.

36. Mirk 2009.

37. Doug Bochsler, interview, May 2010.

38. Karla Newton, [pseudonym], interview, March 2010, Cascade Locks.

39. Dierdre Jefferson, [pseudonym], interview, April 2010, Cascade Locks.

40. Lorraine Harmon [pseudonym], interview, March 2010, Cascade Locks.

41. Kevin Cardozo [pseudonym], interview, March 2010, Cascade Locks.

42. Sierra Club volunteer, interview, May 2010, Portland.

43. House and Graves 2016.

44. Food and Water Watch et al. 2011.

45. Julia DeGraw, interview, September 2019, Portland.

46. House 2015a.

47. Gordon Zimmerman, quoted in House 2015c.

48. DeGraw, interview, September 2019.

49. E. Greene 2015.

50. In 2021, Sams became director of the U.S. National Park Service, the first Indigenous person to hold this office.

51. House 2015b.

52. Pamela Larsen, interview, July 2017, Hood River.

53. Pamela Larsen, interview, May 2019.

54. See, e.g., House 2015d.

55. Cureton 2021.

56. Whitney Kalama, interview, September 2019, Warm Springs.

57. Cureton 2020; McCurdy 2020.

58. New resources have been dedicated to addressing the Warm Springs water crisis, including $3.6 million in state funds and the Chúush Fund, a partnership between the Confederated Tribes of Warm Springs and the foundation Seeding Justice. In 2022, Warm Springs received $24 million from the federal infrastructure bill to completely replace its water system.

59. Foster, interview, May 2018.

60. DeGraw, interview, September 2019.

61. Hood River County Elections Office 2016.

62. Associated Press 2015b.

63. House 2015b.

64. Aurora Del Val, interview, August 2017, Portland.

65. Heidi Jimenez, interview, August 2019, Portland.

66. Larsen, interview, May 2019.

67. Tweed Strategies n.d.; Gumbel 2016.

68. Julia DeGraw, interview, August 2016.

69. Local Water Alliance 2016.

70. Perkins, interview, August 2019.

71. DeGraw, interview, August 2016.

72. Parks 2016.
73. Mulvihill 2016.
74. Godowa-Tufti 2016.
75. Quoted in Carlton 2016.
76. DeGraw, interview, September 2019.
77. KPTV Fox 12 News 2016.
78. City of Cascade Locks 2016.
79. DeGraw, interview, September 2019.
80. Westley, interview, August 2017.
81. The 2016–17 Indigenous-led protests against construction of the Dakota Access Pipeline, near the Standing Rock Sioux Indian Reservation in North and South Dakota.
82. Goudy 2016.
83. K. Brown 2017.
84. DeGraw, interview, September 2019.
85. Ver Valen 2016.
86. DeGraw, interview, September 2019.
87. Del Val, personal communication, August 2019.
88. Nichols 2016.
89. Condon 2017.
90. DeGraw, interview, September 2019.
91. Chronicle Editorial Board 2019.
92. A. Brown 2019.
93. DeGraw, interview, September 2019.
94. Cagle 2020.
95. A. Brown 2020.
96. DeGraw 2016.
97. Larsen, interview, July 2017.
98. Westley, interview, June 2019, Portland.

CHAPTER 6

1. Mike Nagy and Arlene Slocombe, interview, August 2015, Guelph.
2. Mark Goldberg, interview, August 2015, Guelph.
3. C. Butler 2016.
4. Kalmusky 2018.
5. All monetary figures in this chapter are in Canadian dollars.
6. British Columbia allowed bottlers to take water for free until it enacted a fee of $2.25 per million liters in 2015. Québec charges water bottlers $70 per million liters (Leslie 2016a).
7. Case and Connor 2019: 9–10.
8. Wellington Water Watchers n.d.-a.
9. Case and Connor 2019: 4.
10. Lydersen 2008.
11. According to the Council of Canadians (2019), annual bottled water exports from Canada increased 1,460 percent between 2008 and 2018, to 122.9 million liters.

12. GRCA n.d.
13. Case and Connor 2019.
14. Wellington Water Watchers n.d.-b.
15. Libby Carlaw, interview, August 2015, Elora.
16. Ballantyne 2017.
17. Quoted in Ballantyne 2017.
18. Donna McCaw, interview, August 2015, Elora.
19. Jan Beveridge, interview, August 2015, Elora.
20. C. Butler 2016.
21. Donna McCaw, interview, August 2016.
22. Jan Beveridge, interview, August 2016.
23. Don Fisher, interview, August 2016, Elora.
24. Portions of this and the following sections are adapted with permission from Jaffee and Case (2018). I gratefully acknowledge the contribution of Robert A. Case to this material.
25. Spiliotis and Garrote 2015: 522.
26. Joy et al. 2014: 960.
27. Loftus 2009: 953, 959.
28. Reisner 1986.
29. Zwarteveen and Boelens 2014: 151.
30. McCaw, interview, August 2016.
31. Kirk McElwain, interview, August 2016, Elora.
32. Nestlé Waters Canada 2016.
33. James 2015.
34. Peter Gleick, interview, December 2019.
35. Zwarteveen and Boelens 2014.
36. Rod Whitlow, interview, September 2021.
37. Mike Nagy, interview, August 2016, Guelph.
38. CTV News 2016.
39. Nagy, interview, August 2016.
40. Arlene Slocombe, interview, August 2016, Guelph.
41. Nagy, interview, August 2016.
42. Bueckert 2016b.
43. Wynne 2016.
44. Leslie 2016b.
45. Wynne 2016.
46. Leslie 2016.
47. Benzie 2017.
48. Griswold 2017.
49. Nagy and Barlow 2017.
50. Mike Balkwill, interview, November 2019, Guelph.
51. Bueckert 2016a.
52. Mainstreet Research et al. 2018.
53. Davidson 2020.
54. Cribb and Marotta 2019.
55. Cribb and Marotta 2019.
56. Beveridge, interview, August 2016.

57. Cribb and Marotta 2019.

58. The Six Nations are Onondaga, Oneida, Seneca, Mohawk, Cayuga, and Tuscarora (Ho and Miller 2021).

59. Six Nations of the Grand River n.d.

60. Six Nations of the Grand River 2020.

61. Taekema 2021.

62. Human Rights Watch 2016.

63. Public forum, "All Eyes on Nestlé," Wellington Water Watchers, Guelph, November 14, 2019.

64. Shimo 2018.

65. Duignan, Moffat, and Martin-Hill 2022.

66. Duignan, Moffat, and Martin-Hill 2022.

67. Canadian Press 2021.

68. Whitlow, personal communication, September 2021.

69. Story of Stuff Project 2021a.

70. Forum, "All Eyes on Nestlé," November 14, 2019. See also M. Robinson 2019.

71. Environmental Registry of Ontario 2021.

72. Save Our Water 2021.

73. Balkwill, interview, August 2021.

74. Statista 2021b.

75. Statista 2021c.

76. *Guelph Mercury Tribune* 2018.

77. Golder Associates 2021a, 2021b.

78. Rubin 2020.

79. Wellington Water Watchers 2020.

80. Chaudhuri 2021.

81. Chaudhuri 2021.

82. Euromonitor International 2022b.

83. Ewing 2020.

84. M. Peters 2020.

85. Wellington Water Watchers 2021; S&P Global Market Intelligence 2021.

86. Story of Stuff Project 2021b.

87. Balkwill, interview, August 2021.

88. Gerber 2021.

89. Slocombe, interview, August 2021.

90. Balkwill, interview, August 2021.

91. Save Our Water 2021.

92. Balkwill, interview, August 2021.

CHAPTER 7

The chapter's epigraph is from an interview with Lauren DeRusha, December 2019.

1. Franklin Frederick, interview, November 2019, Guelph.

2. Crevoisier 2013.

3. Chatterjee 2013.

4. Crevoisier 2013.
5. Chazan 2018; L'Eau Qui Mord n.d.
6. Bernard Schmitt, interview, November 2019, Guelph.
7. Balkwill 2018.
8. See also Schroering 2021.
9. See also "Troubled Water," an episode of the Netflix documentary series *Rotten*.
10. Michael O'Heaney, interview, August 2021.
11. Mike Balkwill, interview, August 2021.
12. Singh 2021.
13. Mehta 2007: 662.
14. Grossman 2017.
15. Harvey 2003: 148.
16. Parag and Roberts 2009: 633.
17. Barlow 2007: 100–101.
18. Jaffee and Newman 2013b.
19. Gleick 2010: 176.
20. Szasz 2007.
21. Royte 2008: 206.
22. Parag and Roberts 2009: 627.
23. Royte 2008: 207.
24. Clarke 2007: 135–36.
25. Royte 2008: 207.
26. Themba 2022.
27. London et al. 2018.
28. J. Greene 2018: 11.
29. Bond 2005; Laxer and Soron 2006.
30. Vail 2010: 310.
31. Polanyi 1944.
32. Richard Girard, interview, Ottawa, August 2015.
33. Frederick, interview, November 2019.
34. Maude Barlow, interview, December 2019.

CONCLUSION

The chapter's epigraph is from an interview with Maude Barlow, December 2019.

1. Gleick 2010: 175.
2. Barlow 2020.
3. Quoted in Erbentraut 2016.
4. Greens/EFA in the European Parliament 2019: 3.
5. Burros and Warner 2006.
6. Mack and Wrase 2017.
7. Mary Grant, interview, December 2019.
8. Gleick 2013.
9. Berck et al. 2013.
10. Bliss 2015.

11. Michael O'Heaney, interview, August 2021.
12. Metz 2019.
13. Schroeer, Littlejohn, and Wilts 2020: 6.
14. Arumugam 2011.
15. Peggy Case, interview, December 2019.
16. Peter Gleick, interview, December 2019.
17. J. Greene 2018: 11.
18. Solnit 2019: xvii.
19. Linebaugh 2014.

References

Abrams, Lindsay. 2014. "Water Is the New Oil: How Corporations Took Over a Basic Human Right." *Salon,* October 5.

Agua para Todos. n.d. "Agua para Todos, Agua para la Vida." Accessed December 2, 2021. https://aguaparatodos.org.mx/.

Ahlers, Rhodante. 2005. "Gender Dimensions of Neoliberal Water Policy in Mexico and Bolivia: Empowering or Disempowering?" In *Opposing Currents: The Politics of Water and Gender in Latin America,* edited by Vivienne Bennett, Sonia Davila-Poblete, and Maria Nieves Rico, 53–71. Pittsburgh, PA: University of Pittsburgh Press.

———. 2010. "Fixing and Nixing: The Politics of Water Privatization." *Review of Radical Political Economics* 42 (2): 213–30.

American Chemistry Council. 2018. "U.S. Chemical Industry Investment Linked to Shale Gas Reaches $200 Billion." Press release, September 11. www.americanchemistry.com/chemistry-in-america/news-trends/press-release/2018/us-chemical-industry-investment-linked-to-shale-gas-reaches-200-billion.

American Water Works Association. 2012. *Buried No Longer: Confronting America's Water Infrastructure Challenge.* Denver, CO: American Water Works Association.

Antea Group. 2018. *Water and Energy Use Benchmarking Study.* St. Paul, MN: Antea Group.

Arumugam, Nadia. 2011. "EU Bans Bottled Water Claim That Water Prevents Dehydration: Ludicrous or Just?" *Forbes,* December 1.

Arup. 2015. *In-Depth Water Yearbook, 2014–2015.* London: Arup.

ASCE (American Society of Civil Engineers). 2020. *The Economic Benefits of Investing in Water Infrastructure.* Washington, DC: American Society of Civil Engineers.

Associated Press. 2015a. "California Governor Orders Mandatory Water Restrictions." April 1.

———. 2015b. "Hot Water Kills Half of Columbia River Sockeye Salmon." July 27.

———. 2020. "Water Shutoffs Problematic for Many in Detroit." March 28.

Bakker, Karen. 2005. "Neoliberalizing Nature? Market Environmentalism in Water Supply in England and Wales." *Annals of the Association of American Geographers* 95 (3): 542–65.

———. 2007. "Trickle Down? Private Sector Participation and the Pro-poor Water Supply Debate in Jakarta, Indonesia." *Geoforum* 37:855–68.

———. 2010. *Privatizing Water: Governance Failure and the World's Urban Water Crisis.* Ithaca, NY: Cornell University Press.

———. 2013. "Neoliberal versus Postneoliberal Water: Geographies of Privatization and Resistance." *Annals of the Association of American Geographers* 103 (2): 253–60.

———. 2014. "The Business of Water: Market Environmentalism in the Water Sector." *Annual Review of Environment and Resources* 39 (1): 469–94.

Balkwill, Mike. 2018. "Dispatches from Vittel, Day 1." *Wellington Water Watchers Blog,* February 11. www.wellingtonwaterwatchers.ca/vittel-feb11-2019.

Ball, Deborah. 2010. "Bottled Water Pits Nestlé vs. Greens." *Wall Street Journal,* May 25.

Ball, Stephen J. 2004. "Education for Sale! The Commodification of Everything?" King's Annual Education Lecture 2004, London, June 17. www.researchgate.net/publication/267683502_Education_For_Sale_The_Commodification_of_Everything.

Ballantyne, Diane. 2017. "Water Warriors: A Small Community's Struggle to Keep Water Public." Save Our Water, October 10. www.saveourwater.ca/post/water-warriors-a-small-community-s-struggle-to-keep-water-public.

Barlow, Maude. 2007. *Blue Covenant: The Global Water Crisis and the Coming Battle for the Right to Water.* New York: New Press.

———. 2014. *Blue Future: Protecting Water for People and the Planet Forever.* New York: New Press.

———. 2019. *Whose Water Is It, Anyway? Taking Water Protection into Public Hands.* Toronto: ECW Press.

———. 2020. "What Good Is a Single-Use Plastics Ban If It Doesn't Include Water Bottles?" *National Observer,* October 9.

Barlow, Maude, and Tony Clarke. 2002. *Blue Gold: The Fight to Stop the Corporate Theft of the World's Water.* New York: New Press.

Bellinger, David C. 2016. "Lead Contamination in Flint—An Abject Failure to Protect Public Health." *New England Journal of Medicine* 374 (March 24): 1101–3.

Benzie, Robert. 2017. "Fees for Bottlers of Water Jump from $3.71 to $503.71." *Toronto Star,* June 8.

Berck, Peter, Jacob Moe Lange, Andrew Stevens, and Sofia Villas-Boas. 2013. "Measuring Consumer Responses to a Bottled Water Tax Policy." Department of Agricultural and Resource Economics, University of California, Berkeley. https://are.berkeley.edu/~sberto/WaterTaxNov27.pdf.

Beverage Marketing Corporation. 2010. *Bottled Water in the U.S.* New York: Beverage Marketing Corporation.
———. 2013. "Bottled Water Shows Strength Yet Again, New Report from BMC Shows." Press release, April 26. www.beveragemarketing.com/news-detail.asp?id=260.
———. 2017. ""Bottled Water Becomes Number One Beverage in the U.S." Press release, March 10. www.beveragemarketing.com/news-detail.asp?id=438.
———. 2020. "Bottled Water, the Biggest Beverage in the U.S., Grows Again in 2019, Data from Beverage Marketing Corporation Show." Press release, May 19. www.beveragemarketing.com/news-detail.asp?id=609.
———. 2021. "Bottled Water, Unbowed by the Covid-19 Crisis, Grows Again in 2020, Data from Beverage Marketing Corporation Show." Press release, May 18. www.beveragemarketing.com/news-detail.asp?id=654.
Bezner Kerr, Rachel. 2013. "Seed Struggles and Food Sovereignty in Northern Malawi." *Journal of Peasant Studies* 40 (5): 867–97.
Bittner, George D., Chun Z. Yang, and Matthew A. Stoner. 2014. "Estrogenic Chemicals Often Leach from BPA-Free Plastic Products That Are Replacements for BPA-Containing Polycarbonate Products." *Environmental Health* 13 (41): 1–14.
Bliss, Laura. 2015. "The Case for Taxing Bottled Water." Bloomberg, May 28.
Bloomberg News. 2020. "World's Biggest User of Plastic to Curtail Its Use." January 20.
Bond, Patrick. 2005. "Globalisation/Commodification or Deglobalisation/Decommodification in Urban South Africa." *Policy Studies* 26 (3/4): 337–58.
———. 2008. "Macrodynamics of Globalisation, Uneven Urban Development and the Commodification of Water." *Law, Social Justice and Global Development* 10 (2). https://go.gale.com/ps/i.do?id=GALE%7CA187844299&sid=googleScholar&v=2.1&it=r&linkaccess=abs&issn=14670437&p=AONE&sw=w&userGroupName=oregon_oweb&isGeoAuthType=true.
Bonny, Silvie. 2017. "Corporate Concentration and Technological Change in the Global Seed Industry." *Sustainability* 9 (9): 1632–57.
Boreal Water News. 2010. "Global Bottled Water Market to Reach $65.9 Billion by 2012, According to a New Report by Global Industry Analysts, Inc." August 30.
Browdie, Brian. 2018. "Cape Town's Bottlers and Brewers Are Coming under Fire for Guzzling Water in a Drought." *Quartz Africa,* February 23.
Brown, Alex. 2019. "Crystal Geyser Mistakenly Emails Chronicle: Randle Bottling Project Likely 'Dead.'" *The Chronicle,* July 11.
———. 2020. "Washington State Bottled Water Bill Fails, but Congress Scrutinizes Industry." *Pew Charitable Trusts: Stateline Blog,* March 6. www.pewtrusts.org/en/research-and-analysis/blogs/stateline/2020/03/06/washington-state-bottled-water-bill-fails-but-congress-scrutinizes-industry.
Brown, Kate. 2017. "Letter to Kurt Melcher, Director, Oregon Department of Fish & Wildlife." Press release, October 27. https://crag.org/wp-content/uploads/2017/10/10.27.17_ODFW-Letter-Cascade-Locks-FINAL.pdf.
Brown, Ryan Lenora. 2022. "'Day Zero' Water Crisis Looms on South Africa's Eastern Cape." *Washington Post,* June 19.

Bueckert, Kate. 2016a. "Nestlé Says It Wants to Partner with Centre Wellington on Middlebrook Well." CBC News, December 13.

———. 2016b. "Township of Centre Wellington Tried to Buy Elora Well Now Owned by Nestlé." CBC News, August 26.

Burros, Marian, and Melanie Warner. 2006. "Bottlers Agree to a School Ban on Sweet Drinks." *New York Times,* May 4.

Butler, Colin. 2016. "A Rare Look inside Nestlé's Aberfoyle Bottling Plant." CBC News, November 1.

Butler, Lindsey J., Madeleine K. Scammell, and Eugene B. Benson. 2016. "The Flint, Michigan, Water Crisis: A Case Study in Regulatory Failure and Environmental Injustice." *Environmental Justice* 9 (4): 93–97.

Cabernard, Livia, Stephan Pfister, Christopher Oberschelp, and Stefanie Hellweg. 2022. "Growing Environmental Footprint of Plastics Driven by Coal Combustion." *Nature Sustainability* 5 (February): 139–48.

Cagle, Suzie. 2020. "Washington State Takes Bold Step to Restrict Companies from Bottling Local Water." *The Guardian,* February 18.

Cain Miller, Claire. 2016. "Liberals Turn to Cities to Pass Laws and Spread Ideas." *New York Times,* January 26.

California State University. 2022. "CSU Contracts and Procurement." Sec. II.D.3. Last revised September 27. https://calstate.policystat.com/policy/12393471/latest/.

Canadian Press. 2016. "Ontario Premier Wants New Rules for Bottled Water Companies." September 22. www.youtube.com/watch?v=7ZdtcPXOSvA.

———. 2021. "Federal Government Reaches Nearly $8B Deal with First Nations on Drinking Water Suit." July 31.

Carlton, Jim. 2016. "Oregon Ban Wrings Water Bottling Plan." *Wall Street Journal,* May 21–22.

Carpenter, Zoe. 2019. "The Toxic Consequences of America's Plastics Boom." *The Nation,* March 14.

Case, Robert, and Leah Connor. 2019. *History of the Wellington Water Watchers: Part 1: Launching a Local Movement.* Guelph, Ontario: Wellington Water Watchers. https://d3n8a8pro7vhmx.cloudfront.net/wellingtonwaterwatchers/pages/10/attachments/original/1589203650/The_Founding_of_the_Wellington_Water_Watchers.pdf.

Castro, José Esteban. 2007. "Poverty and Citizenship: Sociological Perspectives on Water Services and Public-Private Participation." *Geoforum* 38: 756–71.

———. 2008. "Water Struggles, Citizenship and Governance in Latin America." *Development* 51:72–76.

CBS News. 2007. "Aquafina Labels to Show Source: Tap Water." July 27.

Chapelle, Francis H. 2005. *Wellsprings: A Natural History of Bottled Spring Waters.* New Brunswick, NJ: Rutgers University Press.

Chariton, Jordan, and Jenn Dize. 2021. "How a Flurry of Suspicious Phone Calls Set Investigators on Rick Snyder's Trail." *The Intercept,* January 14.

Chatterjee, Pratap. 2013. "Nestlé Found Guilty of Spying on Swiss Activists." CorpWatch, January 30. www.corpwatch.org/article/nestle-found-guilty-spying-swiss-activists.

Chaudhuri, Saabira. 2020. "Plastic Water Bottles, Which Enabled a Drinks Boom, Now Threaten a Crisis." *Wall Street Journal*, December 12.
———. 2021. "Nestlé to Sell Poland Spring, Pure Life and Other Water Brands for $4.3 Billion." *Wall Street Journal*, February 17.
Chazan, David. 2018. "French Town of Vittel Suffering Water Shortages as Nestle Accused of 'Overusing' Resources." *The Telegraph*, April 26.
Chronicle Editorial Board. 2019. "Our Views: With Email Slip, Crystal Geyser Shows Its True Face." *The Chronicle*, July 19.
CIEL (Center for International Environmental Law). 2019. *Plastic and Climate: The Hidden Costs of a Plastic Planet*. Washington, DC: Center for International Environmental Law.
City of Cascade Locks, Oregon. 2016. "Ballot Measure 14-55 Legal Issues." PDF document no longer available on internet but in author's possession.
City of Flint, Michigan. 2020. "City of Flint Launches Final Push to Get the Lead Out; Service Line Replacement Project Set to Finish by Nov. 30, 2020." Press release, August 13. www.cityofflint.com/city-of-flint-launches-final-push-to-get-the-lead-out-service-line-replacement-project-set-to-finish-by-nov-30-2020/.
Clarke, Tony. 2007. *Inside the Bottle: An Exposé of the Bottled Water Industry*. Rev. ed. Toronto: Canadian Centre for Policy Alternatives.
———. 2008. "Toronto Stood Up to Bottled Water Industry." *Toronto Star*, December 11.
CNN. 2007. "Aquafina Labels to Spell Out Source—Tap Water." July 27.
Colton, Roger. 2020. *The Affordability of Water and Wastewater Service In Twelve U.S. Cities*. Belmont, MA: Fisher, Sheehan and Colton.
Conca, Ken. 2008. "The United States and International Water Policy." *Journal of Environment and Development* 17 (3): 215–37.
Condon, M.B. 2017. "Chamber Too Cozy on Nestlé." *Goldendale Sentinel*, June 13.
Confluence Project. n.d. "Celilo Park—Confluence Project." Accessed July 9, 2021. www.confluenceproject.org/river-site/celilo-park/.
Conlin, Michelle. 2008a. "A Community Goes Up—and Wins a Round—against Nestle." *Business Week*, May 28.
———. 2008b. "A Town Torn Apart by Nestlé." *Business Week*, April 16.
Corporate Accountability. n.d. "About Our Water Campaign." Accessed March 16, 2021. www.corporateaccountability.org/water/about-water-campaign/.
Council of Canadians. 2019. "Landmark Vote by B.C. Municipalities and First Nations Urges B.C. Government to Halt Exports for Bottled Water." Press release, September 27. https://canadians.org/media/landmark-vote-bc-municipalities-and-first-nations-urges-bc-government-halt-exports-bottled/
———. n.d.-a. "Blue Communities." Accessed July 1, 2022. https://canadians.org/bluecommunities.
———. n.d.-b. "Our Story." Accessed December 12, 2022. https://canadians.org/story.
Council on Environmental Health. 2016. "Prevention of Childhood Lead Toxicity." *Pediatrics* 138 (1): 1–15.

Cournoyer, Caroline. 2012. "Cities Tout Municipal Tap Water as Better Than Bottled." *Governing*, March 27.

Cox, Kieran D., Garth A. Covernton, Hailey L. Davies, John F. Dower, Francis Juanes, and Sarah E. Dudas. 2019. "Human Consumption of Microplastics." *Environmental Science and Technology* 53 (12): 7068–74.

Crevoisier, Arnaud. 2013. "Nestlé et Securitas condamnés au civil pour l'infiltration d'ATTAC." *Le Courier*, January 25.

Cribb, Robert, and Stefanie Marotta. 2019. "'We're in a David-and-Goliath Situation': Small Ontario Town Taking on Nestle to Save Its Water." *Toronto Star*, December 17.

CTV News. 2016. "Nestle Seeking 10-Year Renewal of Aberfoyle Water-Taking Permit." April 14. Kitchener, Ontario.

Cureton, Emily. 2020. "Water Crisis Returns to Warm Springs as Virus Cases Rise." OPB News, June 30.

———. 2021. "Warm Springs Treaty Turns 166: 'This Was the Real Story of the West.'" OPB News, June 26.

Cutler, David, and Grant Miller. 2005. "The Role of Public Health Improvements In Health Advances: The Twentieth-Century United States." *Demography* 42 (1): 1–22.

Dangelantonio, Matt. 2015. "Nestlé Waters CEO Isn't Stopping Bottling in California, Says New Tech Will Save Millions of Gallons." *AirTalk*, KPCC, May 14.

Davey, Monica. 2015. "Flint Officials Are No Longer Saying the Water Is Fine." *New York Times*, October 8.

Davidson, Nick. 2020. "The People versus Nestlé's Profiteering of Community Water." *Sierra*, June 7.

Davis, Nicola. 2017. "'They're Just Not Very British': Will Cities Finally Splash Out on Water Fountains?" *The Guardian*, December 5.

———. 2018. "Sadiq Khan Announced 100 New Drinking Fountains for London." *The Guardian*, October 11.

Dawson, Ashley. 2018. "A Water Apartheid." *Washington Post*, July 10.

DeCalma, Justine. 2020. "New York City Is Cracking Down on Plastic Bottles." *The Verge*, February 7.

DeGraw, Julia. 2016. "This is How You Defeat Nestlé." Food and Water Watch, May 24. www.foodandwaterwatch.org/impact/how-you-defeat-nestl%C3%A9 (no longer available).

de Mora e Souza, Marcos. 2018. "MST invade unidade de agua mineral da Nestlé em Minas Gerais." *Valor Economico (O Globo)*, March 20.

DNA India. 2017. "Civic Offices, Educational Institutes to Be Included in Plastic Ban." November 17.

Drewnowski, Adam, Colin D. Rehm, and Florence Constant. 2013. "Water and Beverage Consumption among Adults in the United States: Cross-sectional Study Using Data from NHANES 2005–2010." *BMC Public Health* 13 (1): 1–19.

Driessen, Travis. 2008. "Collective Management Strategies and Elite Resistance in Cochabamba, Bolivia." *Development* 51 (1): 89–95.

Duignan, Sarah, Tina Moffat, and Dawn Martin-Hill. 2022. "Be Like the Running Water: Assessing Gendered and Age-Based Water Insecurity Experiences with Six Nations First Nation." *Social Science and Medicine* 298, April. https://doi.org/10.1016/j.socscimed.2022.114864.

Elliott, Stuart. 2003. "Coca-Cola Tries Selling Sexiness in Promoting Dasani in the Competitive Bottled-Water Market." *New York Times*, June 30.

Ellison, Garrett. 2019. "Why Nestle Pays Next to Nothing for Michigan Groundwater." Michigan Live, April 2.

———. 2020. "U.S. House Democrats Launch Probe into Nestlé Water Bottling." Michigan Live, March 4.

Environmental Registry of Ontario. 2021. "Updating Ontario's Water Quantity Management Framework." Ministry of Environment, Conservation and Parks, March 31. https://ero.ontario.ca/notice/019-1340#proposal-details.

Erbentraut, Joseph. 2016. "Here's Another Reason to Be Worried about Bottled Water." *Huffington Post*, August 22.

Esterl, Mike. 2006. "Dry Hole: Great Expectations for Private Water Fail to Pan Out." *Wall Street Journal*, June 26.

Euromonitor International. 2019. *Bottled Water in the US*. London: Euromonitor International.

———. 2020a. *Bottled Water in Indonesia*. London: Euromonitor International.

———. 2020b. *Bottled Water in Mexico*. London: Euromonitor International.

———. 2021a. *Bottled Water in Canada*. London: Euromonitor International.

———. 2021b. *Bottled Water in the US*. London: Euromonitor International.

———. 2022a. *Bottled Water in the US*. London: Euromonitor International.

———. 2022b. "Bottled Water in World: Datagraphics." December. Available privately through Euromonitor.

Ewing, Jack. 2020. "Nestlé Weighs Sale of Water Unit in Push toward Sustainability." *New York Times*, June 11.

Facundo, Jarod. 2022. "'Benton Harbor Is Not Flint'—It's Worse." *American Prospect*, February 23.

Fasenfest, David. 2016. "Emergency Management in Michigan: Race, Class and the Limits of Liberal Democracy." *Critical Sociology* 42 (3): 331–34.

Fedinick, Kristi Pullen, Steve Taylor, and Michele Roberts. 2019. *Watered Down Justice*. Natural Resources Defense Council, Coming Clean, Environmental Justice Health Alliance. New York: Natural Resources Defense Council. www.nrdc.org/sites/default/files/watered-down-justice-report.pdf.

Felton, Ryan. 2019a. "The FDA Knew the Bottled Water Was Contaminated. The Public Didn't." Consumer Reports, November 21.

———. 2019b. "Should We Break Our Bottled Water Habit?" Consumer Reports, October 9.

———. 2020a. "High Levels of Arsenic Found in US Whole Foods' Bottled Water Brand." *The Guardian*, June 24.

———. 2020b. "How Coke and Pepsi Make Millions From Bottling Tap Water, as Residents Face Shutoffs." Consumer Reports, April 23.

———. 2021. "New Study Finds PFAS in Bottled Water, as Lawmakers Call for Federal Limits." Consumer Reports, June 17.

Fergusson, James. 2015. "The World Will Soon Be at War over Water." *Newsweek,* April 24.

Ferrier, Katherine. 2001. "Bottled Water: Understanding a Social Phenomenon." World Wildlife Fund. *AMBIO: A Journal of the Human Environment* 30 (2): 118–19.

FindTap.com. n.d. "Tap: Stations." Accessed January 13, 2022. https://findtap.com/network.

Five Gyres. 2017. *Plastics B.A.N. List 2.0*. Santa Monica, CA: 5 Gyres.

Foley, Meraiah. 2009. "Bundanoon Journal: Ban on Bottled Water, Apparently a First, Puts a Small Town on a Big Stage." *New York Times,* July 17.

Fonger, Ron. 2015. "Ex-Emergency Manager Says He's Not to Blame for Flint River Water Switch." Michigan Live, October 13.

———. 2021. "Judge Gives Preliminary OK to $641 Million Flint Water Crisis Settlement." Michigan Live, January 21.

———. 2022. "Flint Residents Searching for Water Crisis Accountability after Snyder's Criminal Charges Dismissed." Michigan Live, December 9.

Food and Water Watch. 2009. *All Bottled Up: Nestle's Pursuit of Community Water.* Washington, DC: Food and Water Watch.

———. 2010. *Bottling Our Cities' Tap Water: Share of Bottled Water from Municipal Supplies Up 50 Percent*. Washington, DC: Food and Water Watch.

———. 2011. *Hanging On for Pure Life*. Washington, DC: Food and Water Watch.

———. 2012a. "Private Equity, Public Inequity: The Public Cost of Private Equity Takeovers of U.S. Water Infrastructure." Fact Sheet, August. https://foodandwaterwatch.org/wp-content/uploads/2021/03/private_equity_public_inequity_fs_aug_2012.pdf.

———. 2012b. *Public-Public Partnerships: An Alternative Model to Leverage the Capacity of Municipal Water Utilities*. Washington, DC: Food and Water Watch.

———. 2016a. *America's Secret Water Crisis*. Washington, DC: Food and Water Watch.

———. 2016b. "Campus Campaigns to Ban Bottled Water." Fact Sheet, October. https://foodandwaterwatch.org/wp-content/uploads/2021/03/fs_1610_tbtt-web_0.pdf.

———. 2016c. *The State of Public Water in the United States*. Washington, DC: Food and Water Watch.

———. 2017. "Water Injustice: Economic and Racial Disparities in Access to Safe and Clean Water in the United States." Issue Brief, March. www.foodandwaterwatch.org/wp-content/uploads/2021/04/ib_1703_water-injustice-web.pdf.

———. 2018a. *Take Back the Tap: The Big Business Hustle of Bottled Water.* Washington, DC: Food and Water Watch.

———. 2018b. "The Water Crisis in Martin County, Kentucky." Issue Brief, February. https://foodandwaterwatch.org/wp-content/uploads/2021/03/ib_1802_martincntykywater-web5.pdf.

———. 2022. *The WATER Act: Restoring Federal Support for Clean Water Systems*. Washington, DC: Food and Water Watch.

Food and Water Watch et al. 2011. "Letter to Philip C. Ward, Director, OWRD, Re: Water Rights Transfer Application No. T-11249." Press release, June 30.

Fox, Louis, dir. 2007. *The Story of Stuff*. Berkeley, CA: Free Range Studios.

———, dir. 2010. *The Story of Bottled Water*. U.S.: Story of Stuff Project.

Frederick, Franklin. 2018. *Water: The Warning Coming from South Africa. Dawn News*, February 15.

Gallup. 2017. "In U.S., Water Pollution Worries Highest since 2001." Press release, March 31. http://news.gallup.com/poll/207536/water-pollution-worries-highest-2001.aspx.

Gerber, Leah. 2021. "Blue Triton Keeps Nestlé's Water-Taking Permits." *The Record*, November 17.

Geyer, Roland, Jenna R. Jambeck, and Kara Lavender Law. 2017. "Production, Use, and Fate of All Plastics Ever Made." *Science Advances* 3:1–5.

Gilbert, Jeremy. 2008. "Against the Commodification of Everything: Anti-consumerist Cultural Studies in the Age of Ecological Crisis." *Cultural Studies* 22 (5): 551–66.

Girard, Richard. 2009. "Bottled Water Industry Targets a New Market: The Global South." AlterNet, June 15. www.alternet.org/2009/06/bottled_water_industry_targets_a_new_market_the_global_south/.

Girard, Richard, and Erika Shaker. 2008. "Bottled Up or Tapped Out: Where Have All the Water Fountains Gone?" *Academic Matters: The Journal of Higher Education*, November 18.

Glassman, Jim. 2006. "Primitive Accumulation, Accumulation by Dispossession, Accumulation by 'Extra-economic' Means." *Progress in Human Geography* 30 (5): 608–25.

Gleick, Peter H. 2010. *Bottled and Sold: The Story behind Our Obsession with Bottled Water*. Washington, DC: Island Press.

———. 2013. "Bottled Water Tax." *ScienceBlogs*, May 9. https://scienceblogs.com/significantfigures/index.php/2013/05/09/bottled-water-tax.

Gleick, Peter H., and Heather S. Cooley. 2009. "Energy Implications of Bottled Water." *Environmental Research Letters* 4 (1): 1–6. https://doi.org/10.1088/1748-9326/4/1/014009.

Glenza, Jessica, and Oliver Milman. 2019. "A Hidden Scandal: America's School Students Exposed to Water Tainted by Toxic Lead." *The Guardian*, March 6.

Godowa-Tufti, Kayla. 2016. "Nestlé Water Battle Continues—Tribal Members Speak Out at City Council Meeting." *Last Real Indians*, April 18. https://lastrealindians.com/news/2016/4/18/apr-18-2016-nestle-water-battle-continues-tribal-members-speak-out-at-city-council-meeting-by-kayla-godowa-tufti.

Goff, Matthew, and Ben Crow. 2014. "What Is Water Equity? The Unfortunate Consequences of a Global Focus on 'Drinking Water.'" *Water International* 39 (2): 159–71.

Golder Associates. 2021a. *Nestlé Waters Canada Aberfoyle Site: 2020 Annual Monitoring Report*. Cambridge, Ontario: Golder Associates.

———. 2021b. *Nestlé Waters Canada Erin Site: 2020 Annual Monitoring Report*. Cambridge, Ontario: Golder Associates.

Goldman, Michael. 2005. *Imperial Nature: The World Bank and Struggles for Social Justice in the Age of Globalization.* New Haven, CT: Yale University Press.

———. 2007. "How 'Water for All!' Policy Became Hegemonic: The Power of the World Bank and its Transnational Policy Networks." *Geoforum* 38:786–800.

Gorelick, Marc H., Lindsay Gould, Mark Nimmer, Duke Wagner, Mary Heath, Hiba Bashir, and David C. Brousseau. 2011. "Perceptions about Water and Increased Use of Bottled Water in Minority Children." *Archives of Pediatric and Adolescent Medicine* 165 (10): 928–32.

Goudy, JoDe. 2016. Chairman, Yakama Nation, speech at Rally Against Nestlé, Oregon State Capitol, September 21. https://kboo.fm/media/52698-no-nestle-rally-salem-92116-full-audio.

Grand View Research. 2022. "Bottled Water Market Size Worth $509.2 Billion by 2030." Press release, April.

Grant, Mary. 2022. "Announcing 100+ Sponsors for the WATER Act." Food and Water Watch, June 28. www.foodandwaterwatch.org/2022/06/28/100-sponsors-for-the-water-act/.

GRCA (Grand River Conservation Authority). n.d. "Groundwater Resources." Accessed December 13, 2022. www.grandriver.ca/en/our-watershed/Groundwater-resources.aspx.

Greene, E. Austin, Jr. 2015. "Letter to Governor Kate Brown re: ODFW Cascade Locks/Oxbow Hatchery Transfer." Confederated Tribes of the Warm Springs Reservation of Oregon, Warm Springs, OR. May 12. PDF document in author's possession.

Greene, Joshua. 2018. "Bottled Water in Mexico: The Rise of a New Access to Water Paradigm." *WIREs Water,* April 11, 1–16. https://doi.org/10.1002/wat2.1286.

Greens/EFA in the European Parliament. 2019. "Media Briefing: Trilogue Results of the Revision of the Drinking Water Directive." Press release, December 19. https://123dok.co/document/yd7mkrej-media-briefing-trilogue-results-revision-drinking-water-directive.html.

Griswold, Elizabeth. 2017. "Ontario Government Has Taken Sides against the Bottled Water Industry." *Toronto Star,* April 10.

Grossman, Zoltán. 2017. *Unlikely Alliances: Native Nations and White Communities Join to Defend Rural Lands.* Seattle: University of Washington Press.

Guelph Mercury Tribune. 2018. "Nestlé Lays Off Workers at Its Aberfoyle Water Bottling Plant." November 19.

Gumbel, Andrew. 2016. "Amid Western Drought, Oregon County to Vote on Nestlé Bottling Public Water." *The Guardian,* May 17.

Gurley, Gabrielle. 2021. "Something in the Water." *American Prospect,* March 23.

Hall, David, Emanele Lobina, and Violeta Corral. 2011. *Trends in Water Privatization.* London: Public Services International Research Unit.

Hall, David, Emanuele Lobina, and Robin De la Motte. 2009. *Making Water Privatisation Illegal: New Laws in Netherlands and Uruguay.* London: Public Service International Research Unit.

Hanna-Attisha, Mona, Jenny LaChance, Richard Casey Sadler, and Allison Champney Schnepp. 2016. "Elevated Blood Lead Levels in Children Associated with the Flint Drinking Water Crisis: A Spatial Analysis of Risk and Public Health Response." *AJPH Research* 106 (2): 283–90.

Harris, Leila M. 2013. "Variable Histories and Geographies of Marketization and Privatization." In *Contemporary Water Governance in the Global South,* edited by Leila M. Harris, Jacqueline A. Goldin, and Christopher Sneddon, 118–32. London: Routledge.

Harvey, David. 2003. *The New Imperialism.* New York: Oxford University Press.

———. 2005. *A Brief History of Neoliberalism.* New York: Oxford University Press.

Hawkins, Gay. 2017. "The Impacts of Bottled Water: An Analysis of Bottled Water Markets and Their Interactions with Tap Water Provision." *WIREs Water* 4 (3). https://doi.org/10.1002/wat2.1203.

Hawkins, Gay, Emily Potter, and Kane Race. 2015. *Plastic Water: The Social and Material Life of Bottled Water.* Cambridge, MA: MIT Press.

Haydock, Sophie. 2009. "Banning Bottled Water in Universities." *The Ecologist,* August 11.

Hecht, Peter. 2015. "Bottled Water Companies under Fire amid California's Drought." *Sacramento Bee,* May 14.

Heid, Markham. 2019. "Your Bottled Water Probably Has Plastic in It. Should You Worry?" *Time,* May 29.

Heller, Leo. 2017. *Report of the Special Rapporteur on the Human Rights to Safe Drinking Water and Sanitation on His Mission to Mexico.* New York: United Nations Human Rights Council.

Hemphill, Gary. 2018. "What Is America Drinking? U.S. Market Trends." Paper delivered at the Packaging Conference, Orlando, FL, February 5.

Herr, Alexandria. 2021. "'Long Overdue': The Senate Just Passed $35 Billion for Clean Drinking Water." *Grist,* May 1.

Hirschman, Albert O. 1970. *Exit, Voice, and Loyalty.* Cambridge, MA: Harvard University Press.

Ho, Elaine, and Richelle Miller. 2021. "Indigenous Youth Are Playing a Key Role in Solving Urgent Water Issues." *The Conversation,* March 18.

Holden, Emily, Caty Enders, Niko Kommenda, and Vivian Ho. 2021. "More Than 25m Drink from the Worst US Water Systems, with Latinos Most Exposed." *The Guardian,* February 26.

Holt, Douglas B. 2012. "Constructing Sustainable Consumption: From Ethical Values to the Cultural Transformation of Unsustainable Markets." *Annals of the American Academy of Political and Social Science* 644 (1): 236–55.

Hood River County Elections Office. 2016. "Proof Ballot Content: May Primary Election." March 9. www.google.com/url?sa=t&rct=j&q=&esrc=s&source=web&cd=&cad=rja&uact=8&ved=2ahUKEwjoq5uRmff7AhWaIjQIHSa3ABoQFnoECAsQAQ&url=https%3A%2F%2Fhoodriveror.govoffice2.com%2Fvertical%2FSites%2F%257B4BB5BFDA-3709-449E-9B16-B62A0A0DD6E4%257D%2Fuploads%2FLocal_Candidates_and_Measures_Filed_for_the_May_17th_2016_Primary_Election.pdf&usg=AOvVaw20-j2-h4myRrIMcAu53-D3.

House, Kelly. 2015a. "Bottled Water Wars: Nestle's Latest Move in Cascade Locks Sparks Outcry from Opponents." *The Oregonian,* January 23.

———. 2015b. "Gov. Kate Brown Asks for New Approach to Nestle Water Deal." *The Oregonian,* November 6.

———. 2015c. "ODFW Agrees to New Approach for Nestle Bottled Water Plant in Cascade Locks." *The Oregonian,* April 10.

———. 2015d. "Woman's Five-Day Fast outside Cascade Locks City Hall Targets Nestle Deal." *The Oregonian,* August 19.

House, Kelly, and Mark Graves. 2016. "Draining Oregon." *The Oregonian,* August 26.

Human Rights Watch. 2016. "Make It Safe: Canada's Obligation to End the First Nations Water Crisis." Human Rights Watch Report, June 7. www.hrw.org/report/2016/06/07/make-it-safe/canadas-obligation-end-first-nations-water-crisis.

IBWA (International Bottled Water Association). 2001. "IBWA Code of Advertising Standards." https://bottledwater.org/ibwa-code-of-advertising-standards/.

———. 2015. "Bottled Water Should Be Available as the Healthy Choice Beverage in All of America's National Parks." Press release, April 23.

———. 2020a. "Consumer Reports Article Misleads the Public about Arsenic in Bottled Water." Press release, June 25. https://bottledwater.org/nr/consumer-reports-article-misleads-the-public-about-arsenic-in-bottled-water/.

———. 2020b. "Consumers Want Bottled Water to Be Available Wherever Drinks Are Sold, and If It's Not, Most Will Choose Another Packaged Beverage That Uses Much More Plastic." Press release, January 10. https://bottledwater.org/nr/consumers-want-bottled-water-to-be-available-wherever-drinks-are-sold-and-if-its-not-most-will-choose-another-packaged-beverage-that-uses-much-more-plastic/.

———. 2020c. *2020 IBWA Progress Report.* Alexandria, VA: International Bottled Water Association.

———. 2020d. "What People Need to Know about Bottled Water during the Covid-19 Pandemic." Press release, March 6. https://bottledwater.org/nr/what-people-need-to-know-about-bottled-water-during-the-covid-19-outbreak/.

———. 2022a. "Increased Consumer Demand for Bottled Water as a Healthy Alternative to Other Packaged Drinks." Press release, May 31. www.globenewswire.com/en/news-release/2022/05/31/2453590/0/en/Increased-consumer-demand-for-bottled-water-as-a-healthy-alternative-to-other-packaged-drinks.html.

———. 2022b. *2022 Progress Report.* Alexandria, VA: International Bottled Water Association.

IDB (Inter-American Development Bank). 2011. *Latin America's Other Water Infrastructure.* Washington, DC: IDB.

Industry Week. 2013. "Nestle: Bans Unlikely to Stop Popularity of Bottled Water." April 4.

Ingraham, Christopher. 2016. "This Is How Toxic Flint's Water Really Is." *Washington Post,* January 16.

Izundu, Chi Chi, Mohamed Madi, and Chelsea Bailey. 2022. "Jackson Water Crisis: A Legacy of Environmental Racism?" BBC, September 4.

Jaeger, Andrew Boardman. 2018. "Forging Hegemony: How Recycling Became a Popular but Inadequate Response to Accumulating Waste." *Social Problems* 65:395–415.

Jaffee, Daniel. 2020. "Enclosing Water: Privatization, Commodification, and Access." In *The Cambridge Handbook of Environmental Sociology*, vol. 2, edited by Katharine Legun, Julie Keller, Michael Bell, and Michael Carolan, 303–23. Cambridge: Cambridge University Press.

Jaffee, Daniel, and Robert A. Case. 2018. "Draining Us Dry: Scarcity Discourses in Contention over Bottled Water Extraction." *Local Environment* 23 (4): 485–501. https://doi.org/10.1080/13549839.2018.1431616.

Jaffee, Daniel, and Soren Newman. 2013a. "A Bottle Half Empty: Bottled Water, Commodification, and Contestation." *Organization and Environment* 26 (3): 318–35. https://doi.org/10.1177/1086026612462378.

———. 2013b. "A More Perfect Commodity: Bottled Water, Global Accumulation, and Local Contestation." *Rural Sociology* 78 (1): 1–28. https://doi.org/10.1111/j.1549-0831.2012.00095.x.

James, Ian. 2015. "Bottling Water without Scrutiny." *Desert Sun*, March 8.

Jasechko, Scott, and Debra Perrone. 2021. "Global Groundwater Wells at Risk of Running Dry." *Science* 372:418–21.

Javidi, Ariana, and Gregory Pierce. 2018. "U.S. Households' Perception of Drinking Water as Unsafe and Its Consequences: Examining Alternative Choices to the Tap." *Water Resources Research* 54 (9): 6100–13.

Jeffrey, Kim. 2009. "The Future of Bottled Water." Slides from Power Point presentation, September 18. www.nestle.com/sites/default/files/asset-library/documents/library/presentations/globally_managed_business/future_bottled_water_sep2009_jeffery.pdf.

Johnson, Rhiannon. 2019. "Plan to Ban Single-Use Plastics Has First Nations with Long-Term Drinking Water Advisories Worried." CBC News, June 14.

Joy, K.J., Seema Kulkarni, Dik Roth, and Margreet Zwarteveen. 2014. "Re-politicising Water Governance: Exploring Water Re-allocations in Terms of Justice." *Local Environment* 19 (9): 954–73.

Joyce, Christopher. 2019. "U.S. Recycling Industry Is Struggling to Figure Out a Future without China." National Public Radio, August 20.

Kalmusky, Katie. 2018. "Environmental Groups Call On Ontario Candidates to Put a Cap on Bottled Water." *National Observer*, May 17.

Kaplan, Jennifer. 2016. "Bottled Water to Outsell Soda for First Time This Year." Bloomberg, August 2.

Kaplan, Sarah. 2016. "By 2050, There Will Be More Plastic Than Fish in the World's Oceans, Study Says." *Washington Post*, January 20.

Karunananthan, Meera, and Devlin Tellatin. 2016. "Whose Rights to Water Will the 2030 Agenda Promote?" In *Spotlight on Sustainable Development 2016: Report of the Reflection Group on the 2030 Agenda for Sustainable Development*, 54–59. New York: Global Policy Forum.

Kinchy, Abby, Daniel Lee Kleinman, and Robyn Autry. 2008. "Against Free Markets, Against Science? Regulating the Socio-Economic Effects of Biotechnology." *Rural Sociology* 73 (2): 147–79.

Kingkade, Tyler. 2015. "When the University of Vermont Banned Bottled Water, Students Drank More Unhealthy Beverages." *Huffington Post,* July 14.

Kishimoto, Satoko, Lavinia Steinfort, and Nara Petrovic, eds. 2020. *The Future Is Public: Toward Democratic Ownership of Public Services.* Amsterdam: Transnational Institute.

Klein, Naomi. 2007. *The Shock Doctrine: The Rise of Disaster Capitalism.* Toronto: Knopf Canada.

Kloppenburg, Jack. 2010. "Impeding Dispossession, Enabling Repossession: Biological Open Source and the Recovery of Seed Sovereignty." *Journal of Agrarian Change* 10 (3): 367–88.

Koltrowitz, Silke, and Emma Thomassen. 2020. "Bottled Water Firms Turn On the Taps with Filters, Flavors and Fizz." Reuters, February 4.

Kornberg, Dana. 2016. "The Structural Origins of Territorial Stigma: Water and Racial Politics in Metropolitan Detroit, 1950s-2010s." *International Journal of Urban and Regional Research* 40 (2): 263–83.

KPTV Fox 12 News. 2016. "Plans for Nestle Water Bottling Plant in Cascade Locks Moves Forward Despite Ban." October 18. Portland, OR.

Kretzman, Steve, and Raymond Joseph. 2020. "Coca-Cola and Cape Town's Sweetheart Day Zero Deal." GroundUp, May 8.

Kurland, Nancy B., and Deone Zell. 2011. "Water and Business: A Taxonomy and Review of the Research." *Organization and Environment* 23 (3): 316–53.

Lagos, Marisa. 2014. "S.F. Supervisors Back Ban on Sale of Plastic Water Bottles." *San Francisco Chronicle,* March 5.

Lakhani, Nina. 2020. "Revealed: Millions of Americans Can't Afford Water as Bills Rise 80% in a Decade." *The Guardian,* June 23.

———. 2021. "Vast Coalition Calls on Biden to Impose National Moratorium on Water Shutoffs." *The Guardian,* January 13.

Laville, Sandra. 2020. "Report Reveals 'Massive Plastic Pollution Footprint' of Drinks Firms." *The Guardian,* March 31.

Laville, Sandra, and Matthew Taylor. 2017. "A Million Bottles a Minute: World's Plastic Binge 'as Dangerous as Climate Change.'" *The Guardian,* June 28.

Laxer, Gordon, and Dennis Soron. 2006. *Not for Sale: Decommodifying Public Life.* Peterborough, Canada: Broadview.

L'Eau Qui Mord. n.d. "L'Eau Qui Mord." Collectif Eau 88. Accessed September 26, 2021. www.leauquimord.com/appuis/.

Lederer, Edith. 2010. "Access to Clean Water Is 'Human Right,' Says UN." *The Independent,* July 30.

Lee, Jennifer. 2008. "City Council Shuns Bottles in Favor of Water from Tap." *New York Times,* June 17.

Lemus, J. Jesús. 2019. *El agua o la vida: Otra guerra ha comenzado en Mexico.* Mexico City: Grijalbo.

Leonard, Annie. 2020. "Want to Slow the Climate Crisis? Don't Use Single-Use Plastics." *The Nation,* April 23.

Leslie, Keith. 2016a. "Ontario Proposes Two-Year Hold on New Water-Taking Permits." *Globe and Mail,* October 17.

———. 2016b. "Premier Wynne Wants 'Bigger Look' at Future of Bottled-Water Industry." *Toronto Star,* December 21.

Levin, Sam. 2017. "How San Francisco Is Leading the Way out of Bottled Water Culture." *The Guardian,* June 28.

Linebaugh, Peter. 2014. *Stop, Thief! The Commons, Enclosures, and Resistance.* Oakland, CA: PM Press.

Lobina, Emanuele, Philipp Terhorst, and Vladimir Popov. 2011. "Policy Networks and Social Resistance to Water Privatization in Latin America." *Procedia Social and Behavioral Sciences* 10:19–25.

Local Water Alliance. 2016. "Yes on 14-55: Our Water, Our Future." Campaign mailer, Hood River, OR.

Loftus, Alex. 2009. "Rethinking Political Ecologies of Water." *Third World Quarterly* 30 (5): 953–68.

London, Jonathan, Amanda Fenci, Sara Watterson, Jennifer Jarin, Alfonso Aranda, Aaron King, Camille Pannu, Phoebe Seaton, Laurel Fireston, Mia Dawson, and Peter Nguyen. 2018. *The Struggle for Water Justice: Disadvantaged Unincorporated Communities in California's San Joaquin Valley.* Davis, CA: UC Davis Center for Regional Change. https://regionalchange.ucdavis.edu/sites/g/files/dgvnsk986/files/inline-files/The%20Struggle%20for%20Water%20Justice%20FULL%20REPORT_0.pdf.

Loria, Kevin. 2019. "Most Plastic Products Contain Potentially Toxic Chemicals, Study Reveals." Consumer Reports, October 2.

———. 2020. "What's Gone Wrong with Plastic Recycling." Consumer Reports, April 30.

Luxemburg, Rosa. [1913] 2003. *The Accumulation of Capital.* New York: Routledge.

Lydersen, Kari. 2008. "Bottled Water at Issue in Great Lakes." *Washington Post,* September 29.

Lynch, Jim. 2016. "DEQ: Flint Water Fix Should Have Come by 2014." *Detroit News,* January 21.

Mack, Elizabeth A., and Sarah Wrase. 2017. "A Burgeoning Crisis? A Nationwide Assessment of the Geography of Water Affordability in the United States." *PLoS One* 12 (1). https://doi.org/10.1371/journal.pone.0169488.

Mainstreet Research et al. 2018. "New Poll: Ahead of 2018 Provincial Elections, Ontarians Want to Protect Ontario's Water from Bottled Water Industry." March 22. mainstreetresearch.ca/polls/ (no longer available).

Malkin, Elisabeth. 2012. "Bottled-Water Habit Keeps Tight Grip on Mexicans." *New York Times,* July 16.

Maloney, Jennifer. 2019. "Coke and Pepsi Want to Sell You Bottled Water without the Bottle." *Wall Street Journal,* June 21.

Mapes, Lynda V. 2020. "Salmon People: A Tribe's Decades-Long Fight to Take Down the Lower Snake River Dams and Restore a Way of Life." *Seattle Times,* November 29.

Marini Higgs, Micaela. 2019. "America's New Recycling Crisis, Explained by an Expert." Vox, April 2.

Martin, Glen. 2008. "Liquid Gold." *California Lawyer,* November, 22–41. www.bhfs.com/Templates/media/files/news/LiquidGold.CaLaw.Nov08.pdf.

Marx, Karl. 1867. *Capital: A Critique of Political Economy.* Chicago: Charles H. Kerr.

Masalai Blog. 2009. "'Majority World'—A New Word for a New Age." February 11. https://masalai.wordpress.com/2009/02/11/majority-world-a-new-word-for-a-new-age/.

Mascarenhas, Michael, and Lawrence Busch. 2006. "Seeds of Change: Intellectual Property Rights, Genetically Modified Soybeans and Seed Saving in the United States." *Sociologia Ruralis* 46 (2): 122–38.

Mason, Sherri A., Victoria G. Welch, and Joseph Neratko. 2018. "Synthetic Polymer Contamination in Bottled Water." *Frontiers in Chemistry* 6 (407): 1–11.

Matheny, Keith, and Paul Egan. 2016. "Nestlé Bottled-Water Company Seeks to Take More Michigan Water." *Detroit Free Press,* November 20.

McCurdy, Christen. 2020. "Broken Pipes, Broken Promises." *The Oregonian,* July 15.

McDonald, David A. 2016. "To Corporatize or Not to Corporatize (and If So, How?)." *Utilities Policy* 40:107–14.

McSpirit, Stephanie, and Caroline Reid. 2009. "Residents' Perceptions of Tap Water and Decisions to Purchase Bottled Water." *Society and Natural Resources* 24 (5): 511–20.

Meehan, Katie, Jason R. Jurjevich, Nicholas M. J. W. Chun, and Justin Sherrill. 2020. "Geographies of Insecure Water Access and the Housing–Water Nexus in US Cities." *Proceedings of the National Academy of Sciences* 117 (46): 28700–28707.

Mehta, Lyla. 2007. "Whose Scarcity? Whose Property? The Case of Water in Western India." *Land Use Policy* 24:654–63.

———. 2016. "Why Invisible Power and Structural Violence Persist in the Water Domain." *IDS Bulletin* 47 (5). Institute of Development Studies. https://doi.org/10.19088/1968-2016.165.

Melo Zurita, Maria de Lourdes, Dana C.Thomsen, Timothy F. Smith, Anna Lyth, Benjamin L. Preston, and Scott Baum. 2015. "Reframing Water: Contesting H2O within the European Union." *Geoforum* 65:170–78.

Mercille, Julien, and Enda Murphy. 2015. "Conceptualising European Privatisation Processes after the Great Recession." *Antipode* 48 (3): 685–704.

Mestre Rodríguez, Eduardo. 2019. "Presupuesto para el sector agua: Lo que debería ocuparnos." *El Universal,* October 16.

Metz, Barbara. 2019. "Lessons Learned: The Refillable Quota in Germany." Reuse Conference, Brussels, September 24. Deutsche Umwelthilfe. www.reloopplatform.org/wp-content/uploads/2019/10/190924_Reuse_Deutsche_Umwelthilfe_Metz.pdf.

Milman, Oliver. 2016. "Millions Exposed to Dangerous Lead Levels in US Drinking Water, Report Finds." *The Guardian,* June 28.

———. 2022. "US Government to Ban Single-Use Plastic in National Parks." *The Guardian,* June 8.

Mintel. 2018. *Bottled Water—US—February 2018: What's Next?* London: Mintel Group.

———. 2019. *Asda's Refill Points Add to Threat to Bottled Water Sales.* London: Mintel Group.

———. 2020. *Still and Sparkling Waters—US—February 2020.* Report. London: Mintel Group.

———. 2021. *Still and Sparkling Water: U.S., March 2021.* London: Mintel Group.

———. 2022. *Still and Sparkling Water, US 2022.* London: Mintel Group.

Mirk, Sarah. 2009. "Debating for Dollars: Lobbyists Spent $18 Million Influencing Oregon." *Portland Mercury,* September 10.

Montero Contreras, Delia P. 2016. "El consumo de agua embotellada en la Ciudad de México desde una perspectiva institucional." *Agua y Territorio* 7:35–49.

———. 2019. *Instituciones y actores: Un enfoque alternativo para entender el consumo de agua embotellada en México.* Mexico City: UAM; Tirant Humanidades.

Montuori, P., E. Jover, M. Morgantini, J.M. Bayona, and M. Triassi. 2008. "Assessing Human Exposure to Phthalic Acid and Phthalate Esters from Mineral Water Stored in Polyethylene Terephthalate and Glass Bottles." *Food Additives and Contaminants:* 511–18.

Mosbergen, Dominique. 2019. "Why Southeast Asia Is Flooded with Trash from America and Other Wealthy Nations." *Huffington Post,* March 8.

MST (Movimento Sem Terra). 2018. "Carta de São Lourenço—Fora Nestlé! A água é nossa!" Press release, March 20. https://mst.org.br/2018/03/20/carta-de-sao-lourenco-fora-nestle-a-agua-e-nossa/.

Mulvihill, Patrick. 2016. "Cascade Locks Council Votes against 14-55." *Hood River News,* April 13.

Murdock, Riley, Mila Murray, and Isis Simpson-Mersha. 2020. "'The Money Is Not Justice': Flint Mothers Express Relief, Skepticism about Water Crisis Settlement." Michigan Live, August 20.

Nagy, Mike, and Maude Barlow. 2017. "Time for Ontario to Protect Its Water Supplies." *The Record* (Kitchener, Ontario), May 11.

Narain, Vishal. 2014. "Whose Land? Whose Water? Water Rights, Equity and Justice in a Peri-urban Context." *Local Environment* 19 (9): 974–89.

NCSL (National Conference of State Legislatures). 2020. "State Beverage Container Deposit Laws." March 13. www.ncsl.org/research/environment-and-natural-resources/state-beverage-container-laws.aspx.

Nelson, Paul. 2017. "Citizens, Consumers, Workers, and Activists: Civil Society during and after Water Privatization Struggles." *Journal of Civil Society* 13 (2): 202–21.

Nestlé. 2020. "Nestlé Reports Full-Year Results for 2019." Press release, February 13. www.nestle.com/media/pressreleases/allpressreleases/full-year-results-2019.

Nestlé Waters Canada. 2016. "Nestlé in Canada." Fact sheet, September. https://dokumen.tips/documents/nestle-in-canada-proud-to-nestle-in-canada-5-pure-facts-about-nestle.html?page=2.

———. 2017. *The Facts about Bottled Water.* N.p.: Nestlé Waters. www.nestle.com/sites/default/files/asset-library/documents/about_us/ask-nestle/british-columbia/nestle-waters-canada-facts.pdf.

Nestlé Waters North America. 2009. "Nestlé Waters North America Withdraws McCloud Project Proposal." Press release, September 10. www.caltrout.org/wp-content/uploads/2021/03/091009-McCloudWithdrawal-FinalPR.pdf.

———. 2018. "Bottled Water Donations to Flint, Michigan." TV advertisement, Nestle Waters North America. www.youtube.com/watch?v=REn5QclOcXA.

Nichols, Rodger. 2016. "City Hears Outcries against Nestlé." *Goldendale Sentinel,* November 8.

NRDC (Natural Resources Defense Council). 1999. *Bottled Water: Pure Drink or Pure Hype?* Washington, DC: Natural Resources Defense Council. www.nrdc.org/sites/default/files/bottled-water-pure-drink-or-pure-hype-report.pdf.

———. 2019. "New Drinking Water Report: Communities of Color More Likely to Suffer Drinking Water Violations for Years." Press release, September 24. www.nrdc.org/media/2019/190924.

NYC Mayor's Office of Sustainability. 2020. "Water Management." https://www1.nyc.gov/site/sustainability/initiatives/water-management.page.

Oaten, James, and Som Patidar. 2020. "Delhi Is Facing a Water Crisis. Ahead of Day Zero, the City's Residents Have Turned to the Mafia and Murder." ABC News, February 8.

Olivera, Oscar, and T. Lewis. 2004. *Cochabamba!: Water War in Bolivia.* Cambridge, MA: South End Press.

O'Mahoney, Ash. 2020. "The Rise and Fall." *The Grocer* (U.K.), March 21.

Opel, Andy. 1999. "Constructing Purity: Bottled Water and the Commodification of Nature." *Journal of American Culture* 22 (4): 67–76.

O'Reilly, Kathleen. 2011. "'They Are Not of This House': The Gendered Costs of Drinking Water's Commodification." *Economic and Political Weekly* 46 (18): 49–55.

Oreskes, Naomi, and Erik M. Conway. 2010. *Merchants of Doubt: How a Handful of Scientists Obscured the Truth on Issues from Tobacco Smoke to Climate Change.* New York: Bloomsbury.

Overstreet, Kim. 2021. "Beverage Manufacturers Respond to Consumer Trends on Plastic Packaging." *Packaging World,* June 17.

Pacheco-Vega, Raúl. 2019. "Human Right to Water and Bottled Water Consumption: Governing at the Intersection of Water Justice, Rights and Ethics." In *Water Politics,* edited by F. Sultana and A. Loftus, 113–28. London: Routledge.

Parag, Y., and J.T. Roberts. 2009. "A Battle against the Bottles: Building, Claiming, and Regaining Tap-Water Trustworthiness." *Society and Natural Resources* 22 (7): 625–36.

Parker, Laura. 2014. "The Best Way to Deal With Ocean Trash." *National Geographic,* April 16.

———. 2019. "How the Plastic Bottle Went from Miracle Container to Hated Garbage." *National Geographic,* August 23.

Parks, Bradley W. 2016. "Nestle Funds PAC Opposing Measure to Block Cascade Locks Bottling Plant." OPB News, May 14.

Pauli, Benjamin. 2019. *Flint Fights Back: Environmental Justice and Democracy in the Flint Water Crisis.* Cambridge, MA: MIT Press.

———. 2020a. "The Flint Water Crisis." *WIREs Water* 7 (3): 1–14.

———. 2020b. "The Long Road Out of Crisis: (Re)Building Trust in Flint's Public Water from Poisoning to Pandemic." In *Public Water and Covid-19: Dark Clouds and Silver Linings,* edited by David McDonald, Susan Spronk and Daniel Chavez, 311–27. Kingston, Ontario: Municipal Service Project.

Pechlaner, Gabriela, and Gerardo Otero. 2010. "The Neoliberal Food Regime: Neoregulation and the New Division of Labor in North America." *Rural Sociology* 75 (2): 179–208.

Peck, Jamie. 2015. *Austerity Urbanism: The Neoliberal Crisis of American Cities.* New York: Rosa Luxemburg Stiftung.

Pempetzoglou, Maria, and Zoi Patergiannaki. 2017. "Debt-Driven Water Privatization: The Case of Greece." *European Journal of Multidisciplinary Studies* 5 (1): 102–11.

Peters, Adele. 2015. "The Bottled Water Industry Is Fighting to Keep Plastic Bottles in National Parks." *Fast Company,* July 20.

Peters, Maren. 2020. "Bottle Empty at Nestlé: Dismantling the Water Business." SRF 1 TV, July 31.

PET Planet. 2022. "New US PET Recycling Coalition." *PET Planet,* June 15.

Pew Charitable Trusts and Systemiq Ltd. 2020. *Breaking the Plastic Wave: A Comprehensive Assessment of Pathways towards Stopping Ocean Plastic Pollution.* Washington, DC: Pew Charitable Trusts.

Pierce, Gregory. 2015. "Beyond the Strategic Retreat? Explaining Urban Water Privatization's Shallow Expansion in Low- and Middle-Income Countries." *Journal of Planning Literature* 30 (2): 119–31.

Pierce, Gregory, and Silvia Gonzalez. 2017. "Mistrust at the Tap? Factors Contributing to Public Drinking Water (Mis)perception across US Households." *Water Policy* 19:1–12.

Pierre-Louis, Kendra. 2015. "We Don't Trust Drinking Fountains Anymore, and That's Bad for Our Health." *Washington Post,* July 8.

Polanyi, Karl. 1944. *The Great Transformation: The Political and Economic Origins of Our Time.* Boston: Beacon Press.

Polaris Institute. 2014. "Bottled Water Maps." www.polarisinstitute.org/bottled_water_maps.

Powers, Lucas. 2015. "California Drought's Newest Battlefront Is Bottled Water." CBC News, May 18.

Prasetiawan, Teddy, Anindrya Nastiti, and Barti Setiani Muntalif. 2017. "'Bad' Piped Water and Other Perceptual Drivers of Bottled Water Consumption in Indonesia." *WIREs Water* 4 (4): 1–12.

Primo Water. n.d. "Primo Water and Water Dispensers." Accessed December 16, 2022. https://primowater.com/walmart-offer/.

Pulido, Laura. 2016. "Flint, Environmental Racism, and Racial Capitalism." *Capitalism Nature Socialism* 27 (3): 1–16.

Race, Kane. 2012. "'Frequent Sipping': Bottled Water, the Will to Health and the Subject of Hydration." *Body and Society* 18 (3–4): 72–98.

Radonic, Lucero, and Cara E. Jacob. 2021. "Examining the Cracks in Universal Water Coverage: Women Document the Burdens of Household Water Insecurity." *Water Alternatives* 14 (1): 60–78.

Raman, K. Ravi. 2010. "Transverse Solidarity: Water, Power, and Resistance." *Review of Radical Political Economics* 42 (2): 251–68.

Ranganathan, Malini. 2014. "Paying for Pipes, Claiming Citizenship: Political Agency and Water Reforms at the Urban Periphery." *International Journal of Urban and Regional Research* 38 (2): 590–608.

Rankin, Jennifer. 2019. "European Parliament Votes to Ban Single Use Plastics." *The Guardian,* March 27.

Readfearn, Graham. 2018. "WHO Launches Health Review after Microplastics Found in 90% of Bottled Water." *The Guardian,* March 15.

Rector, Josiah. 2016. "Neoliberalism's Deadly Experiment: In Michigan, Privatization and Free-Market Governance Has Left 100,000 People without Water." *Jacobin,* October 21.

Refill. n.d. "Life with Less Plastic, Made Easy." Accessed March 26, 2021. www.refill.org.uk/.

Rehm, Colin D., Matthieu Maillot, Florent Vieux, Pamela Barrios, and Adam Drewnowski. 2020. "Who Is Replacing Sugar-Sweetened Beverages with Plain Water? Analyses of NHANES 2011–16 Data." *Current Developments in Nutrition* 4:557.

Reisner, Marc. 1986. *Cadillac Desert: The American West and Its Disappearing Water.* New York: Viking Penguin.

Roberts, Adrienne. 2008. "Privatizing Social Reproduction: The Primitive Accumulation of Water in an Era of Neoliberalism." *Antipode* 40 (4): 535–60.

Robertson, Derek. 2020. "Flint Has Clean Water Now. Why Won't People Drink It?" *Politico,* December 23.

Robinson, Joanna. 2013. *Contested Water: The Struggle against Water Privatization in the United States and Canada.* Cambridge, MA: MIT Press.

Robinson, Mike. 2019. "Protesters Demand Nestle Cease and Desist Water-Taking." *Wellington Advertiser,* June 13.

Rodwan, John G., Jr. 2011. "Bottled Water 2010: The Recovery Begins: U.S. And International Developments and Statistics." *Bottled Water Reporter,* April-May, 11–17.

———. 2018. "Bottled Water 2017: Staying Strong." *Bottled Water Reporter,* July/August, 12–20.

———. 2019. "Significant, but Slower, Growth for Bottled Water in 2018." *Bottled Water Reporter,* July/August, 10–18.

———. 2021. "Bottled Water 2020: Continued Upward Movement." *Bottled Water Reporter,* August 5, 10–19.

Rose, Joan. 2019. "The US Drinking Water Supply Is Mostly Safe, but That's Not Good Enough." *The Conversation,* May 29.

Rosinger, Asher Y., Kirsten A. Herrick, Amber Y. Wutich, Jonathan S. Yoder, and Cynthia L. Ogden. 2018. "Disparities in Plain, Tap and Bottled Water

Consumption among US Adults: National Health and Nutrition Examination Survey (NHANES), 2007–2014." *Public Health Nutrition* 21 (8): 1455–64.
Ross, Gavin. 2020. *Bottled Water Production in the US*. New York: IBISWorld.
Rouda, Harley, and Rashida Tlaib. 2020. Letter to Fernando Mercé, President and CEO of Nestlé Waters North America. March 3. U.S. House of Representatives, Committee on Oversight and Reform, Environment Subcommittee. https://oversightdemocrats.house.gov/sites/democrats.oversight.house.gov/files/2020-03-03.Rouda%20Tlaib%20to%20Merce-Nestle%20re%20Bottled%20Water%20FINAL.pdf.
Royte, Elizabeth. 2008. *Bottlemania: Big Business, Local Springs, and the Battle over America's Drinking Water.* New York: Bloomsbury.
Rubin, Josh. 2020. "Environmentalists Cheer as Nestle Sells Canadian Water Business." *Toronto Star,* July 2.
Ruiters, Greg. 2007. "Contradictions in Municipal Services in Contemporary South Africa: Disciplinary Commodification and Self-Disconnections." *Critical Social Policy* 27 (4): 487–508.
Rushe, Dominic. 2014. "Blow to Detroit's Poorest as Judge Rules Water Shutoffs Can Continue." *The Guardian,* September 29.
Salzman, James. 2012. *Drinking Water: A History*. New York: Overlook.
S&P Global Market Intelligence. 2021. "Triton Water Launches $1.4B, 2-Part Bond Offering." *S&P Global Market Intelligence,* March 15.
Save Our Water. 2021. "The Middlebrook Well Story: Is It Over?" Webinar, April 22. www.saveourwater.ca/post/the-middlebrook-well-story-is-it-over-not-by-a-long-shot.
Sax, Leonard. 2010. "Polyethylene Terephthalate May Yield Endocrine Disruptors." *Environmental Health Perspectives* 118 (4): 445–48.
Schlosberg, Deia, dir. 2019. *The Story of Plastic.* New York: Outcast Films.
Schnell, Urs, dir. 2012. *Bottled Life: Nestlé's Business with Water.* Zurich: DokLab.
Schroeer, Anne, Matt Littlejohn, and Henning Wilts. 2020. *Just One Word: Refillables.* Washington, DC: Oceana. https://oceana.org/wp-content/uploads/sites/18/3.2.2020_just_one_word-refillables.pdf.
Schroering, Caitlin. 2021. "Constructing Another World: Solidarity and the Right to Water." *Studies in Social Justice* 15 (1): 102–28.
Scroll.in. 2017. "Maharashtra to Ban Plastic Bottles in All Government Offices, Schools and Colleges from March." November 17.
Shimo, Alexandra. 2018. "While Nestlé Extracts Millions of Litres from Their Land, Residents Have No Drinking Water." *The Guardian,* October 4.
Shiva, Vandana, ed. 2016. *Seed Sovereignty, Food Security: Women in the Vanguard of the Fight against GMOs and Corporate Agriculture*. Berkeley, CA: North Atlantic Books.
Shoup, Mary Ellen. 2017a. "Beverage Marketing Corporation: Bottled Water Will Continue as 'Undisputed' Top Player of Beverage Market." *Beverage Daily,* May 1.
———. 2017b. "Bottled Water Surpasses Soda in Consumption with 86% Purchase Rate among Americans." *Beverage Daily,* March 13.

Singh, Maanvi. 2021. "Drought-Hit California Moves to Halt Nestlé from Taking Millions of Gallons of Water." *The Guardian,* April 27.
Six Nations of the Grand River. 2020. *Six Nations of the Grand River: Land Rights, Financial Justice, Creative Solutions*. Six Nations: Six Nations Lands and Resources Department. www.sixnations.ca/LandsResources/SNLands-LandRightsBook-FINALyr2020.pdf.
———. n.d. "About: Six Nations of the Grand River." Accessed September 10, 2021. www.sixnations.ca/about.
Sloan, Carrie. 2016. "How Wall Street Caused a Water Crisis in America's Cities." *The Nation,* March 11.
Snider, Annie. 2021. "Biden Says Bipartisan Deal Will Solve the Country's Lead Problem. It Won't." Politico, August 2.
Snitow, Alan, Deborah Kaufman, and Michael Fox. 2007. *Thirst: Fighting the Corporate Theft of Our Water.* San Francisco: Wiley and Sons.
Solnit, Rebecca. 2019. *Hope in the Dark: Untold Histories, Wild Possibilities.* 3rd ed. Chicago: Haymarket Books.
Sparrow, Jeff. 2015. "Wipe Right: Toilet App Looie Forces Movement of 'Sharing Economy' towards Privatisation." *The Guardian,* August 21.
Spiliotis, M., and L. Garrote. 2015. "A Fuzzy Multicriteria Categorization of Water Scarcity in Complex Water Resources Systems." *Water Resource Management* 29:521–39.
Spronk, Susan. 2015. "Roots of Resistance to Urban Water Privatisation in Bolivia: The 'New Working Class,' the Crisis of Neoliberalism, and Public Services." In *Crisis and Contradiction: Marxist Perspectives on Latin America in the Global Political Economy,* edited by S. J. Spronk and J. R. Webber, 29–51. Leiden, Netherlands: Brill.
Spronk, Susan, and Jeffery R. Webber. 2007. "Struggles against Accumulation by Dispossession in Bolivia: The Political Economy of Natural Resource Contention." *Latin American Perspectives* 34 (2): 31–47.
Statista. 2020. "Sales of Nestlé Waters Worldwide from 2010 to 2019, by Region." www.statista.com/statistics/268906/sales-of-nestle-waters-worldwide-by-region/.
———. 2021a. "Carbonated Soft Drink (CSD) All-Channel Sales Volume in the United States from 2010 to 2020." No longer available on internet; a copy is in the author's possession.
———. 2021b. "Per Capita Consumption of Packaged Water in Canada from 2010 to 2022 (in Liters)." www.statista.com/statistics/1121097/bottled-water-per-capita-consumption-canada/.
———. 2021c. "Share of Past-Day Bottled Water Consumers in Canada from 2011 to 2020." www.statista.com/statistics/452645/share-of-past-day-bottled-water-consumers-in-canada/.
———. 2022a. "Consumption of Packed Beverages Worldwide in 2021, by Beverage Type (in Billion Liters)." www.statista.com/statistics/232924/global-consumption-of-packed-beverages-by-beverage-tpye/.
———. 2022b. *Non-alcoholic Beverages in the United States.* www.statista.com/study/10629/non-alcoholic-beverages-and-soft-drinks-in-the-united-states-statista-dossier/. Hamburg: Statista.

Stephenson, John B. 2009. *Bottled Water: FDA Safety and Consumer Protections Are Often Less Stringent Than Comparable EPA Protections for Tap Water.* Report. Washington, DC: U.S. General Accounting Office. www.gao.gov/assets/gao-09-610.pdf.

Story of Stuff Project, dir. 2018. *A Tale of Two Cities: Flint, Evart, and the Fight for Michigan's Water.* United States: Leighton Woodhouse.

———. 2021a. "Nestlé's Troubled Waters." Webinar, March 18. www.storyofstuff.org/nestle/.

———. 2021b. "Public Statement on the Sale of Nestlé Waters." February. www.storyofstuff.org/blog/public-statement-on-the-sale-of-nestle-waters/.

Subramaniam, Mangala, and Beth Williford. 2012. "Contesting Water Rights: Collective Ownership and Struggles against Privatization." *Sociology Compass* 6 (5): 413–24.

Sultana, Farhana, and Alex Loftus, eds. 2012. *The Right to Water: Politics, Governance, and Social Struggles.* London: Earthscan.

Sunder, Kalpana. 2021. "How to Stop Another 'Day Zero.'" *BBC Future Planet,* January 5.

Sustainable Practices. n.d. "Bottle Ban." Accessed July 1, 2022. https://sustainablepracticesltd.org/bottle-ban.

Swanson, Emily. 2016. "AP-GfK Poll: About Half of Americans Confident in Tap Water." Associated Press, March 6.

Switzer, David, and Manuel P. Teodoro. 2017. "Class, Race, Ethnicity, and Justice in Safe Drinking Water Compliance." *Social Science Quarterly* 99 (2): 524–35.

Swyngedouw, Erik. 2005. "Dispossessing H2O: The Contested Terrain of Water Privatization." *Capitalism Nature Socialism* 16 (1): 81–98.

Sydney Water. n.d. "Find a Water Refill Station." Accessed September 6, 2022. www.sydneywater.com.au/water-the-environment/what-you-can-do/find-a-water-refill-station.html.

Szasz, Andrew. 2007. *Shopping Our Way to Safety: How We Changed from Protecting the Environment to Protecting Ourselves.* Minneapolis: University of Minnesota Press.

Taekema, Dan. 2021. "Six Nations Elected Chief Stresses Unity, Reiterates Call for Moratorium on Development." CBC News, April 27.

Teodoro, Manuel P., and Robin R. Saywitz. 2020. "Water and Sewer Affordability in the United States: A 2019 Update." *AWWA Water Science* 2 (2): e1176.

Thara, Kaveri. 2017. "In Troubled Waters: Water Commodification, Law, Gender, and Poverty in Bangalore." *Gender and Development* 25 (2): 253–68.

Themba, Makani. 2022. "Jim Crow Infrastructure and the Jackson, Miss., 'Water Crisis.'" *The Nation,* September 6.

Thomas, Daniel. 2020. "People Still Want Plastic Bottles, Says Coca-Cola." BBC, January 21.

Thomas, Gregory, and Eduardo Medina. 2019. "No Fly Zone: SFO Bans Sale of Plastic Water Bottles." *San Francisco Chronicle,* August 2.

TOMRA. 2021. "Bottle Bill States and How They Work." September 8. www.tomra.com/en/discover/reverse-vending/feature-articles/bottle-bill-states-and-how-they-work.

Toner, J. P. 2021. "Beating Back the Ban: Communicating the Truth about Bottled Water." *Bottled Water Reporter* 61 (4).

Tweed Strategies. n.d. "Tweed Strategies." Accessed July 16, 2021. www.tweedstrategies.com/.

2030 Water Resources Group. n.d. "Governance—2030 Water Resources Group—World Bank Group." Accessed December 21, 2022. www.2030wrg.org/who-we-are/.

UCLA (University of California, Los Angeles). 2020. "Policy 809: Single-Use Plastics." Administrative Policies and Procedures. www.adminpolicies.ucla.edu/APP/Number/809.0.

UCLA Sustainability. n.d.. "Single-Use Plastics Policy: UCLA Will Be Plastic-Free Campus by 2023." Accessed March 23, 2021. www.sustain.ucla.edu/single-use-plastic-policy/.

UN Convention to Combat Desertification. 2022. *Drought in Numbers 2022*. New York: United Nations.

UNDP (United Nations Development Program). 2006. *Human Development Report 2006: Beyond Scarcity: Power, Poverty and the Global Water Crisis*. New York: United Nations Development Program.

UN Human Rights Office of the High Commissioner. 2014. "Detroit: Disconnecting Water from People Who Cannot Pay—An Affront to Human Rights, Say UN Experts." Press release, June 25. www.ohchr.org/en/press-releases/2014/06/detroit-disconnecting-water-people-who-cannot-pay-affront-human-rights-say.

United Nations. 1948. *Universal Declaration of Human Rights*. New York: United Nations.

———. 1992. *Dublin Statement on Water and Sustainable Development*. New York: United Nations.

———. 2010. *General Assembly Resolution 64/292: The Human Right to Water and Sanitation*. New York: United Nations General Assembly.

———. 2018. *Sustainable Development Goal 6: Synthesis Report on Water and Sanitation*. United Nations.

UN News. 2020. "Three Billion People Globally Lack Handwashing Facilities at Home: UNICEF." October 15.

UN-Water. 2018. *Sustainable Development Goal 6: Synthesis Report 2018 on Water and Sanitation*. New York: United Nations.

Upadhyay, Aishwarya. 2019. "To Reduce Plastic Waste and Save Water, South Delhi Municipal Corporation Bans Plastic Water Bottles in the Office." NDTV.com, July 25.

U.S. Census Bureau. n.d. "U.S. Census QuickFacts: Flint, Michigan." Accessed October 22, 2019.

Vail, John. 2010. "Decommodification and Egalitarian Political Economy." *Politics and Society* 38 (3): 310–46.

Van Esterik, Penny. 2013. "The Politics of Breastfeeding: An Advocacy Update." In *Food and Culture: A Reader,* 3rd ed., edited by Carole Counihan and Penny Van Esterik, 510–30. New York: Routledge.

Varghese, Shiney. 2007. *Privatizing U.S. Water.* Minneapolis: Institute for Agriculture and Trade Policy.

Vasquez, M., J. Carter, and P. Valko, 2015. "Bottled Water Ban: Update 2015." Washington University, St. Louis, Office of Sustainability. https://gallery.mailchimp.com/7066ff9764a24116835 2699e8/files/160118_Bottle_Brief_Final.pdf.

Ver Valen, Dian. 2016. "Updated: Waitsburg Mayor Walt Gobel Resigns." *Walla Walla Union Bulletin,* August 2.

Vidal, John. 2006. "Big Water Companies Quit Poor Countries." *The Guardian,* March 22.

Vieux, Florent, Matthieu Maillot, Colin D. Rehm, Pamela Barrios, and Adam Drewnowski. 2020. "Trends in Tap and Bottled Water Consumption among Children and Adults in the United States: Analyses of NHANES 2011–16 Data." *Nutrition Journal* 19 (10). https://doi.org/10.1186/s12937-020-0523-6.

Villanueva, Cristina M., Marianna Garfí, Carles Milà, Sergio Olmos, Ivet Ferrer, and Cathryn Tonne. 2021. "Health and Environmental Impacts of Drinking Water Choices in Barcelona, Spain: A Modelling Study." *Science of the Total Environment* (795): 148884.

Viscusi, W. Kip, Joel Huber, and Jason Bel. 2015. "The Private Rationality of Bottled Water Drinking." *Contemporary Economic Policy* 33 (3): 450–67.

Volcovici, Valerie. 2022. "U.S. Plastic Recycling Rate Drops to Close to 5%—Report." Reuters, May 5.

Wagenhofer, Erwin, dir. 2005. *We Feed the World.* Documentary film. Austria: Katharina Bogensberger and Helmut Grasser.

Wait, Isaac W., and William A. Petrie. 2017. "Comparison of Water Pricing for Publicly and Privately Owned Water Utilities in the United States." *Water International* 42 (8): 967–80.

Walter, Carolin Tina, Michelle Kooy, and Indrawan Prabaharyaka. 2017. "The Role of Bottled Drinking Water in Achieving SDG 6.1: An Analysis of Affordability and Equity from Jakarta, Indonesia." *Journal of Water, Sanitation and Hygiene for Development* 7 (4): 642–50.

Water Is Life Alliance. n.d. "Member Groups—Water Is Life Alliance." Accessed February 10, 2021. https://waterislifealliance.net/about-us/.

Watts, Jonathan. 2018. "Cape Town Faces Day Zero: What Happens When the City Turns Off the Taps?" *The Guardian,* February 3.

Wellington Water Watchers 2020. "All Eyes on Nestlé: Is This Really Goodbye?" Webinar, July 13. www.youtube.com/watch?v=IiA1g-LHUFg.

———. 2021. "Blue Triton Fact Sheet." https://d3n8a8pro7vhmx.cloudfront.net/wellingtonwaterwatchers/pages/1391/attachments/original/1623027515/BlueTriton_FACT_SHEET.pdf?1623027515.

———. n.d.-a. "About Us—Wellington Water Watchers." Accessed September 13, 2021. http://wellingtonwaterwatchers.ca/about-us/.

———. n.d.-b. "Hillsburgh." Accessed September 2, 2021. www.wellingtonwaterwatchers.ca/hillsburgh.

Westerhoff, P., Panjai Prapaipong, Everett Shock, and Alice Hillaireau. 2008. "Antimony Leaching from Polyethylene Terephthalate (PET) Plastic Used for Bottled Drinking Water." *Water Research* 42 (3): 551–56.

WEYI TV. 2022. "Ice Mountain to Provide Bottled Water to Flint through 2022." *Mid-Michigan Now,* June 6.

Whiteford, Linda, and Scott Whiteford, eds. 2005. *Globalization, Water, and Health: Resource Management in Times of Scarcity.* Oxford: James Currey.

WHO (World Health Organization). 2019. *Microplastics in Drinking-Water.* Geneva: World Health Organization.

WHO (World Health Organization) and UNICEF (United Nations Children's Fund). 2017. *Progress on Drinking Water, Sanitation and Hygiene, 2017.* Geneva: World Health Organization and United Nations Children's Fund.

———. 2021. *Progress on Household Drinking Water, Sanitation and Hygiene, 2000–2020: Five Years into the SDGs.* Geneva: WHO and UNICEF.

Wilkins, Matt. 2018. "More Recycling Won't Solve Plastic Pollution." *Scientific American Blog,* July 6.

Wong, Venessa. 2017. "Almost No Plastic Bottles Get Recycled into New Bottles." Buzzfeed, April 13.

World Bank. 2005. *Infrastructure Development: The Roles of the Public and Private Sectors; World Bank Group's Approach to Supporting Investments in Infrastructure.* Washington, DC: World Bank.

World Economic Forum. 2016. *The New Plastics Economy: Rethinking the Future of Plastics.* Geneva: World Economic Forum.

Young, Angelo. 2019. "Coca-Cola, Pepsi Highlight the 20 Corporations Producing the Most Ocean Pollution." *USA Today,* June 17.

Zdanowicz, Christina. 2016. "Flint Family Uses 151 Bottles of Water per Day." CNN, March 7.

Zhang, Xue, Rivas Gonzáles, M. Grant, and Mildred E. Warner. 2022. "Water Pricing and Affordability in the U.S.: Public vs. Private Ownership." *Water Policy* 24 (3): 500–516.

Zhang, Xue, and Mildred E. Warner. 2021. "The Relationship between Water Shutoffs and COVID Infections and Deaths." Food and Water Watch and Cornell University, report, March. www.foodandwaterwatch.org/2021/03/24/the-relationship-between-water-shutoffs-and-covid-19/.

Zwarteveen, Margreet Z., and Rutgerd Boelens. 2014. "Defining, Researching and Struggling for Water Justice: Some Conceptual Building Blocks for Research and Action." *Water International* 39 (2): 143–58.

Index

Note: *fig.* refers to figures.

Founded in 1893,
UNIVERSITY OF CALIFORNIA PRESS
publishes bold, progressive books and journals on topics in the arts, humanities, social sciences, and natural sciences—with a focus on social justice issues—that inspire thought and action among readers worldwide.

The UC PRESS FOUNDATION
raises funds to uphold the press's vital role as an independent, nonprofit publisher, and receives philanthropic support from a wide range of individuals and institutions—and from committed readers like you. To learn more, visit ucpress.edu/supportus.